Lecture Notes in Civil Engineering

Volume 255

Lecture Notes in Civil Engineering (LNCE) publishes the latest developments in Civil Engineering—quickly, informally and in top quality. Though original research reported in proceedings and post-proceedings represents the core of LNCE, edited volumes of exceptionally high quality and interest may also be considered for publication. Volumes published in LNCE embrace all aspects and subfields of, as well as new challenges in, Civil Engineering. Topics in the series include:

- Construction and Structural Mechanics
- Building Materials
- Concrete, Steel and Timber Structures
- Geotechnical Engineering
- Earthquake Engineering
- Coastal Engineering
- Ocean and Offshore Engineering; Ships and Floating Structures
- Hydraulics, Hydrology and Water Resources Engineering
- Environmental Engineering and Sustainability
- Structural Health and Monitoring
- Surveying and Geographical Information Systems
- Indoor Environments
- Transportation and Traffic
- Risk Analysis
- Safety and Security

To submit a proposal or request further information, please contact the appropriate Springer Editor:

- Pierpaolo Riva at pierpaolo.riva@springer.com (Europe and Americas);
- Swati Meherishi at swati.meherishi@springer.com (Asia—except China, Australia, and New Zealand);
- Wayne Hu at wayne.hu@springer.com (China).

All books in the series now indexed by Scopus and EI Compendex database!

Muhammad Adly Rahandi Lubis
Editor

Proceedings of the 10th International Symposium for Sustainable Humanosphere

ISSH 2021

Springer

Editor
Muhammad Adly Rahandi Lubis
Research Center for Biomass
and Bioproducts
National Research and Innovation Agency
Cibinong, Indonesia

ISSN 2366-2557 ISSN 2366-2565 (electronic)
Lecture Notes in Civil Engineering
ISBN 978-981-19-1421-8 ISBN 978-981-19-1422-5 (eBook)
https://doi.org/10.1007/978-981-19-1422-5

This Springer imprint is published by the registered company Springer Nature Singapore Pte Ltd.
The registered company address is: 152 Beach Road, #21-01/04 Gateway East, Singapore 189721, Singapore

Preface

Proceedings of the 10th International Symposium for Sustainable Humanosphere was held virtually on September 13–14, 2021. The forum was organized by the JSPS Alumni Association of Indonesia (JAAI) and Research Center for Biomaterials of Indonesian Institute of Sciences Indonesia (LIPI), and supported by JSPS Bangkok Office, Research Institute for Sustainable Humanosphere (RISH), Kyoto University, IPB University, University of Padjajaran, University of Jambi, University of Islam Indonesia, University of Lambung Mangkurat, and the University of North Sumatera.

As a scientific community, we aim to design and assess effective solutions for a sustainable future as one of the actions we can actively contribute to the sustainable development goals and climate change mitigation. Therefore, our step-up action is to gather all the researchers, scientists, academia, and practitioners in the field of *biomass conversion, bio-based smart materials, forest and environment, atmospheric science, and socio-economy* in a symposium entitled "Proceedings of the 10th International Symposium for Sustainable Humanosphere."

The event showcased a "Frontier Innovation of Lignocellulose and Humanosphere Sciences Towards Sustainable Development Goals and Climate Change Mitigation" tagline. Together with climate change mitigation, science, technology, and innovation are two thematic-ongoing activities alongside other sectors such as urbanization, transportation, water, ocean, and energy. In a global context, the bio-economy strategy connects the sustainable use of renewable biological resources for food, bio-based products, and bioenergy while maintaining the safety and restoration of biodiversity, ecosystems, and natural habitat.

This conference proceedings contain original research articles and review articles in forest and environment, enhancing mitigation for climate change and socio-economy in sustainable development goals. In total, 27 articles were accepted for publication after the peer-review process. Together, the papers presented here represent a set of high-quality contributions to the literature on frontier innovation of lignocellulose and humanosphere sciences towards sustainable development goals and climate change mitigation.

Finally, we want to thank everyone who contributed to this effort, including paper authors, session presenters, track chairs, program committee members, volunteers, and sponsors. Without their support, the event would not have been as successful.

Cibinong, Indonesia
January 2022

Muhammad Adly Rahandi Lubis

Editorial Board

Managing Editors

Contents

Forest and Environment (Forest Product Technology, Forest Management)

Response of Fertilizer Type and Urea Dosage Application to Support Sorghum Growth in Ultisols

I Made Sudiana, Azra Zahrah Nadhirah Ikhwani, Atit Kanti, Toga Pangihotan Napitupulu, Indriati Ramadhani, Arwan Sugiharto, Muhamad Amin, Ismu Purnaningsih, Idris, Arief Nurahmadiyanto, Reni Lestari, Rizki Estiningtyas, and Masaru Kobayashi

Abstract Marginal land in Indonesia is about 24 millions which is dominated by dry land and most of it is Ultisols. This soil has the potential to be planted with sorghum which is drought tolerant. However, the low nutritional content of Ultisols requires fertilization to increase the productivity. The objective of this research was to evaluate the effect of fertilizers type and urea dose on plant growth and soil biology in sorghum cultivation for production of biomass energy in soil. This experiment used completely by randomized design with 25 combinations of soil treatment on two factors—the

Authors 'Azra Zahrah Nadhirah Ikhwani, Atit Kanti, Toga Pangihotan Napitupulu' are contribution equally.

I. M. Sudiana (✉) · A. Z. N. Ikhwani · T. P. Napitupulu · A. Sugiharto · Idris
Research Center for Applied Microbiology, Research Organization for Life Sciences and Environment, National Research and Innovation Agency, Cibinong, Bogor, Indonesia
e-mail: imad003@brin.go.id

A. Kanti · I. Ramadhani
Research Center for Biosystematics and Evolution, Research Organization for Life Sciences and Environment, National Research and Innovation Agency, Cibinong, Bogor, Indonesia

M. Amin
Indonesian Cereals Research Institute (ICERI), Maros, Indonesia

I. Purnaningsih
Directorate of Scientific Collection Management, Deputy for Research and Innovation Infrastructure, National Research and Innovation Agency, Cibinong, Bogor, Indonesia

A. Nurahmadiyanto · R. Lestari
Research Center for Applied Botany, Research Organization for Life Sciences and Environment, National Research and Innovation Agency, Cibinong, Bogor, Indonesia

R. Estiningtyas
Soil Sciences, Faculty of Agriculture, Brawijaya University, Malang, Indonesia

M. Kobayashi
Graduate School of Agriculture, Kyoto University, Kyoto, Japan

M. A. Rahandi Lubis (ed.), *Proceedings of the 10th International Symposium for Sustainable Humanosphere*, Lecture Notes in Civil Engineering 255,
https://doi.org/10.1007/978-981-19-1422-5_1

type of fertilizer consisting of compost (P1), compost plus inoculant (P2), slow-release fertilizer (P3), and the three combinations (P4); and different urea doses. The agronomic observation showed the highest values which were significantly different in the P4 treatment. This treatment also gave significant results on the total population of soil bacteria and fungi, and increased the C-organic content. However, the urea dose treatment did not give significant results which also there is no positive interaction with the type of fertilizer treatment. Increased production of sorghum biomass with higher soil fertility Ultisols was achieved when biofertilizers and a small amount of synthetic fertilizers were applied.

Keywords Compost application · Fertilizer · Soil treatment · Sorghum · Urea dosage

1 Introduction

In order to achieve the sustainable development goal especially goal no 1, 2 and 13 (zero hunger, no poverty, and mitigation of climate change), we have to optimize the use of land resources particularly the marginal land which are greatly available in Indonesia. The marginal land does have a lower soil quality, lower productivity due to some of the limiting factors. The limiting factors such as topography sloping, the dominance of the parent material, the content of nutrients, organic material, moisture content, and pH unfovarobale for plant growth. Other issue is accumulation of the metal element that is poison for the plants [1]. Marginal land area in Indonesia reached 157,246,565 hectares. However, the potential of land that can be used for new agricultural is about 91,904,643 hectares, or about 58.4% only. The land that still 'sleeping' is certainly cannot be used for cultivation. However, those marginal land or often called sub-optimal land can be optimized for planting food crops, which is expected to be able to support national food security.

The sub-optimal land area in Indonesia is around 78.2% consisting of dry land, swamp land, and peat land. Dry land dominates for 77.2% of the total sub-optimal land and about 68.98% of the total dry land has the potential to be optimized for agriculture [2]. This sub-optimal dry land is divided into acid and non-acid which is dominated by acid dry land with a percentage of 84.68% and the widest type of soil is from the order Ultisols which reaches 45,794,000 ha or 44.55% of acid dry land [3, 4]. The nutrient content of Ultisols soil is low because basic washing takes place intensively, besides that the content of organic matter is low because the decomposition process runs fast and partly is carried out by erosion. Even with such characteristics, Ultisols soil has a high potential for the development of dry land agriculture. However, it does require special treatment to use Ultisols soil for crop plantations [5].

Management of dry land to be used as agricultural land has been carried out in various parts of the world, including Indonesia. Several agricultural food commodities are known to have the potential to be cultivated on sub-optimal dry land with certain treatments. One of these potential crops is sorghum which is known as a drought-resistant and grows better on marginal lands than other cereals [6]. Sorghum in Indonesia itself has the potential as one of the commodities in food diversification in fulfilling calories and protein as an alternative to rice, which also means that it can contribute to food security [7]. Apart from being food for humans, sorghum can also be used as animal feed and bioenergy [8]. However, the use of sorghum in Indonesia has not been fully carried out because of several obstacles in terms of its economic benefits.

In the development of sorghum in Ultisols soil, it is necessary to provide fertilizer to increase the growth of sorghum. The use of fertilizers could increase nutrient availability and diversity of soil microbes. Application of chemical fertilizer and compost increase the growth of soybean plants on Ultisols [9]. In addition, fertilizing with a type of slow release fertilizer allows nutrients available for a longer time, and more effective in providing nitrogen sources for optimal growth of plant [10]. Compost fertilizer is an organic fertilizer which functions to increase the cation exchange capacity (CEC) of the soil, increasing the availability of soil nutrients and enhance soil microbial activity [11]. Composting on Ultisols soil is known to increase dry biomass of sorghum stover by 1.55 times and 1.68 times with the addition of zeolite [12].

Combination application of compost, chemical fertilizer and plant growth promoting rhizobacteria (PGPR) could increase nutrient availability and stimulate plant growth effectively [13]. Application of PGPR also create better plant microbes interaction through production of various growth hormones, vitamins and various organic acids which play an important role in root growth [14]. Urea is a fertilizer containing a single N. Nitrogen is an essential macronutrient needed for large plant growth. Plants take nitrogen in the form of NH_4^+ and NO_3^- but this nitrate is immediately reduced to ammonium [15]. Addition of several types of fertilizers and urea doses are expected to improve soil biological properties and enhance plant growth. As with the study of micro-dose of N and P fertilizers on semi-arid land with acidic pH in Ethiopia showed an increase in sorghum production and productivity [16]. This study aims to determine the effect of adding fertilizer types and urea doses on the biological properties of dry land Ultisols soil and the growth of sorghum.

2 Material and Methods

2.1 *Research Site*

This research was conducted in February to May 2018 which was carried out in the Laboratory and Greenhouse of Environmental Microbiology, Microbiology, Biology Research Center of the Indonesian Institute of Sciences (LIPI), Cibinong, West Java.

2.2 *Research Materials*

The Ultisols soil and compost used was obtained from the Cibinong Botanical Garden ecology park, Indonesian Institute of Sciences (*Lembaga Ilmu Pengetahuan Indonesia*, LIPI). Plus Compost made from compost added with inoculants in a ratio of 1:1 and slow release fertilizer (SRF) obtained from Research Center for Biology—LIPI. The sorghum seeds used were 'super 2' variety developed by Indonesian Cereals Research Institute (ICERI, Maros Indonesia). Other fertilizers used were urea (Petro, Petrokimia Gresik, Indonesia), SP-36 (Petro, Petrokimia Gresik, Indonesia), and KCl (Petro, Petrokimia Gresik, Indonesia). Other materials used in this study were $K_2Cr_2O_7$ (ACS, Merck, Germany), H_2SO_4 (ISO, Merck, Germany), nutrient agar (NA) and potato dextrose agar (PDA) media (microbiology, Merck, Germany), and chloroform (ACS, Merck, Germany).

2.3 *Soil Physico-Chemical Analysis*

The Ultisols soil used before being treated with fertilizer was analyzed for its physicochemical character following the method previously [17].

2.4 *Design Treatment*

The research stages included soil preparation with compost, plus compost, slow release fertilizer (SRF) and urea fertilizer treatment, planting, maintenance, observation of plant growth, harvesting and soil analysis in the laboratory. Fertilization treatment in sorghum growing media was carried out using two factors, namely the type of fertilizer (P) and the dose of urea (U). The type of fertilizer used were 90 g compost (P1), 60 g plus compost (P2), 45 g SRF (P3), and their combination (P4) in 15 kg of soil. Treatment of compost and plus compost mixed with soil before planting. The addition of SRF was carried out one week after planting (1 WAP). Treatment doses of urea used were 25% (U1), 50% (U2), 75% (U3), and 100% (U4).

The soil without fertilizer was also observed (P0 and U0). Thus, the total treatment was 25 combinations (see Table 1). After one and four weeks of planting, this fertilizer treatment was added with half a dose of SP 36 (P_2O_5) at a dose of 1.4 g and 0.42 g KCl fertilizers. Analysis to determine the effect of treatment was carried out by agronomic observations (plant height, number of leaves and plant biomass) and soil analysis to the end of the vegetative phase (8 WAP). The data was processed by grouping it into two factors that were used (P and U) to see the effect of each of these factors (see Table 1).

Table 1 Experimental design of sorghum growing media on Ultisols soil with fertilizer treatment

Sample code	Factor	
	Type of fertilizer	Urea dose
P0U0[1,a]	P0: without fertilizer	U0: without urea
P0U1[1,b]		U1: 25% dose of urea
P0U2[1,c]		U2: 50% dose of urea
P0U3[1,d]		U3: 75% dose of urea
P0U4[1,e]		U4: 100% dose of urea
P1U0[2,a]	P1: compost	U0: without urea
P1U1[2,b]		U1: 25% dose of urea
P1U2[2,c]		U2: 50% dose of urea
P1U3[2,d]		U3: 75% dose of urea
P1U4[2,e]		U4: 100% dose of urea
P2U0[3,a]	P2: plus compost	U0: without urea
P2U1[3,b]		U1: 25% dose of urea
P2U2[3,c]		U2: 50% dose of urea
P2U3[3,d]		U3: 75% dose of urea
P2U4[3,e]		U4: 100% dose of urea
P3U0[4,a]	P3: SRF	U0: without urea
P3U1[4,b]		U1: 25% dose of urea
P3U2[4,c]		U2: 50% dose of urea
P3U3[4,d]		U3: 75% dose of urea
P3U4[4,e]		U4: 100% dose of urea
P4U0[5,a]	P4: Combination of compost, plus compost, and SRF	U0: without urea
P4U1[5,b]		U1: 25% dose of urea
P4U2[5,c]		U2: 50% dose of urea
P4U3[5,d]		U3: 75% dose of urea
P4U4[5,e]		U4: 100% dose of urea

Type of fertilizer (P) treatment measurement is denoted by numbers: 1 = P0, 2 = P1, 3 = P2, 4 = P3, and 5 = P4

Treatment of urea dose (U) measurement is denoted by letters: a = U0, b = U1, c = U2, d = U3, and e = U4

2.5 *Soil Analysis*

Soil Microbial Respiration

Respiration measurements were carried out using the Gray Wolf AdvanceSense Environment Test Meter Serial 00,255. Soil samples of 100 g were put in a plastic bag. Data was measured at one minute intervals for ten minutes with ten repetitions [18].

Bacterial and Fungi Population

The total microbial population was calculated using the serial dilution technique which was calculated through colony forming units (CFU). Bacterial population was carried out on NA and fungi on PDA media.

Microbial Biomass Carbon

The method used was Chloroform Fumigation Incubation (CFE). Soil samples that passed a 2 mm sieve were incubated for 36 h with 30 ml of alcohol-free $CHCl_3$ in desiccator. The same thing was done on the same soil subsample, but without fumigation and defumigation treatment. The amount of microbial biomass was measured by the Walkley and Black method which was then calculated by the formula:

$$\text{Microbial biomass C} = (\text{c-organic fumigation}) - (\text{c-organic non-fumigation}) \times 2.64$$

pH H_2O

Soil samples that passed a 2 mm sieve were shaken with distilled water in a bottle at a ratio of 1:5 for 30 min. The pH of soil solution was then measured with pH meter.

C-Organic

Soil sample 0.5 g that passed a 0.5 mm sieve was put into a 100 ml volumetric flask and added 5 ml of 1 N $K_2Cr_2O_7$, then shaken. A total of 7.5 ml of concentrated H_2SO_4 was added to it and then shaken and allowed to stand for 30 min. The solution is then diluted with distilled water and left overnight. The absorbance of the clear solution was measured by a spectrophotometer with a wavelength of 561 nm [19]. Calculation of C-organic in the soil is carried out using the formula:

$$\text{Organic Content}(\%) = \text{ppm curve} \times 10/50 \times \text{correction factor}$$

Data Analysis

The data that has been obtained will be tested by the F test level of 5%. If there is a treatment effect, DMRT (Duncan Multiple Range Test) is tested with a level of 5% with Genstat (version 12, VSNi, England UK).

3 Results and Discussion

The Ultisols soil used had a silty texture with very little sand and almost no clay, although most of this type of soil tends to have a clay texture [20–22], but there was also dominated by a silt texture [23]. The characteristics of this Ultisols soil have an acidic pH [20, 24, 25], although the one used in this study is slightly acidic with a value close to 6 which means it is close to a neutral pH. As is known, Ultisols soil is lacking in nutrients for plant growth [20, 26, 27]. As measured in this study, the C-organic content was low and the cation exchange capacity was quite low as well (see Table 2).

In order to improve the limitation of Ultisols, one strategy is to increase the presence of organic content in the soil. Beside increasing the availability of carbon, the organic content can also improve soil structure, increase the cation exchange capacity value and the water retention ability. Because the lack of nutrient content in the Ultisols, the improvement could be done by addition of soil fertilizer, with purpose to increase the amount of nutrients in the soil. One benefit of Ultisols is its deep cross section which make it a suitable media for plants. Therefore, the improvement of Ultisols soil fertility for sorghum growing media was attempted by applying fertilizer in this study with two treatment factors, type of fertilizer (P) and dose of urea (U). Responses were studied on agronomic observations and soil analysis.

Agronomic observation of sorghum plant height showed that there were significant differences in the treatment of fertilizer types (P) and urea dose (U) compared to

Table 2 The physical and chemical characteristics of soil in Ecopark LIPI Cibinong

Parameters	Methods/Instruments	Value
Texture:	Pipette	
Sand		9.21%
Silt		90.76%
Clay		0.01%
pH H_2O	pH meter	5.7
C-Organic	Walkley and black	1.66%
Cation exchange capacity	Methylene blue titration	11.56
Bulk density	Cylinder	1.06
Particle density	Pycnometer	2.31
Water retention	Pressure plate apparatus and caoline box	
PF1		47.82
PF2		34.36
PF2.54		20.77
PF4.2		9.41
Available water	Gravimetry	11.36
Porosity	(1-BD/PD) × 100%	54.11

their controls (P0 and U0). However, on the urea dose factor, there was no significant difference between the doses given. Combination of compost, plus compost, and SRF (P4) has the highest plant height. Organic fertilizers containing nutrients and microbes contained in plus compost, so it can increase the efficiency of nutrient extraction [28]. As for the highest U factor is the 75% urea dose (U3) (see Fig. 1).

The number of leaves did not show any significant difference between the types of fertilizers given, but it was quite significant compared to the control (P0). The largest number is still the same as in the plant height shown in P4. Meanwhile, in the urea dose treatment, there was a significant difference at the 50% (U2) dose which was the lowest value. The highest number of leaves was shown at a dose of 100% (U4) whose value was not significantly different from the control (U0) (see Fig. 2).

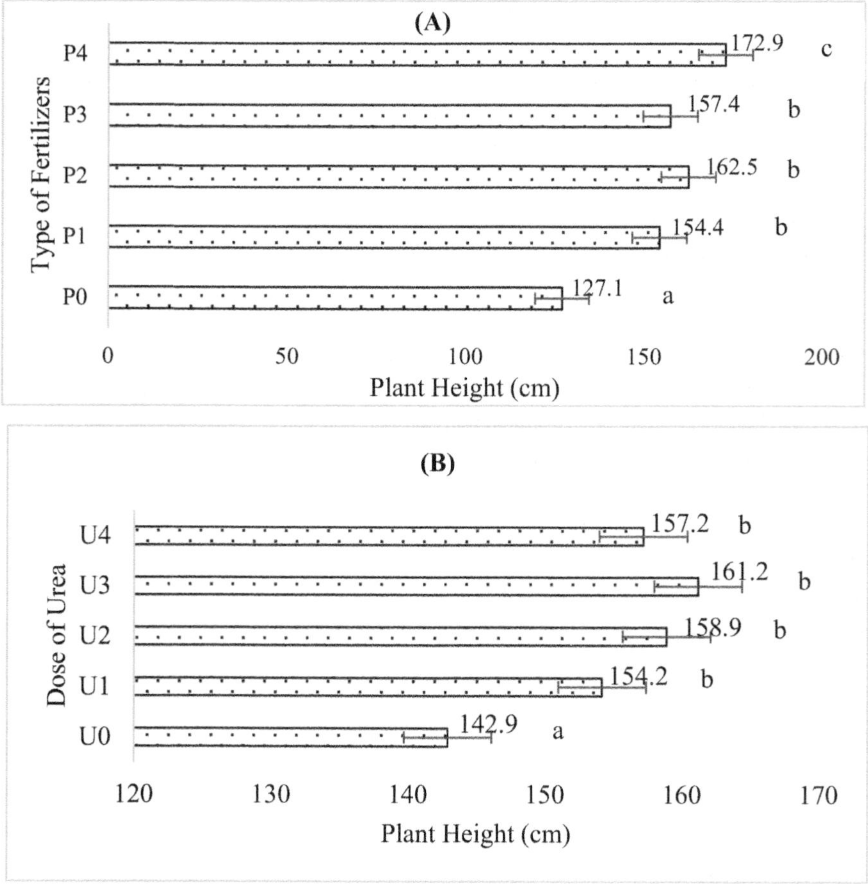

Fig. 1 Plant height at 8 weeks after planting. Figures followed by the same letters show no significant difference according to the DMRT test level of 5%; P0: without fertilizer; P1: compost fertilizer; P2; plus compost fertilizer; P3: SRF; P4: combination (compost, plus compost, SRF); U0: without urea; U1: urea dose 25%; U2: urea dose of 50%; U3: urea dose 75%; U4: urea dose 100%

Organic matter is known to increase nutrient availability for plants, especially N elements [29].

The biomass of sorghum plants measured from the wet and dry weight of the stover and root showed the same thing in the type of fertilizer treatment (P) with the two agronomic responses mentioned earlier. There is a significant difference compared to control (P0) which is the lowest value, while the highest weight is shown in P4. The urea dose treatment showed a significant difference in the highest value obtained

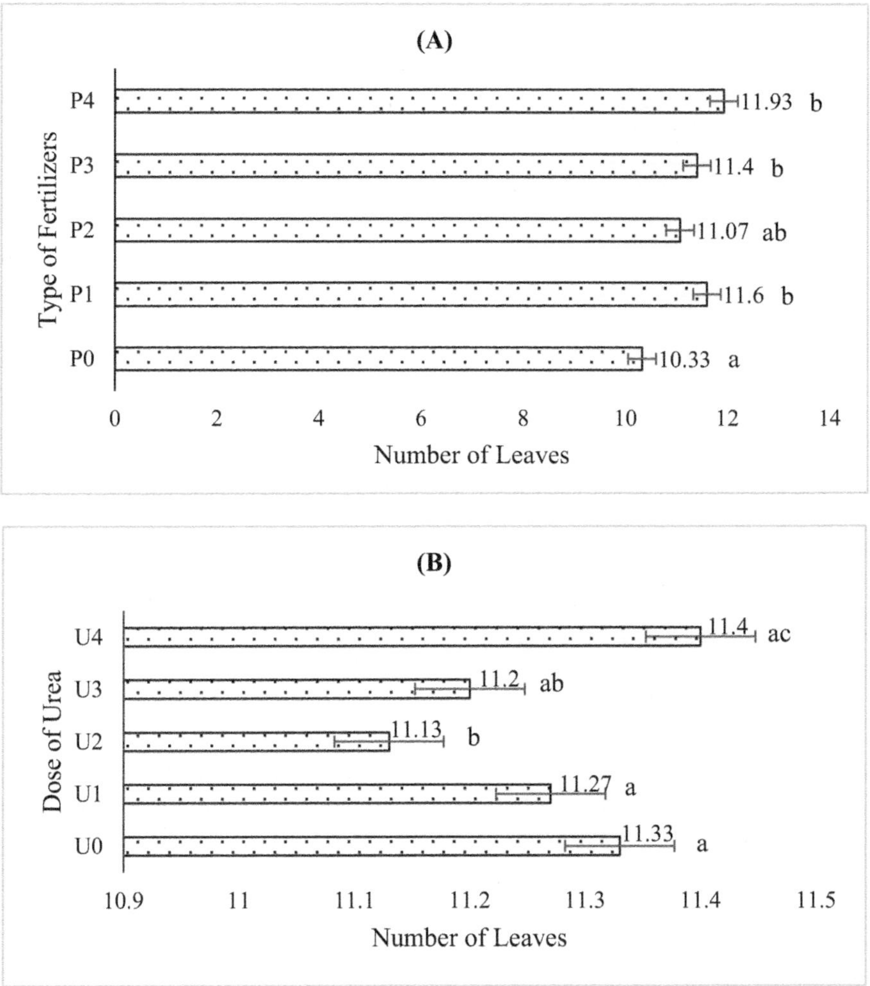

Fig. 2 Number of leaves at 8 weeks after planting. Figures followed by the same letters show no significant difference according to the DMRT test level of 5%; P0: without fertilizer; P1: compost fertilizer; P2; plus compost fertilizer; P3: SRF; P4: combination (compost, plus compost, SRF); U0: without urea; U1: urea dose 25%; U2: urea dose of 50%; U3: urea dose 75%; U4: urea dose 100%

at a dose of 25% (U1), while the other doses were not significantly different from the control (U0) where the lowest value was a dose of 50% (U2) (see Figs. 3, 4, 5, and 6).

Provision of fertilizer and dosage of urea can provide sufficient nutrients for plants, so that it can increase the weight of plant stover. Adequate nitrogen availability can

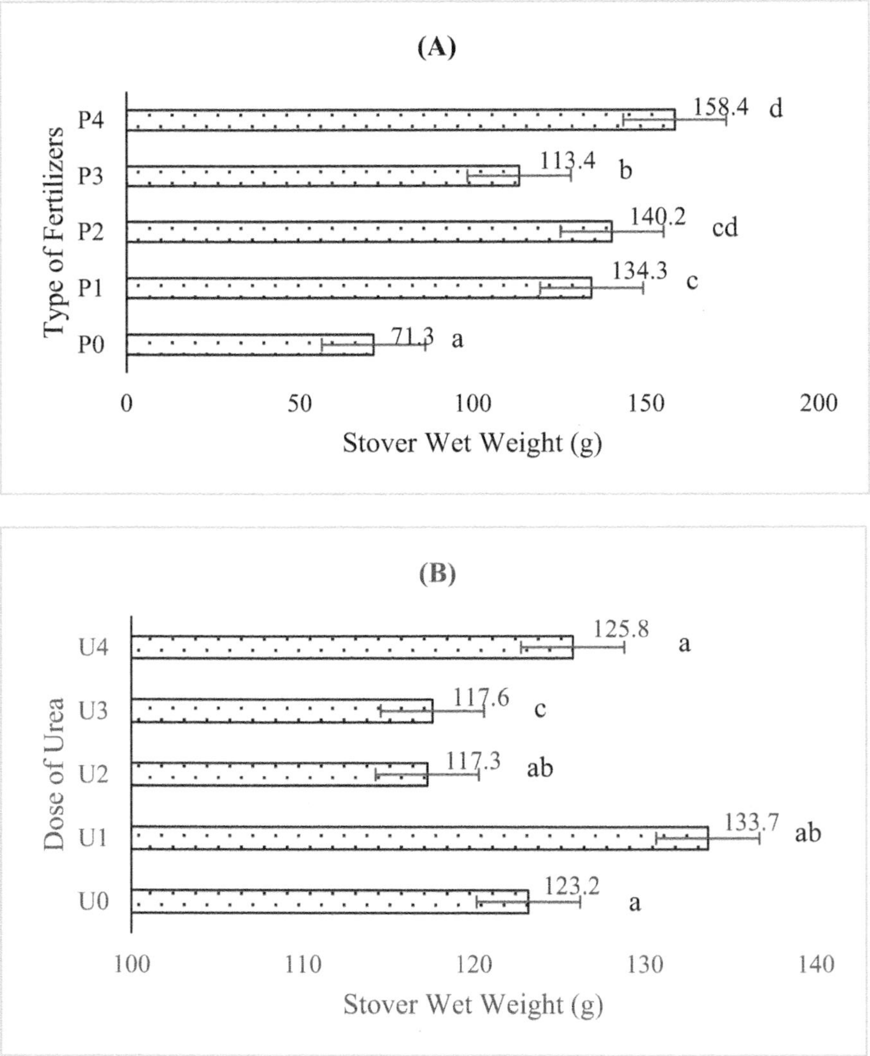

Fig. 3 Brackish wet weight in fertilizer type treatment and urea dosage. Figures followed by the same letters show no significant difference according to the DMRT test level of 5%; P0: without fertilizer; P1: compost fertilizer; P2; Plus compost fertilizer; P3: SRF; P4: combination (compost, plus compost, SRF); U0: without urea; U1: urea dose 25%; U2: urea dose of 50%; U3: urea dose 75%; U4: urea dose 100%

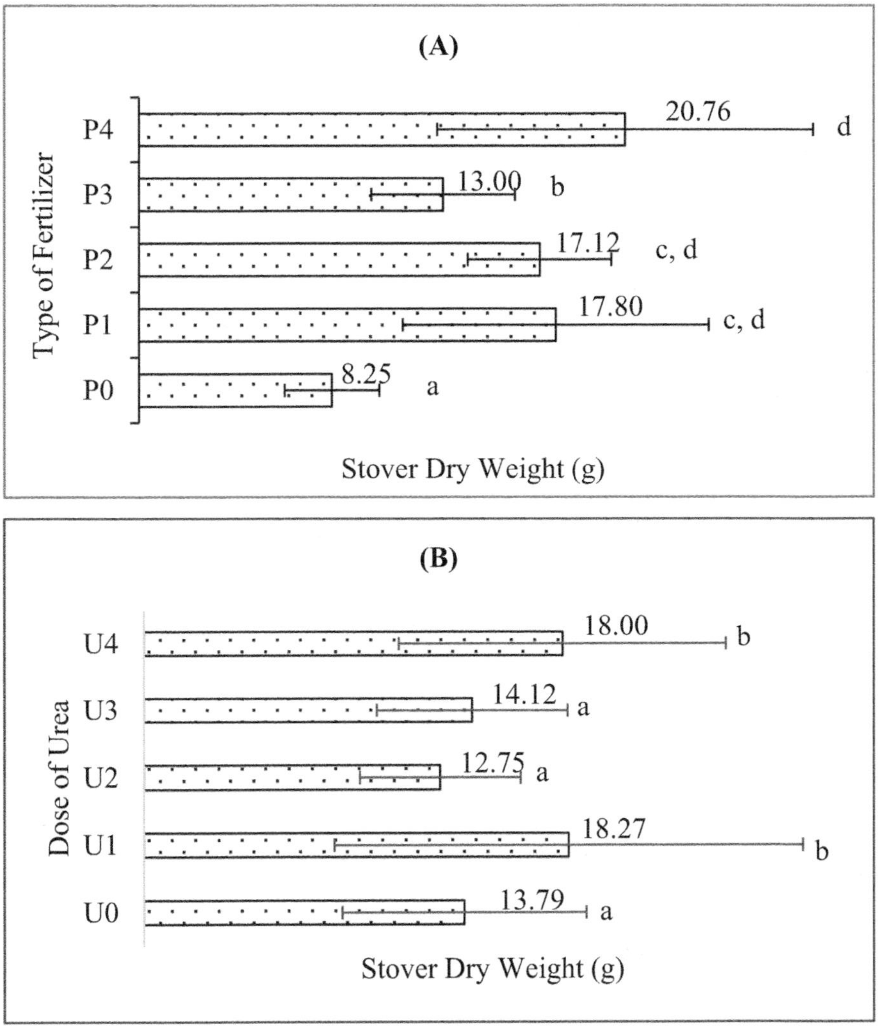

Fig. 4 Brackish dry weight in fertilizer type treatment and urea dosage. Figures followed by the same letters show no significant difference according to the DMRT test level of 5%; P0: without fertilizer; P1: compost fertilizer; P2; plus compost fertilizer; P3: SRF; P4: combination (compost, compost plus, SRF); U0: without urea; U1: urea dose 25%; U2: urea dose of 50%; U3: urea dose 75%; U4: urea dose 100%

improve vegetative growth perfectly, in this condition will affect plants to enter the generative growth phase [30]. The results in root biomass is in accordance earlier study on the use of urea fertilizer and manure that can provide an increase in total root length and root dry weight compared to controls [31]. Organic matter can reduce soil density so as to facilitate the development of roots in nutrient absorption. The

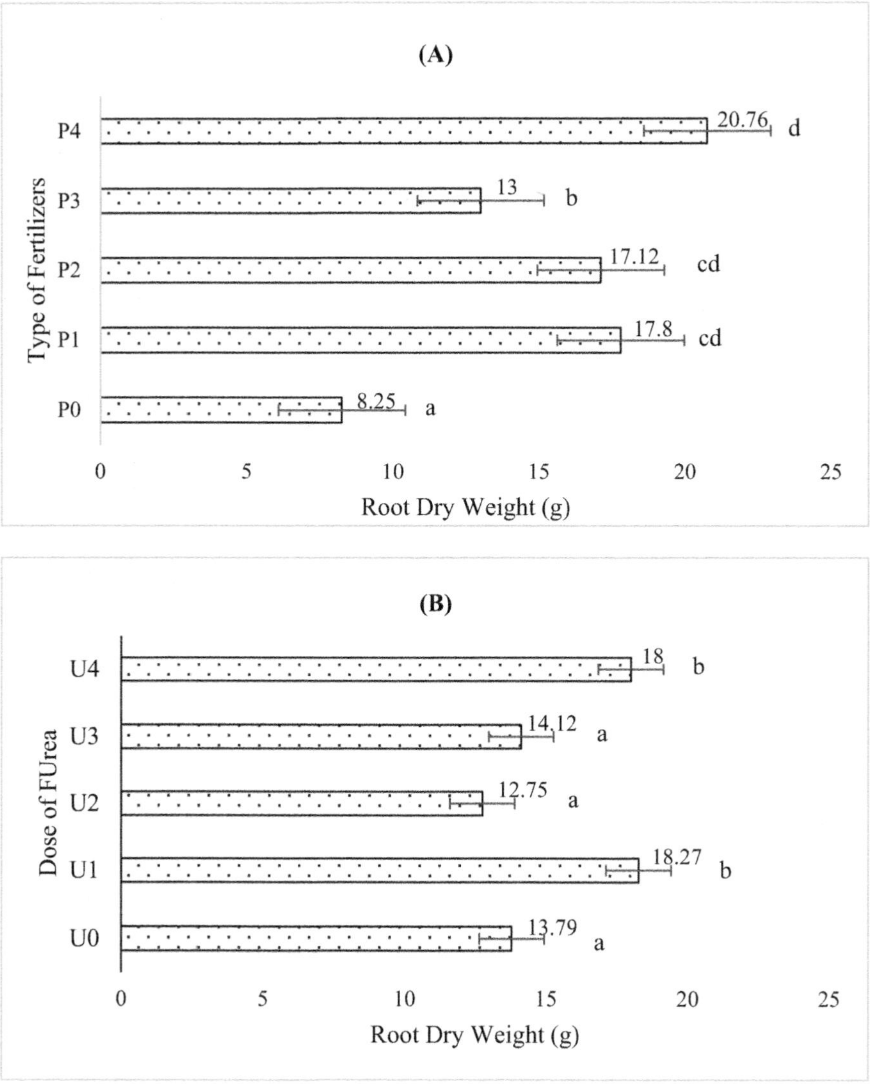

Fig. 5 Dry root weight in fertilizer type treatment and urea dosage. Figures followed by the same letters show no significant difference according to the DMRT test level of 5%; P0: without fertilizer; P1: compost fertilizer; P2; plus compost fertilizer; P3: SRF; P4: combination (compost, plus compost, SRF); U0: without urea; U1: urea dose 25%; U2: urea dose of 50%; U3: urea dose 75%; U4: urea dose 100%

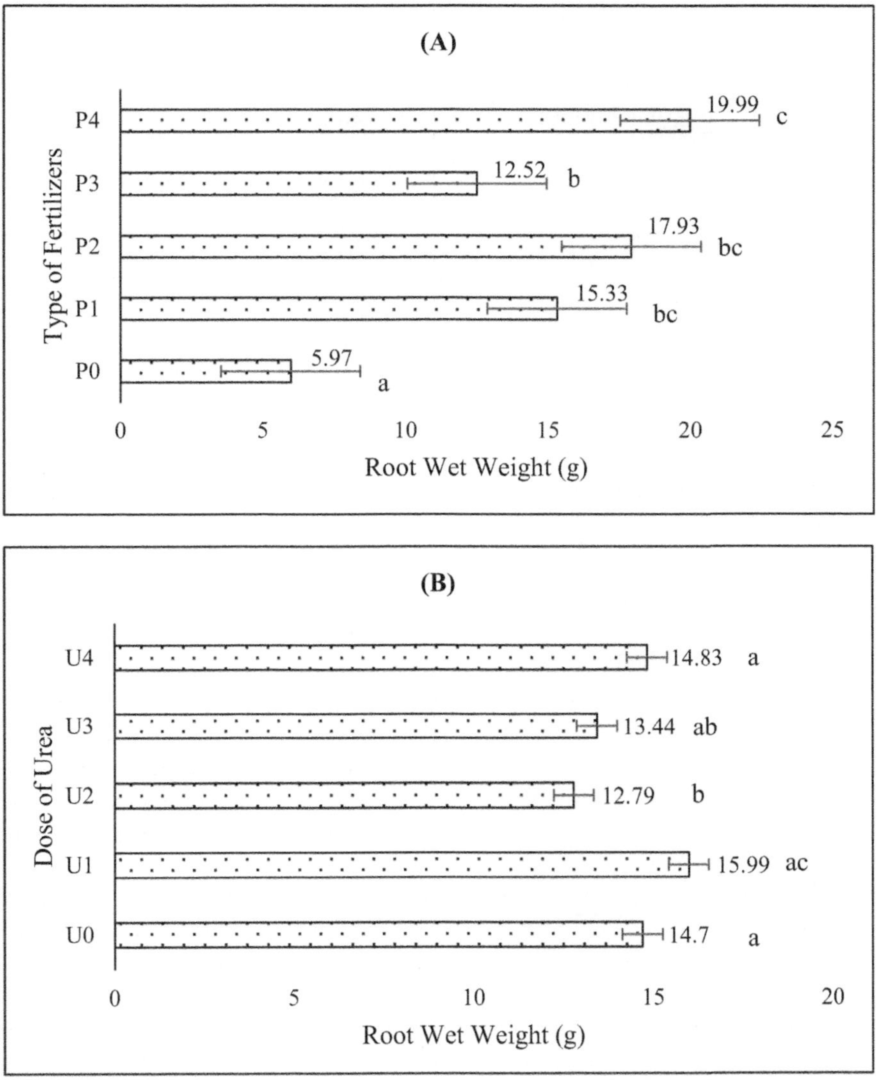

Fig. 6 Root wet weight in fertilizer type treatment and urea dosage. Figures followed by the same letters show no significant difference according to the DMRT test level of 5%; P0: without fertilizer; P1: compost fertilizer; P2; plus compost fertilizer; P3: SRF; P4: combination (compost, plus compost, SRF); U0: without urea; U1: urea dose 25%; U2: urea dose of 50%; U3: urea dose 75%; U4: urea dose 100%

rapid growth of roots will benefit plants in adhering to the soil and absorption of water and nutrients in the soil [32].

Analysis of soil treated with fertilizer as a growing medium for sorghum showed the highest soil microbial respiration which was not significantly different from the control (P0 and U0). The highest value was shown in the type of compost fertilizer (P1) and 75% urea dose (U3). Meanwhile, the lowest value is owned by SRF fertilizer (P3) and urea dose of 25% (U1) (see Fig. 7).

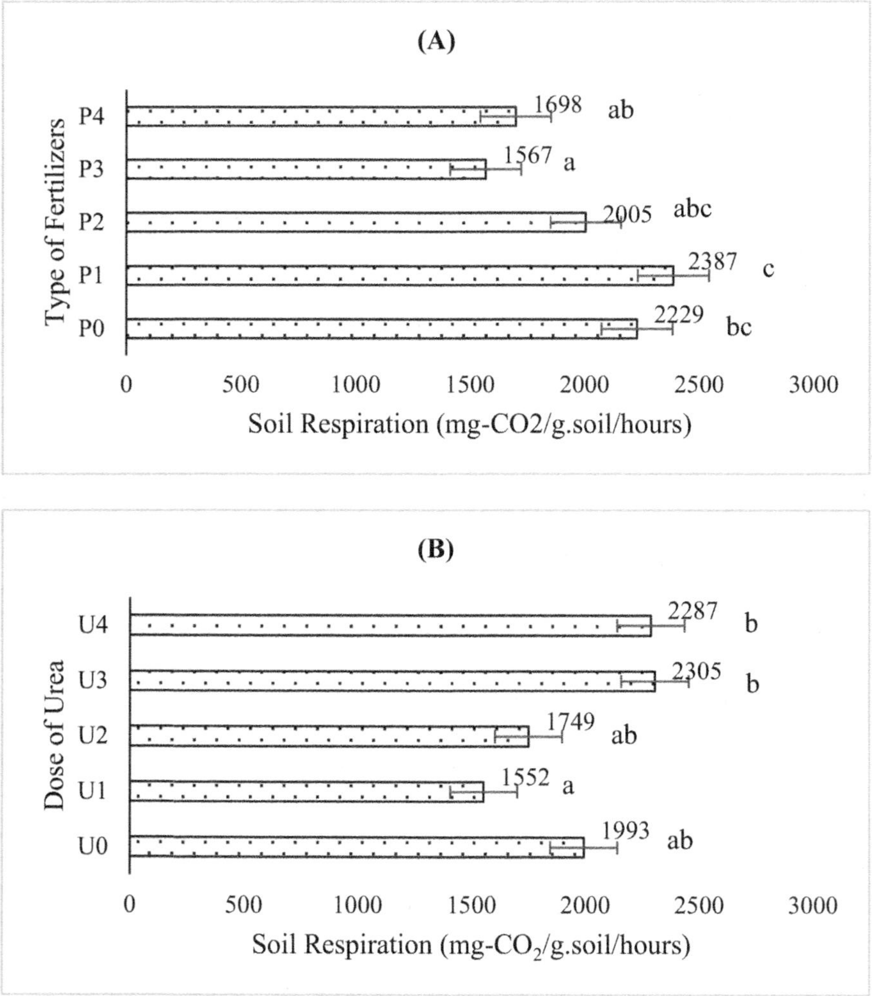

Fig. 7 Respiration in fertilizer type treatment and urea dosage. Figures followed by the same letters show no significant difference according to the DMRT test level of 5%; P0: without fertilizer; P1: compost fertilizer; P2; plus compost fertilizer; P3: SRF; P4: combination (compost, compost plus, SRF); U0: without urea; U1: urea dose 25%; U2: urea dose of 50%; U3: urea dose 75%; U4: urea dose 100%

The addition of compost containing synthetic fertilizer can increase the reproduction of soil microbes. Organic material is a food source for microbes, so microbial activity occurs in the soil. Addition of organic fertilizer has a positive effect on CO_2 release because the C compoent is more readily available [33]. In the treatment with slow release fertilizer application there was no significant increase in treatment of untreated soil. This is suspected, there is not much carbon content in slow release fertilizer. Soil microbes obtain energy from the process of reforming carbon-containing materials [34]. With the availability of sufficient energy sources, the soil microbes will be able to carry out activities optimally, which among other things increases the availability of nutrients for plants.

The type of fertilizers treatment has very significant effect on the number of bacterial and fungal populations in the soil, while the dose of urea has no significant effect. The P0 type of fertilizer treatment (without fertilizer) had the lowest bacterial population, while the P4 treatment (combination of compost, plus compost, SRF) had the highest bacterial population. The 50% dose of urea (U2) and 75% (U3) treatment had the lowest bacterial and fungal population, respectively, while the 100% urea dose (U4) treatment had the highest bacterial and fungal population (see Figs. 8 and 9).

The P4 treatment includes not only compost but also beneficial microbes and other nutrients contained in SRF. A high microbe population is suspected because of the addition of compost and inoculants. Compost has organic compounds which are food sources for microorganisms in the soil. Level of microbial sensitivity toward humidity and light intensity also will affect the microbial population present in the soil [35]. The availability of organic matter has a positive effect on soil fertility which is utilized by microorganisms in the process of decomposition of organic and inorganic compounds, including the nutrient cycle [36]. Fungi are microfauna which play a role in remodeling organic matter in nature into simple elements. Giving organic fertilizer can increase the population of bacteria and fungi, and the administration of chemical fertilizers does not have a significant impact on the population of bacteria and fungi. The application of inorganic fertilizers that are not accompanied by organic fertilizers tends to reduce the population of soil bacteria.

C microbial biomass reflects the level of organic C in the soil and shows the amount of substrate available for plant growth. Addition of fertilizer type had significant effect on C microbial biomass, while the addition of urea dose did not have a significant effect. The result of this C microbial biomass is similar to the C-organic of the soil. The treatment of P0 (without fertilizer) has the lowest C microbial biomass and C-organic, while the treatment P2 and P4 is the highest. The treatment of U0 (without urea) had the lowest of thus C, whereas the treatment of U2 (urea dose 50%) had the highest C (see Figs. 10 and 11).

C microbial biomass acts as an indicator of soil microbial activity, the greater the amount of microbial C biomass indicates that microbial activity in the soil is good. C microbial biomass is influenced by soil organic matter which is a source of nutrients for microbes in metabolic activity. Microbial biomass depends on the availability of organic matter in microbial activity [37]. Organic matter has a positive influence on soil fertility because organic matter is a food source for microorganisms

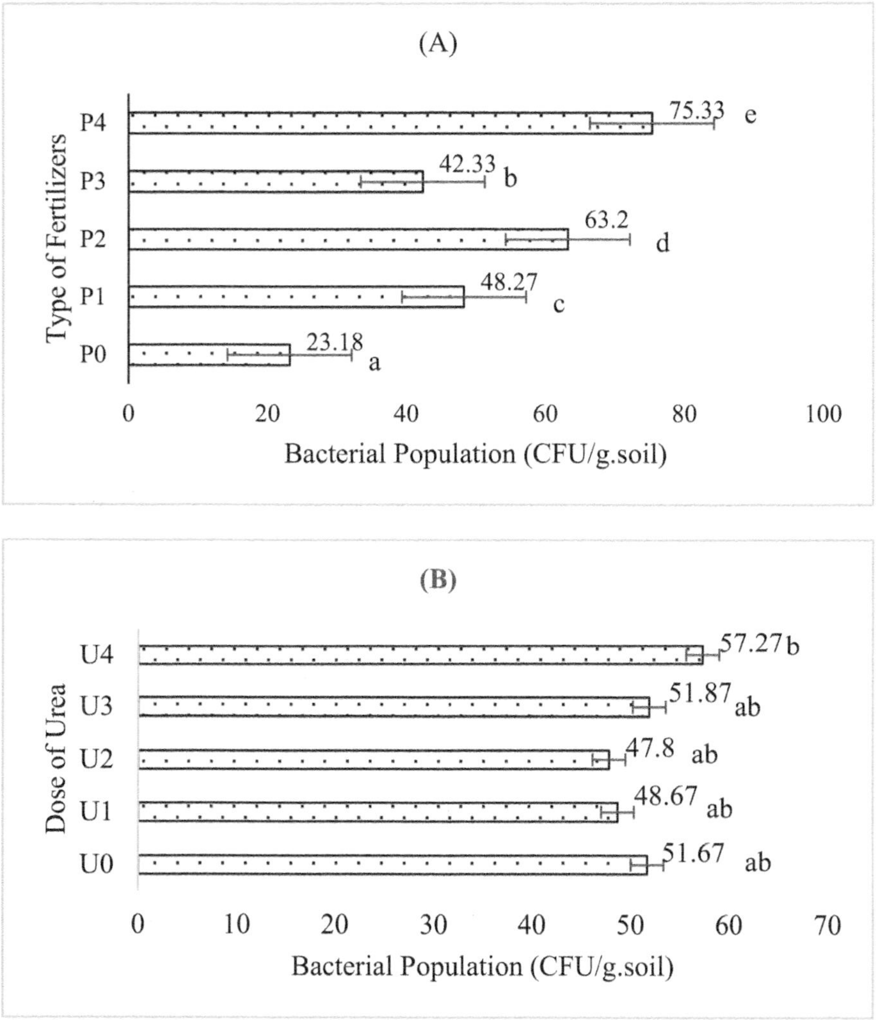

Fig. 8 Number of bacterial populations in fertilizer type treatment and urea dosage. Figures followed by the same letters show no significant difference according to the DMRT test level of 5%; P0: without fertilizer; P1: compost fertilizer; P2; plus compost fertilizer; P3: SRF; P4: combination (compost, plus compost, SRF); U0: without urea; U1: urea dose 25%; U2: urea dose of 50%; U3: urea dose 75%; U4: urea dose 100%

in the process of decomposition, nutrient cycle and decomposition of organic and inorganic compounds [38]. The provision of organic matter can increase the content of c-organic soil, so that an increase in soil organic c-can affect soil properties to be better both physically, chemically and biologically [39].

The addition of fertilizer types had a very significant effect on soil pH, while the addition of urea doses did not have a significant effect on soil pH. The interaction

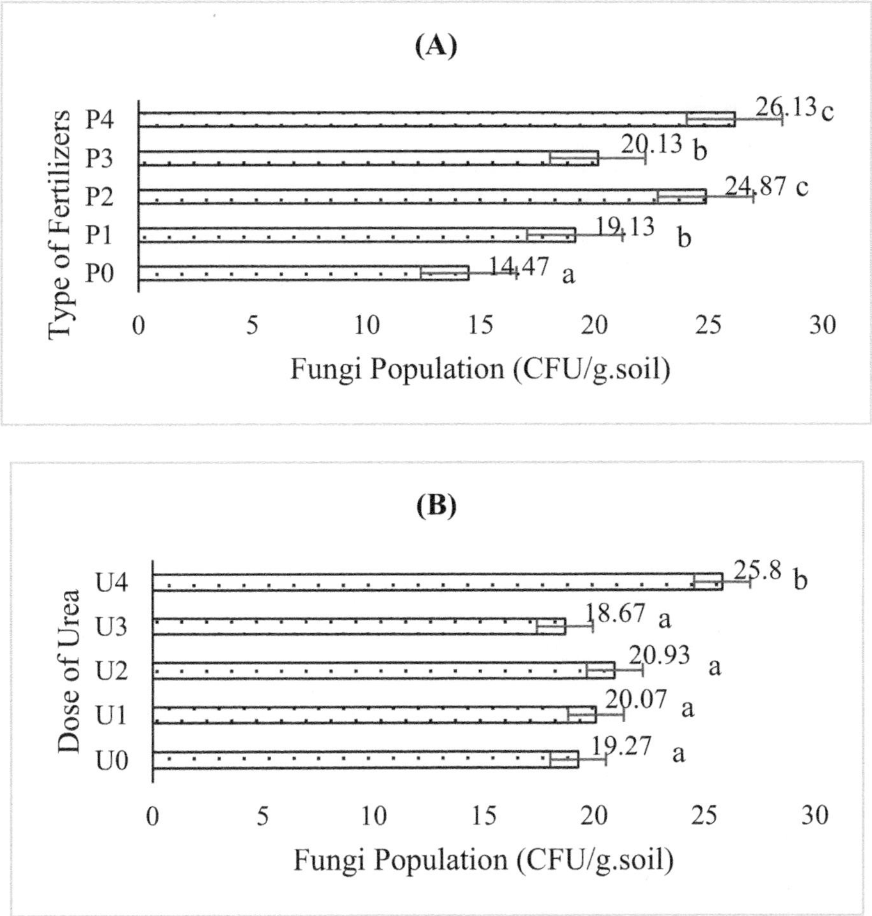

Fig. 9 Number of fungi populations in fertilizer type treatment and urea dosage. Figures followed by the same letters show no significant difference according to the DMRT test level of 5%; P0: without fertilizer; P1: compost fertilizer; P2; plus compost fertilizer; P3: SRF; P4: combination (compost, plus compost, SRF); U0: without urea; U1: urea dose 25%; U2: urea dose of 50%; U3: urea dose 75%; U4: urea dose 100%

between the two factors had a significant effect on soil pH. The treatment of P0 (without fertilizer) has the lowest soil pH, while the treatment P2 (compost plus) has the highest soil pH. The effect of compost as an addition to organic matter on soil pH can increase or decrease depending on the level of maturity of organic matter and the type of soil. The treatment of urea dosage did not have a significant effect on soil pH. The treatment of U4 (100% urea dose) has the lowest soil pH, while treatment U0 (without urea) has the highest soil pH (see Fig. 12). Organic fertilizers can improve soil fertility compared to inorganic fertilizers because in addition to the process of releasing nutrients gradually there are also other ingredients such as organic carbon,

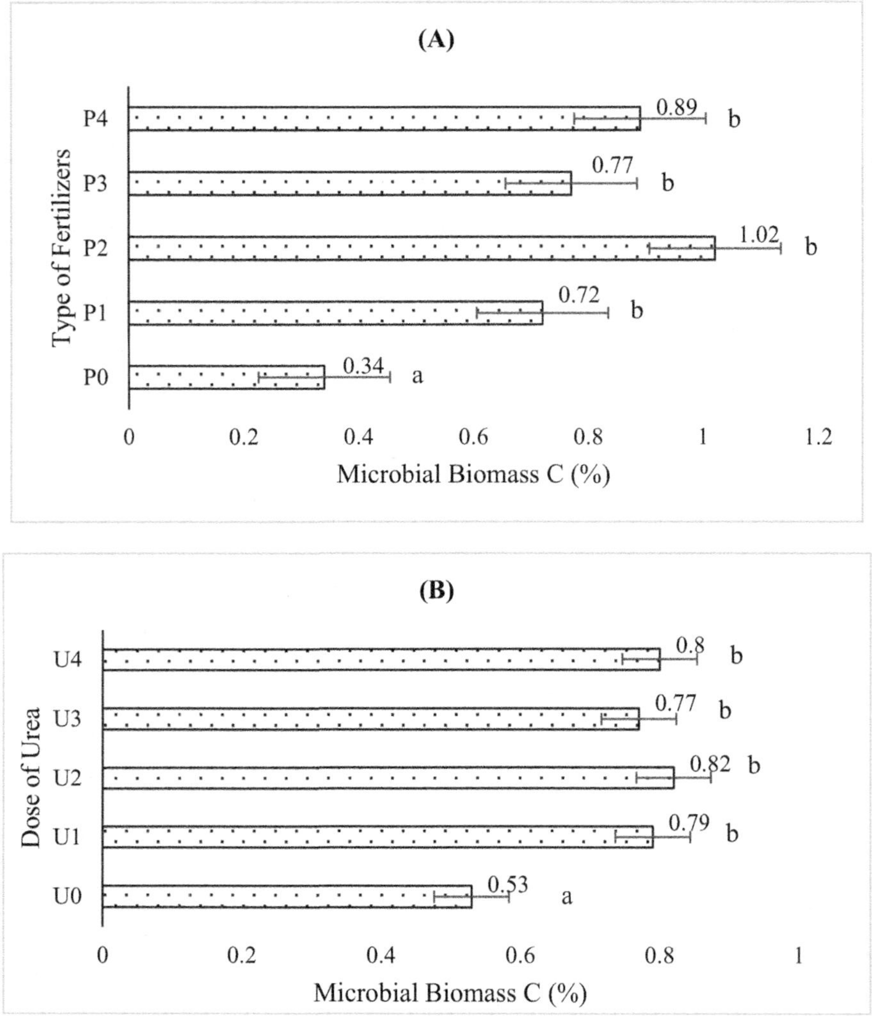

Fig. 10 Microbial biomass C in fertilizer type treatment and urea dosage. Figures followed by the same letters show no significant difference according to the DMRT test level of 5%; P0: without fertilizer; P1: compost fertilizer; P2; plus compost fertilizer; P3: SRF; P4: combination (compost, plus compost, SRF); U0: without urea; U1: urea dose 25%; U2: urea dose of 50%; U3: urea dose 75%; U4: urea dose 100%

nitrogen, phosphate, potassium, calcium which will increase the pH value of the soil [40]. A neutral pH value can make it easier for plants to absorb nutrients contained in the soil [41].

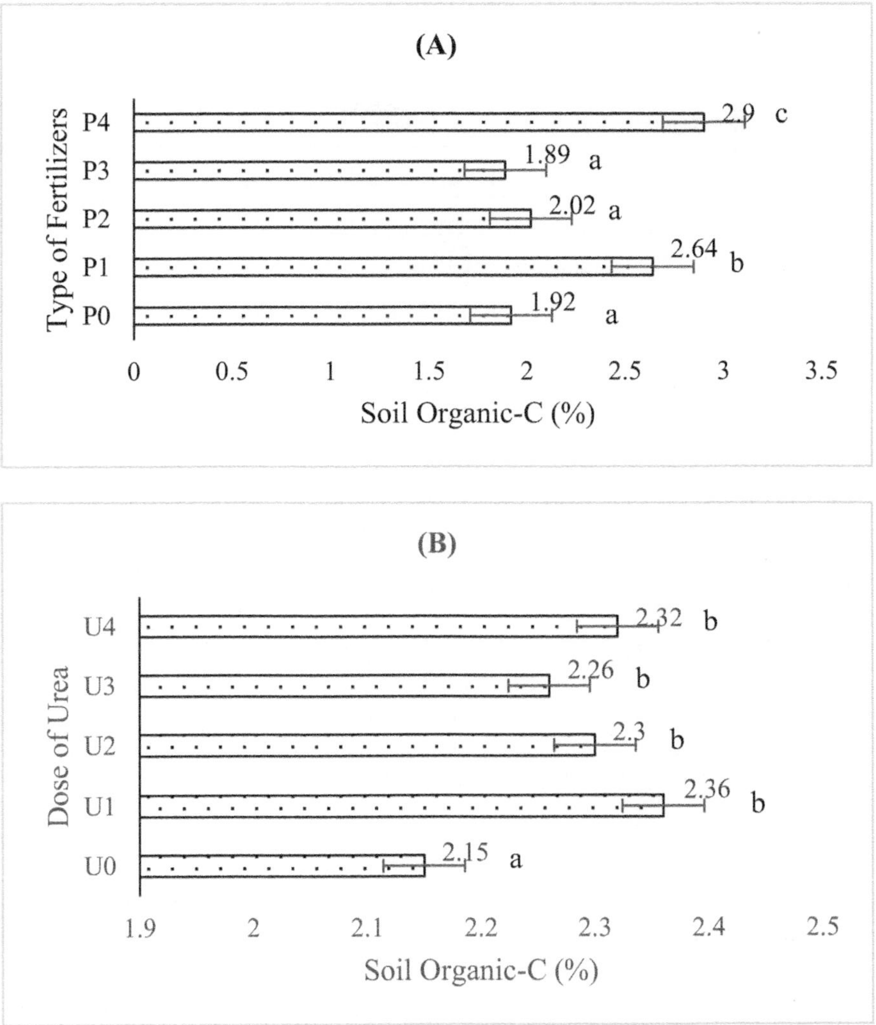

Fig. 11 C-Organic in fertilizer type treatment and urea dosage. Figures followed by the same letters show no significant difference according to the DMRT test level of 5%; P0: without fertilizer; P1: compost fertilizer; P2; plus compost fertilizer; P3: SRF; P4: combination (compost, plus compost, SRF); U0: without urea; U1: urea dose 25%; U2: urea dose of 50%; U3: urea dose 75%; U4: urea dose 100%

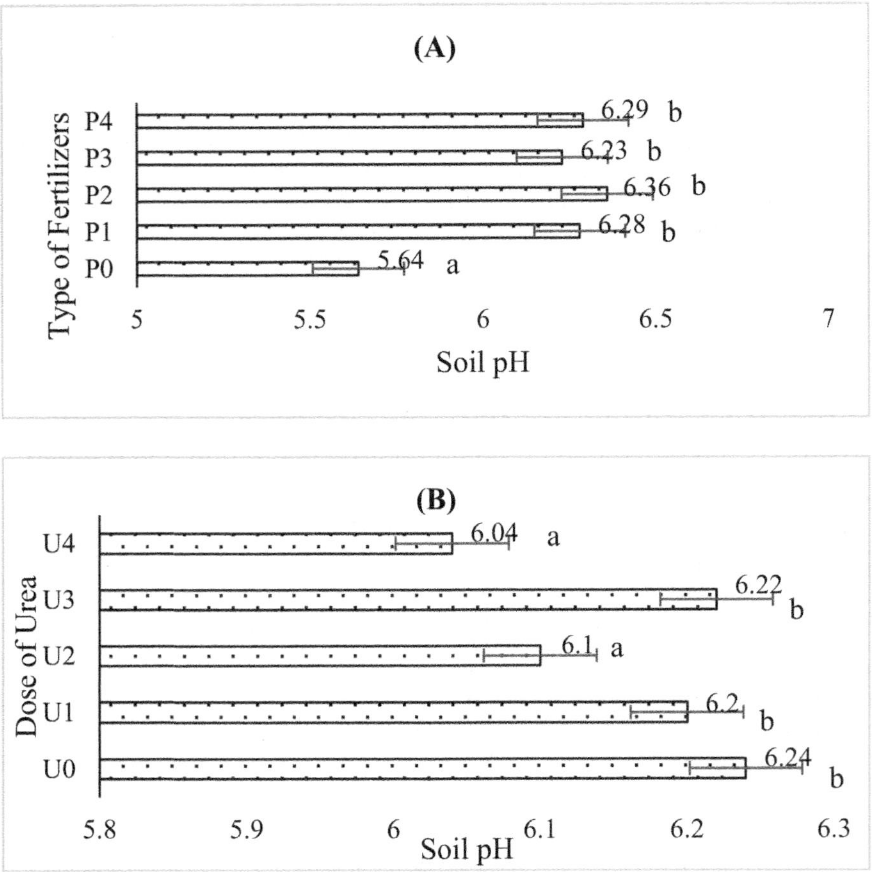

Fig. 12 Soil pH in fertilizer type treatment and urea dosage. Figures followed by the same letters show no significant difference according to the DMRT test level of 5%; P0: without fertilizer; P1: compost fertilizer; P2; plus compost fertilizer; P3: SRF; P4: combination (compost, plus compost, SRF); U0: without urea; U1: urea dose 25%; U2: urea dose of 50%; U3: urea dose 75%; U4: urea dose 100%

4 Conclusion

The application of fertilizer types and urea doses has a significant effect on increasing the growth of sorghum plants, plant biomass and soil biological properties. The significant interaction between compost, urea and microorganism was on the weight of stovers. The treatment also affect the soil microbial respiration, bacterial and fungal population. Application of fertilizer, compost and beneficial microbes are necessary to increase biomass production of sorghum, hence supporting the production of biomass energy toward achievement of sustainable development goals.

References

1. Millaleo R, Reyes-Diaz M, Ivanov A, Mora M, Alberdi M (2010) Manganese as essential and toxic element for plants: transport, accumulation and resistance mechanisms. J Soil Sci Plant Nutr 10(4):470–481. https://doi.org/10.4067/s0718-95162010000200008
2. Mulyani A, Nursyamsi D, Harnowo D (2017) Potensi dan tantangan pemanfaatan lahan suboptimal untuk tanaman aneka kacang dan umbi [Potency and challenges for the use of suboptimal land for development of legumes and tuber crops]. In: Rahmianna AA, Harnowo D, Sholihin, Nugrahaeni N, Taufiq A, Suharsono Yusnawan E, Ginting E, Hermanto FR (eds) Hasil Penelitian Tanaman Aneka Kacang dan Umbi 2016. Pusat Penelitian dan Pengembangan Tanaman Pangan, Bogor, pp 16–30
3. Mulyani A, Hikmatullah, Subagyo H (2004) Karakteristik dan potensi tanah masam lahan kering di Indonesia. In: Prosiding Simposium Nasional Pendayagunaan Tanah Masam. Pusat Penelitian dan Pengembangan Tanah dan Agroklimat, Bogor, pp 1–32
4. Subagyo H, Suharta N, Siswanto AB (2004) Tanah-tanah pertanian di Indonesia. In: Adimihardja A, Amien LI, Agus F, Djaenudin D (eds) Sumberdaya Lahan Indonesia dan Pengelolaannya. Pusat Penelitian dan Pengembangan Tanah dan Agroklimat, Bogor, pp 21–66
5. Wicke B, Sikkema R, Dornburg V, Faaij A (2011) Exploring land use changes and the role of palm oil production in Indonesia and Malaysia. Land Use Policy 28(1):193–206. https://doi.org/10.1016/j.landusepol.2010.06.001
6. Saballos A (2008) Development and utilization of sorghum as a bioenergy crop. In: Vermerris W (ed) Genetic improvement of bioenergy crops. Springer, New York, pp 211–248
7. Irawan B, Sutrisna N (2011) Prospek pengembangan sorgum di Jawa Barat mendukung diversifikasi pangan [Prospect of sorghum development in West Java to support food diversification]. Forum Penelitian Agro Ekonomi 29(2):99–113
8. Fujii K, Hartono A, Funakawa S, Uemura M, Sukartiningsih, Kosaki T (2011) Acidification of tropical forest soils derived from serpentine and sedimentary rocks in East Kalimantan, Indonesia. Geoderma 160(3–4):311–323. https://doi.org/10.1016/j.geoderma.2010.09.027.
9. Gascho GJ (2005) Soil fertility decline in the tropics: with case studies on plantations. soil science [Internet]. Ovid Technologies (Wolters Kluwer Health) 170(2):149–51. https://doi.org/10.1097/00010694-200502000-00009
10. Wu L, Jacobson AD, Chen H-C, Hausner M (2007) Characterization of elemental release during microbe–basalt interactions at T = 28 °C. Geochim Cosmochim Acta 71(9):2224–2239. https://doi.org/10.1016/j.gca.2007.02.017
11. Peng X, Zhu QH, Xie ZB, Darboux F, Holden NM (2016) The impact of manure, straw and biochar amendments on aggregation and erosion in a hillslope Ultisol. CATENA 138:30–37. https://doi.org/10.1016/j.catena.2015.11.008
12. Napitupulu TP, Sudiana IM (2019) The use of compost and zeolite on improving growth of sorghum cultivated in Ultisols. In: The 3rd SATREPS Conference on 2018. LIPI, Bogor, pp 167–172
13. Antonopoulou G, Gavala HN, Skiadas IV, Angelopoulos K, Lyberatos G (2008) Biofuels generation from sweet sorghum: fermentative hydrogen production and anaerobic digestion of the remaining biomass. Biores Technol 99(1):110–119. https://doi.org/10.1016/j.biortech.2006.11.048
14. Nunes WAGDA, Menezes JFS, Benites VDM, Lima Junior SAD, Oliveira ADS (2015) Use of organic compost produced from slaughterhouse waste as fertilizer in soybean and corn crops. Scientia Agricola 72(4):343–350. https://doi.org/10.1590/0103-9016-2014-0094
15. Ni B, Liu M, Lü S, Xie L, Wang Y (2011) Environmentally friendly slow-release nitrogen fertilizer. J Agric Food Chem 59(18):10169–10175. https://doi.org/10.1021/jf202131z
16. Sebnie W, Mengesha M, Girmay G, Feyisa T, Asgedom B, Beza G, Dejene D (2020) Evaluation of micro-dosing fertilizer application on sorghum (*Sorghum bicholor* L) production at Wag-Lasta Areas of Amhara Region Ethiopia. Scientific Reports 10:6889. https://doi.org/10.1038/s41598-020-63851-6

17. Jackson ML (2005) Soil chemical analysis: advanced course, 2nd edn. Parallel Press, Wisconsin
18. Rahmansyah M, Suliasih (2016) Karakter respirasi dan mineralisasi karbon organik pada sampel tanah dikoleksi dari Pulau Bangka [Respiration and organic carbon mineralization character in soil samples collected from Bangka Island]. Berita Biologi 15(1):27–37
19. Eviati S (2009) Analisis kimia tanah, tanaman, air, dan pupuk. Petunjuk teknis, 2nd edn. Balai Penelitian Tanah, Bogor
20. West LT, Beinroth FH, Sumner ME, Kang BT (1998) Ultisols: characteristics and impacts on society. Adv Agron 63:179–236
21. Yulnafatmawita, A (2014) Physical characteristics of Ultisols and the impact on soil loss during soybean (*Glycine* max Merr) cultivation in a wet tropical area. Agrivita 36(1):57–64
22. Purwanto S, Gani RA, Suryani E (2020) Characteristics of Ultisols derived from basaltic andesite materials and their association with old volcanic landforms in Indonesia. Sains Tanah—Journal of Soil Science and Agroclimatology 17(2):135–143
23. Prasetyo BH, Suharta N, Subagyo H, Hikmatullah (2001) Chemical and mineralogical properties of Ultisols of Sasamba Area, East Kalimantan. Indonesian J Agric Sci 2(2):37–47
24. Ifansyah H (2013) Soil pH and solubility of aluminum, iron, and phosphorus in Ultisols: the roles of humic acid. J Tropic Soils 18(3):203–208
25. Li J, Masud MM, Li Z, Xu R (2015) Amelioration of an Ultisol profile acidity using crop straws combined with alkaline slag. Environ Sci Pollut Res 22(13):9965–9975. https://doi.org/10.1007/s11356-015-4176-5
26. Prasetyo BH, Subardja D, Kaslan B (2005) Ultisols bahan volkan andesitik: Diferensiasi potensi kesuburan dan pengelolaannya. Jurnal Tanah dan Iklim 23:1–12
27. Taisa R, Maulida D, Salam A, Niswati A (2019) Improvement of soil chemical properties and growth of maize due to biochar application on Ultisol. J Tropic Soils 24(3):99–109
28. Zhong W, Gu T, Wang W, Zhang B, Lin X, Huang Q, Shen W (2009) The effects of mineral fertilizer and organic manure on soil microbial community and diversity. Plant Soil 326(1–2):511–522. https://doi.org/10.1007/s11104-009-9988-y
29. Jones C, Engel R, Olson-Rutz K (2019) Soil acidification in the semiarid regions of North America's Great Plains. Crops & Soils 52(2):28–56. https://doi.org/10.2134/cs2019.52.0211
30. Tan XW, Ikeda H, Oda M (2000) The absorption, translocation, and assimilation of urea, nitrate or ammonium in tomato plants at different plant growth stages in hydroponic culture. Sci Hortic 84(3–4):275–283. https://doi.org/10.1016/s0304-4238(99)00108-9
31. Limon-Ortega A, Govaerts B, Sayre KD (2008) Straw management, crop rotation, and nitrogen source effect on wheat grain yield and nitrogen use efficiency. Eur J Agron 29(1):21–28. https://doi.org/10.1016/j.eja.2008.01.008
32. Smith SE, Smith FA (2011) Roles of arbuscular mycorrhizas in plant nutrition and growth: new paradigms from cellular to ecosystem scales. Annu Rev Plant Biol 62(1):227–250. https://doi.org/10.1146/annurev-arplant-042110-103846
33. Pérez-Piqueres A, Edel-Hermann V, Alabouvette C, Steinberg C (2006) Response of soil microbial communities to compost amendments. Soil Biol Biochem 38(3):460–470. https://doi.org/10.1016/j.soilbio.2005.05.025
34. Bolan NS, Kunhikrishnan A, Choppala GK, Thangarajan R, Chung JW (2012) Stabilization of carbon in composts and biochars in relation to carbon sequestration and soil fertility. Sci Total Environ 424:264–270. https://doi.org/10.1016/j.scitotenv.2012.02.061
35. Ward BB (2002) How many species of prokaryotes are there? Proc Natl Acad Sci 99(16):10234–10236. https://doi.org/10.1073/pnas.162359199
36. Tkacz A, Cheema J, Chandra G, Grant A, Poole PS (2015) Stability and succession of the rhizosphere microbiota depends upon plant type and soil composition. ISME J 9(11):2349–2359. https://doi.org/10.1038/ismej.2015.41
37. Kemmitt SJ, Lanyon CV, Waite IS, Wen Q, Addiscott TM, Bird NRA, O'Donnel AG, Brookes PC (2008) Mineralization of native soil organic matter is not regulated by the size, activity or composition of the soil microbial biomass—a new perspective. Soil Biol Biochem 40(1):61–73. https://doi.org/10.1016/j.soilbio.2007.06.021

38. Chen G, Zhu H, Zhang Y (2003) Soil microbial activities and carbon and nitrogen fixation. Res Microbiol 154(6):393–398. https://doi.org/10.1016/s0923-2508(03)00082-2
39. Berg G (2009) Plant–microbe interactions promoting plant growth and health: perspectives for controlled use of microorganisms in agriculture. Appl Microbiol Biotechnol 84(1):11–18. https://doi.org/10.1007/s00253-009-2092-7
40. Wu SC, Cao ZH, Li ZG, Cheung KC, Wong MH (2005) Effects of biofertilizer containing N-fixer, P and K solubilizers and AM fungi on maize growth: a greenhouse trial. Geoderma 125(1–2):155–166. https://doi.org/10.1016/j.geoderma.2004.07.003
41. Ding Y, Liu Y, Liu S, Li Z, Tan X, Huang X, Zeng G, Zhou L, Zheng B (2016) Biochar to improve soil fertility: a review. Agronomy Sustain Dev 36:36. https://doi.org/10.1007/s13593-016-0372-z

Enhanced Adhesion and Reduced Formaldehyde Emission of Urea-Formaldehyde Resin Bonded Plywood Panel Using Intercalated Montmorillonite and Laponite Nanoclay

Apri Heri Iswanto, Muhammad Adly Rahandi Lubis, and Sumit Manohar Yadav

Abstract The crystal formation in urea–formaldehyde (UF) resin of low formaldehyde/urea (F/U) mole ratio generates a poor adhesion at the expense of low formaldehyde emission. Therefore, this study investigated the way to prevent the crystallization in cured UF resin by mixing 5% of nanoclay during the addition reaction of resin's synthesis. Montmorillonite (MN) and laponite (LN) were used as the nanoclay. X-ray diffraction (XRD) and Fourier Transform Infrared (FTIR) results showed that the chemical structure of pristine nanoclay-UF resins was still crystal, while modified MN-UF resin was partially amorphous. In addition, modified LN-UF resin was still crystal, similar with that of pristine nanoclay-UF resin. The neat UF resin had a crystallinity of about 54.1% after curing. But, after the addition of nanoclay during resin's synthesis, the crystallinity of cured UF resin remarkably decreased to 29.3% using modified MN and 47.5% using modified LN. This eventually improved the adhesion and reduced formaldehyde emission of plywood bonded with UF resin. These results showed that the prevention of crystallization in UF resin could be done by mixing 5% modified MN during the synthesis, which subsequently improved the adhesion strength by 33.6% (1.37 MPa) and lowered the formaldehyde emission by 56.1% (0.73 mg/L). The modified MN was intercalated in UF resin polymers during synthesis and eventually partially blocked the crystal formation in cured UF resins. This could facilitate the formation of three-dimensional cross-linking structures in

A. H. Iswanto
Department of Forest Product, Faculty of Forestry, Universitas Sumatera Utara, Medan, Indonesia

M. A. R. Lubis (✉)
Research Center for Biomass and Bioproducts, National Research and Innovation Agency, Cibinong, Indonesia
e-mail: marl@biomaterial.lipi.go.id

S. M. Yadav
Department of Forest Products and Utilization, Forest College and Research Institute, Hyderabad, India

M. A. Rahandi Lubis (ed.), *Proceedings of the 10th International Symposium for Sustainable Humanosphere*, Lecture Notes in Civil Engineering 255,
https://doi.org/10.1007/978-981-19-1422-5_2

cured UF resin and led to a better adhesion strength and reduced formaldehyde emission of plywood bonded modified UF resin.

Keywords Adhesion strength · Formaldehyde emission · Nanoclay · Plywood · UF resin adhesive

1 Introduction

Urea–formaldehyde (UF) resin is dominantly used as wood adhesive in the production of wood-based panels such as plywood, particleboard and medium density fiberboard [1]. UF resin possesses numerous industrial desirable features like low pressing time for curing, high reactivity, clear glue line, and comparatively low cost [2]. However, there is also some limitations such as formaldehyde emission (FE) owing to its lower water resistance, which must be tackling for continuing industrial application of UF resin [3, 4]. The main cause for the FE and low moisture resistance of UF resin bonded wood based panels is the reversibility of the amino-methylene linkages in UF resin over a time [3–5].

A way for minimizing FE of UF resin bonded panels is by lowering the formaldehyde to urea (F/U) molar ratio for the synthesis of UF resin. But, UF resin with low F/U mole ratio has slower curing rate and low adhesion strength in wood-based panels bonded with UF resin. In addition, the existence of crystal structure in lo molar ratio UF resin has been reported, showing when the F/U mole ratio was lowered from 1.6 to 1.0, UF resin exhibited the crystallinity [6, 7]. The crystallinity of the cured UF resin is depended on the F/U mole ratio and the mixture of minimum energy in its molecular forms [8]. A study found the crystalline region in colloidal particles of UF resin, revealing the colloidal aggregates were built and finally changed to super clusters by coalescence during aging [9]. It was also reported that low F/U mole ratio cured UF resin possessed greater amount of hydrogen bonds thus resulted in a large amount of closely packed spherical structures than those of high F/U mole ratio [10]. As lowering the F/U mole ratio of UF resin generally results in low adhesion strength, wood adhesive scientist found that the responsible chemical species for adhesion to cellulose are hydroxymethyl oligomers. The hydroxymethyl content decreases by lowering the F/U mole ratio and therefore has a negative effect on adhesion strength.

The application of UF resin bonded wood based panels could be extended with a significant reduction in the FE in line with a drastic enhancement in the durability and hydrolytic stability of UF resin. At the present time, utilization of nanotechnology has opened a new opportunity to develop new adhesives with improved properties. A simultaneous improvement of different properties of material can be achieved by incorporating nanoparticles into a polymer matrix [11, 12]. The development of nanoparticles-reinforced adhesive materials is presently one of the most investigated areas in materials science and engineering [13]. The exceptional properties of nanoparticles have led to widespread research in this area. For example, the adhesion strength of plywood bonded with UF resin can be enhanced with addition of nanoclay in UF resin, and further reduction in the FE of plywood panels has also been reported [14, 15]. Nanoclay, a nanoparticle with a structure of stacked platelets and one dimension of the platelet in the nano-meter scale, has a high aspect ratio and specific surface when intercalated or exfoliated. If nanoclay is dispersed properly, its nano-size can provide a significant amount of interface between the clay and the resin and may restrict the formation of crystals in UF resin [14, 15].

However, utilization of layered silicate in nanoclay can be effective when the layers are intercalated or exfoliated instead aggregated. Pristine monmorillonite (MN) and laponite (LN) nanoclay are two kinds of nanoclay that have high cation exchange capacity (CEC), up to 150 and 140 mg equivalent/100 g, respectively [16, 17]. High CEC provides greater ion exchange of nanoclay with metal ions for the modification of layered silicate, allowing modification of the interlayer space to achieve better compatibility with the UF resin polymer. This process separates the nanoclay platelets and reduces the sizes of nanoclay for intercalation or exfoliation, and also enables the nanoclay to be more compatible with UF resin polymer [18].

The main objective of this work was to enhance the adhesion strength and reduce FE of UF resin bonded plywood by intercalating either pristine or modified MN and LN during the synthesis of the resin. In this study, ion exchange method using Copper (II) chloride has been used to improve the compatibility of pristine MN and LN for the modification of UF resin during synthesis. The influence of pristine and modified MN and LN addition during the UF resin synthesis and their effect on the performance of modified UF resin bonded plywood panels were also investigated. X-ray diffraction (XRD) and Fourier Transform Infrared (FTIR) analysis were used to study the structural and chemical properties of modified nanoclay and modified UF resins. Three-ply plywood panels were prepared from modified UF resins for the evaluation of adhesion strength and FE according to Japanese Agriculture Standard (JAS) No. 233 [19].

2 Experimental Section

2.1 *Materials*

Pristine montmorillonite (MN) and laponite (LN) nanoclay were supplied by BYK (BYK-Chemie GmbH, Wesel, Germany). Copper (II) chloride ($CuCl_2$) and Sodium chloride (NaCl) were analytical grade purchased from Merck (Merck, Darmstadt, Germany). The synthesis of UF resins was done by using technical grade of urea granules (99%) and formalin (37%). The pH level was adjusted using aqueous solutions of both formic acid (20%) and sodium hydroxide (20%) during the resin's synthesis. Sengon (*Paraserianthes falcataria* (L.) Nielsen) wood veneers, sizing 250 × 250 × 2 mm, was bought from local market (Cibinong, West Java, Indonesia) and was used to fabricate plywood.

2.2 Modification of Pristine Nanoclay

Copper (II) chloride was used as a transition metal ion (TMI) to modify the pristine nanoclay [18]. Initially, the pristine nanoclay was washed by water before TMI treatment. Approximately 80 g of pristine nanoclay was placed in 1600 ml water and the suspension was vigorously stirred for 24 h at temperature of 27 ± 2 °C. The dispersion turned into viscous slurry due to swelling of nanoclay, upon stirring for a few hours. The slurry was kept in 1 ml of NaCl solution for 48 h, to remove the exchangeable ion contaminations. The slurry was then filtered using vacuum pump and dried in an oven for 12 h at temperature of 80 °C. Using same solvent, the dried clay was placed in 0.30 M of TMI solutions for the ion exchange process. The suspension was stirred for 36 h at temperature of 27 ± 2 °C. After the TMI treatment, the slurry was filtered and dried for 12 h at 80 °C. The dried clay then powdered using a mortar and pestle.

2.3 Intercalation of Nanoclay During the Synthesis of UF Resin

In this study, low molar ratio UF resins with 1.0 F/U mole ratio was in-situ modified using MN and LN by following alkaline-acid two-step reaction (Fig. 1). Briefly, formaldehyde solution was placed in the glass reactor, and set up the temperature control of the heating mantle at 90 °C. As the temperature increases, the pH was adjusted to 8.0 using aqueous solutions of sodium hydroxide. When the temperature reached at 40 °C, urea-1 was poured into the reactor. After reaching 90 °C, approximately 5% nanoclay based on the theoretical solids content of UF resin was dispersed into the reactor [15, 16, 19, 20]. The temperature then was controlled at 90 °C for 60 min to allow the methylolation reaction. The next step of UF resins synthesis consists of the condensation reaction. The formic acid was added to bring the pH down to 4.6–5.0, and the condensation reaction was carried out until it reached a target viscosity of around 250 mPa.s. The urea-2 then was added into the reactor and kept for 30 min to dissolve. After that, the reaction was cooled to room temperature (27 ± 2 °C) and the pH was adjusted to 8.0. UF resin then was stored in a plastic bottle and kept for 24 h prior to characterization and application. As a control, neat UF resin was synthesized using similar method without addition of nanoclay during the synthesis.

2.4 Characterization of Nanoclay and Modified UF Resin

The non-volatile solids content of neat and modified UF resin was determined by measuring the weight of the resin before and after drying. Approximately 1 g of

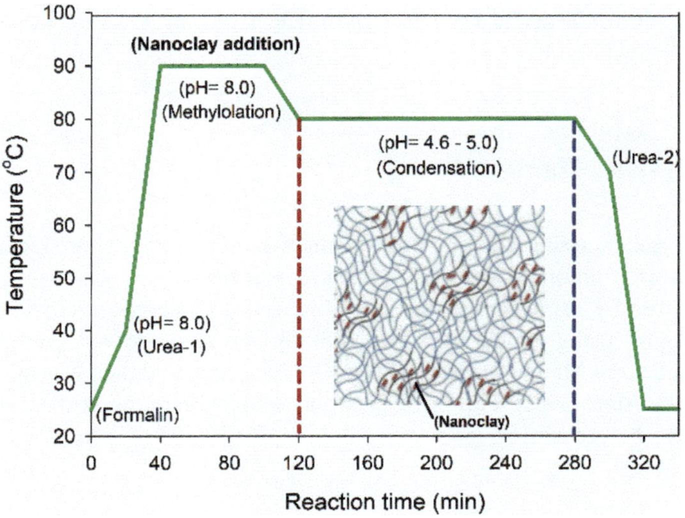

Fig. 1 Partially intercalated of nanoclay during the synthesis of UF resin

modified UF resin was poured in disposable aluminium foil, and then dried at 105 °C in an oven for 3 h. The measurement was conducted in triplicate.

The gelation time of neat and modified UF resin was measured by adding 3% of ammonium chloride hardener (based on UF resin solids content) in about 10 g of resin sample. The sample then was heated using gel time meter (Techne GT-6, Coleparmer, United States) at 100 °C. The measurement was carried out with three replications.

A rotational rheometer (Rheolab QC, Anton Paar, Austria) was used to measure the viscosity of neat and modified UF resin. The viscosity of the resin was measured at 25 °C using a cyclone cylinder type spindle No. 27 at 60 rpm. The viscosity measurement was conducted in triplicate.

Prior to pH measurement of neat and modified UF resin, the digital pH meter (Orion Star™ A211, ThermoFisher Scientific, MA, United States) was calibrated using buffer solution having pH 4.0 and 10.0. Around 50 g of resin was poured into a 100 ml beaker for the pH measurement.

Functional groups nanoclay, neat UF and modified UF resin was investigated using an UATR–FTIR spectrometer (Spectrum Two, Perkin Elmer, United States) in the range of 400–4000 cm^{-1} with a resolution of 2 cm^{-1}. An average of 16 scans were used for scanning the background and sample. The spectrum of neat and modified UF resin was then normalized using a min–max normalization.

In this study, X-Ray Diffractometer (XRD, Rigaku Miniflex II, Tokyo, Japan) is used to analyse the structure of nanoclay, neat UF and modified UF resin. Before analysis, samples were powdered using a grinder and dried in an oven at 105 °C for 3 h. The XRD patterns nanoclay, neat and modified UF resin were obtained at room temperature and λ of 0.15406 radiation. The diffractograms were obtained in the range of 1–50° of 2θ and at a step of 1°/min. The d-spacing was calculated according to Bragg's law [11, 18].

2.5 Plywood Manufacturing

Plywood panels, sizing 250 mm × 250 mm × 6 mm, were prepared using Sengon wood veneer. Ammonium chloride (3% based on resin solids content) was mixed as a hardener with UF resins before glue spreading. The glue spread of UF resin adhesive was 170 g/m^2 using double glue line. Initially, the plywood was cold-pressed for 20 min under 0.8 MPa of pressure at room temperature, and then was hot pressed at 120 °C under 0.8 MPa for 4 min. The plywood samples were conditioned at 27 ± 2 °C for 2 weeks before cutting and testing.

2.6 Evaluation of Plywood Properties

Adhesion strength of UF resin bonded plywood was determined by means of tensile shear strength (TSS) according to JAS No. 233 [19]. Approximately twelve specimens were cut into 25 mm × 80 mm × 6 mm. Dry and wet TSS values of the specimens were measured using a universal testing machine (UTM 50 kN, Shimadzu, Japan) at a crosshead speed of 2 mm/min. The wood failure analysis was performed using the TSS samples right after the analysis.

The FE values were determined by placing ten plywood samples (150 mm × 50 mm × 6 mm) together with 300 ml of distilled water in a desiccator at 20 ± 2 °C for 24 h. After that, 25 mL of the solution was pipetted into 100-mL flasks and mixed with 25 mL acetyl acetone–acetic acid ammonium solutions. The mixtures were heated for 10 min at 65 ± 2 °C in a water bath and then cooled to room temperature. The FE was measured using UV–visible spectrophotometry (UV-1800, Shimadzu, Japan) at a wavelength of 412 nm.

3 Results and Discussion

3.1 Characteristics of Nanoclay

XRD technique was employed to investigate the layer confirmation of pristine and modified nanoclay. The nanoclay structure may be identified by observing the position and intensity of the basal reflections from the distributed silicate layers. Figure 2 shows the XRD patterns of pristine MN and LN and their modification. Theoretically, the diffraction peak (asterisk *) of pristine nanoclay appears at 2θ of 7.86° for MN and 7.28° for LN. The *d*-spacing of MN and LN was 1.14 nm and 1.21 nm. In contrast, after the modification of pristine nanoclay with transition metal ion, the *d*-spacing values has increased. The XRD data obtained for the modified nanoclay showed that the diffraction peak (asterisk) moved towards left. After modification, the diffraction peak of modified MN shifted to 2θ of 5.02° with increased in *d*-spacing value from 1.14 to 1.76 nm, while the 2θ of modified LN shifted to 6.90° with increased in *d*-spacing value from 1.21 to 1.28 nm. An increase in *d*-spacing of modified nanoclay indicates that the copper ions (Cu^{2+}) have been intercalated into the interlayer of pristine nanoclay, and the higher *d*-spacing of modified nanoclay might support the exfoliation or intercalation the nanoclay structure during the synthesis of UF resins. Several published work reported a similar observation, showing that the modification of nanoclay using ion exchange method could enhance the *d*-spacing [14, 15, 18, 21]. This study showed that the modified MN has greater *d*-spacing value compared to modified LN. This could be due to that the MN has higher cation exchange capacity (CEC) than the LN, therefore the MN have more cations for ion exchange with Cu^{2+}. The MN has up to 150 mg equivalent/100 g of CEC while the LN has around 140 mg equivalent/100 g of [16, 17].

Functional groups analysis of pristine and modified nanoclay was done using FTIR spectroscopy. FTIR spectra of pristine and modified nanoclay are displayed in Fig. 3. Pristine nanoclay has a typical peak at 3640 cm^{-1} and 1625 cm^{-1}, which were assigned to O–H of Si–OH or Al–OH of nanoclay and their intra-molecular hydrogen bonds [14, 15]. After the modification, several new peaks appeared in the modified nanoclay, particularly in modified MN. These peaks were detected at 2920 cm^{-1}, 2830 cm^{-1} and 1468 cm^{-1} were assigned to the C–H stretching, $-CH_3$ vibration, and $-CH_2-$ vibration of modified MN. No particular alteration of functional groups was

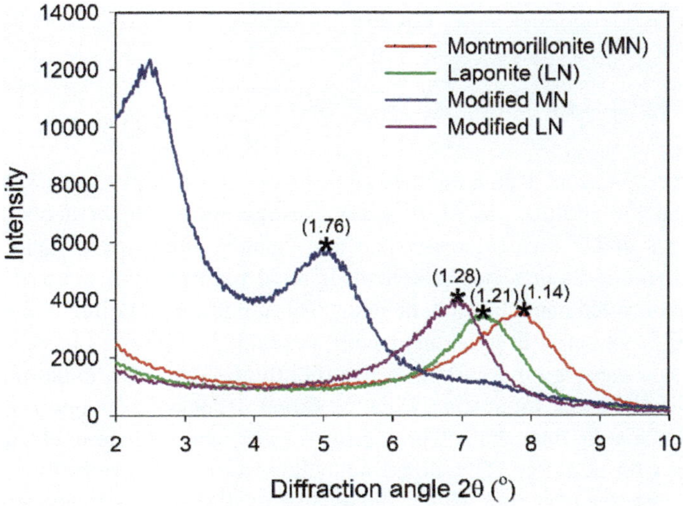

Fig. 2 XRD diffractograms of pristine and modified nanoclay

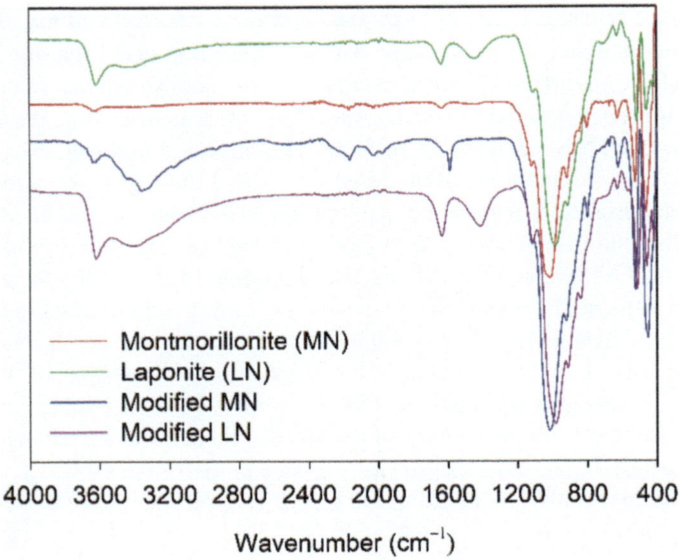

Fig. 3 FTIR spectra of pristine and modified nanoclay

observed in modified LN, showing that the modification did not significantly change the functional groups. This results supported the XRD diffractograms that showed the modified MN has greater *d*-spacing value compared to modified LN. This could be due to that the MN has higher CEC than the LN, therefore the MN have more cations for ion exchange with Cu^{2+}. The MN has up to 150 mg equivalent/100 g of CEC while the LN has around 140 mg equivalent/100 g of [16, 17].

3.2 Properties of Modified UF Resin

The properties of neat UF and modified UF resins are presented in Table 1. The values showed that the addition of 5% of either pristine and modified nanoclay during the synthesis of UF resin increased the non-volatile solids content and viscosity. Increased in non-volatile solids content of UF resin might possibly due to the physical interaction between nanoclay and the resin. The solids content of uF resin increased by 4.0–5.0%, which is in agreement with the published work [14, 15]. Nanoclay is known as a rheology modifier, thus incorporation of nanoclay could increase the viscosity of UF resin. Interestingly, the modified nanoclay increased the viscosity of UF resin higher than that of pristine nanoclay. The viscosity increased up to 100 mPa·s with addition of modified nanoclay, wihle it was only 20 mPa·s with addition of pristine nanoclay. This indicated that modification of nanoclay enhanced the *d*-spacing which expanded the nanoclay interlayer gallery for intercalation during the synthesis of UF resin. The intercalation of nanoclay eventually increased the viscosity of UF resin [18, 21]. By contrast, the pristine nanoclay just dispersed in UF resin during the synthesis. This could result in aggregation of pristine nanoclay after the synthesis of UF resin, thus the viscosity did not increase remarkably. Gelation time of UF resin was slower with addition of pristine nanoclay compared with that of modified nanoclay during the synthesis. Gelation time increased by around 100 s with incorporation of pristine nanoclay, while it was reduced by 30 s with addition of modified nanoclay. As mentioned, modification of nanoclay enhanced the *d*-spacing which swelled the nanoclay for intercalation during the synthesis of UF resin. The intercalation of nanoclay eventually reduced the gelation time of UF resin [18, 21]. By contrast, the pristine nanoclay just dispersed in UF resin during the synthesis. This could result in aggregation of pristine nanoclay in UF resin, which could increase the gelation time.

The FTIR spectra of neat and modified UF resin with pristine and modified nanoclay are shown in Fig. 4. FTIR band assignment was interpreted according to published works [14, 15]. In general, no additional of new peak was observed in the spectra, indicating that nanoclay and UF resin were interacted via physical reaction. The absorption bands at 3325 cm^{-1}, 2940 cm^{-1}, and 1645 cm^{-1} were attributed to the N–H of primary aliphatic amines, C–H stretching, and C=O of primary amides, respectively, which were detected in the neat and modified UF resin. The hydrogen bonds occurred from the reaction between C=O and N–H in UF resin, causing the C–N of the secondary amide bond at 1530 cm^{-1} of neat UF to have a lower vibration frequency in the FTIR spectra. Similar observation was found in published works [14, 15]. The addition of either pristine or modified nanoclay might have hindered the formation of hydrogen bonds in UF resin, resulting in a higher C–N bond frequency, from 1530 to 1540 cm^{-1}. The splitting of hydrogen bonds might occur due to the intercalation or exfoliation of nanoclay in UF resin polymer during the synthesis. Intercalated nanoclay extended the polymer chains and inserted UF resin polymer into the gallery space between silicate layers [20, 22].

The XRD diffractometer was used to analyse any changes in the crystalline structure modified UF resin (Fig. 5). The XRD analysis results showed that the crystalline patterns of neat UF resin had four crystalline peaks (asterisk, $*^1$, $*^2$, $*^3$ and $*^4$) at 21.6°, 24.0°, 31.1° and 40.3°, respectively. Incorporation of pristine nanoclay during the synthesis of UF resin did not altered the crystalline structure of the resin, while addition of modified nanoclay changed the structure particularly with addition of

Table 1 Properties of neat UF resin and modified UF resin with montmorillonite and laponite

Type of nanoclay	Non-volatile solids content (%)	Gelation time (s)	Viscosity (mPa·s)	pH
Neat UF	60.6 ± 0.8	220.0 ± 5.0	251.4 ± 3.0	7.9
MN-UF	65.1 ± 0.2	300.0 ± 4.0	274.1 ± 7.0	7.9
LN-UF	65.4 ± 0.3	310.0 ± 6.0	259.9 ± 9.0	7.9
Modified MN-UF	64.2 ± 0.9	190.0 ± 5.0	371.6 ± 5.0	7.9
Modified LN-UF	64.3 ± 0.6	200.0 ± 4.0	354.7 ± 6.0	7.9

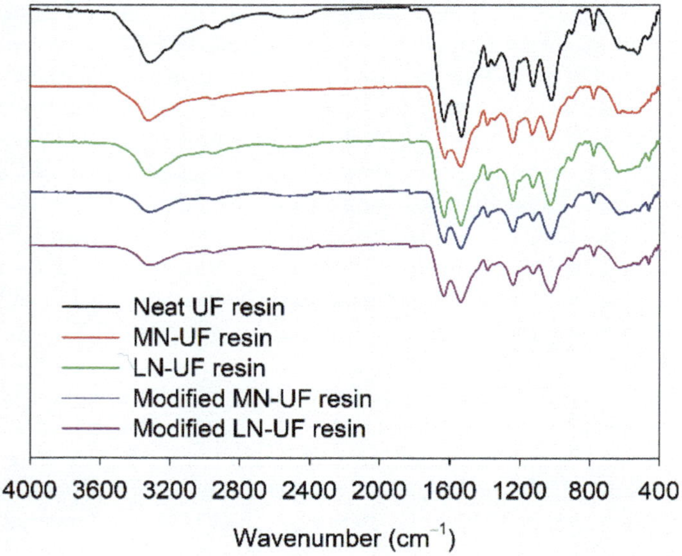

Fig. 4 Typical FTIR spectra of neat and modified UF resin mixed with nanoclay

5% modified MN. In particular, the new peaks that occurred at a lower 2θ range of 3.8°~6.2°, indicating the presence of nanoclay. This resulted in an increase in the interlayer gallery of *d*-spacing of nanoclay in the modified UF resins. Modification of nanoclay further increased the *d*-spacing of nanoclay in the modified UF resins. The modified MN-UF resin has a *d*-spacing value of 2.32 nm, which increased by 18.4% from 1.96 nm. By contrast, the modified LN-UF resin has a *d*-spacing value of 1.61 nm, that increased by 9.5% from 1.47 nm. This indicated that the interlayer gallery had been expanded due to the intercalation of nanoclay in the modified UF resin during the synthesis. Published works reposted similar observation on modification of UF resin using TMI and octadecylamine (ODA) modified nanoclay [14, 15].

Deconvolution method using a Gaussian curve-fitting function was performed to quantify the crystallinity of modified UF resin (Fig. 6a). The typical fit-fitting curve showed that UF resin had four crystalline peaks and one amorphous peak. The Gaussian curve-fitting function helped the calculation of crystallinity by dividing total area under the crystalline regions with total area of crystalline and amorphous [14, 15]. The result showed that neat UF resin has crystallinity of around 54.1%. Addition of pristine nanoclay during the synthesis of UF resin slightly decreased the crystallinity to 51.8% for MN-UF resin and 52.0% for LN-UF resin (Fig. 6b). No significant reduction in crystallinity of UF resin with addition of pristine nanoclay. This was probably due to the interlayer gallery of pristine nanoclay still closely pack together, therefore the intercalation of UF resin polymer into the interlayer gallery of nanoclay did not occur. By contrast, addition of modified nanoclay remarkably decreased the crystallinity of UF resin, particularly by adding modified MN during

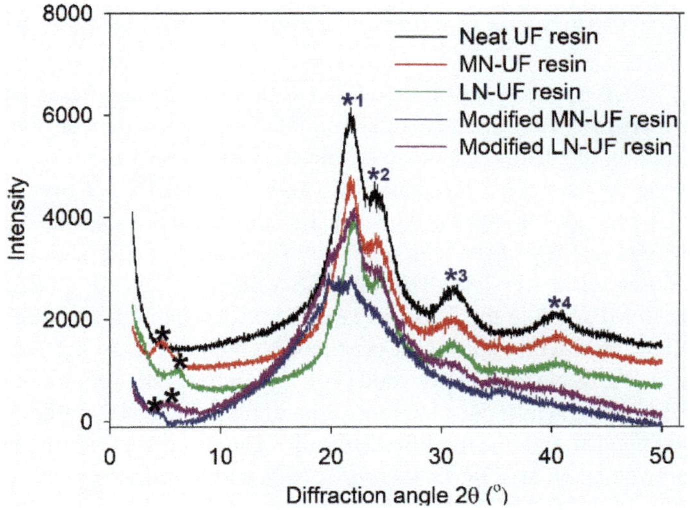

Fig. 5 XRD analysis of neat and modified UF resin mixed with nanoclay

the synthesis of the resin. The crystallinity of modified UF resin went down to 29.3% with addition of modified MN, while it was only decreased to 47.5% with addition of modified LN. As mentioned, the modified MN has greater *d*-spacing value compared to the modified LN. The MN has higher CEC than that of LN, therefore the MN have more cations for ion exchange with Cu^{2+}. The MN has up to 150 mg equivalent/100 g of CEC while the LN has around 140 mg equivalent/100 g of [18, 19].

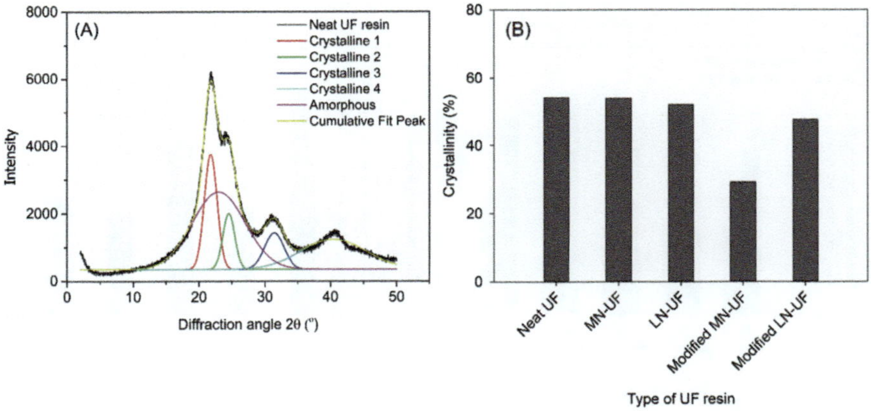

Fig. 6 Crystallinity measurement of neat and modified UF resin mixed with nanoclay. **a** typical deconvolution method using Gaussian Function, **b** crystallinity of the resin

3.3 Performance of Plywood

Properties of UF bonded plywood modified with nanoclay are presented in Fig. 7. Adhesion strength of plywood was determined by means of tensile shear strength (TSS). Addition of pristine nanoclay during the synthesis of UF resin just slightly enhanced the dry and wet TSS values by 1.0–5.0%. Dry TSS of plywood bonded with neat UF resin was 0.98 MPa, while the TSS increased to 1.04 MPa with addition of pristine MN and 0.99 MPa with addition of pristine LN (Fig. 7a). Similar trend was also observed in wet TSS. The wet TSS value of plywood bonded with neat UF was 0.50 MPa, while the TSS increased to 0.53 MPa with addition of pristine MN and 0.51 MPa with addition of pristine LN. None of them met the minimum requirement of JAS No. 233 of plywood [19]. By contrast, incorporation of modified nanoclay during the synthesis of UF resin remarkably enhanced the adhesion strength of plywood bonded with the modified UF resin. The result showed that the dry TSS value of plywood increased by 33.6% to 1.32 MPa with addition of modified MN, and increased by 18.3% to 1.17 MPa with addition of modified LN. Similar observation was found in wet TSS. The wet TSS value of plywood bonded with modified MN-UF resin was 0.79 MPa and around 0.71 MPa with addition of modified LN. This remarkable improvement was probably due to lower crystallinity of modified MN-UF resin compared to others (Fig. 6b). It was stated that the modified MN has greater *d*-spacing value compared to the modified LN. The MN has higher CEC than that of LN, therefore the MN have more cations for ion exchange with Cu^{2+}. The MN has up to 150 mg equivalent/100 g of CEC while the LN has around 140 mg equivalent/100 g of [18, 19]. This result was in agreement with the published work that reported an improvement in adhesion strength of UF-bonded plywood modified with TMI and ODA modified nanoclay [14, 15].

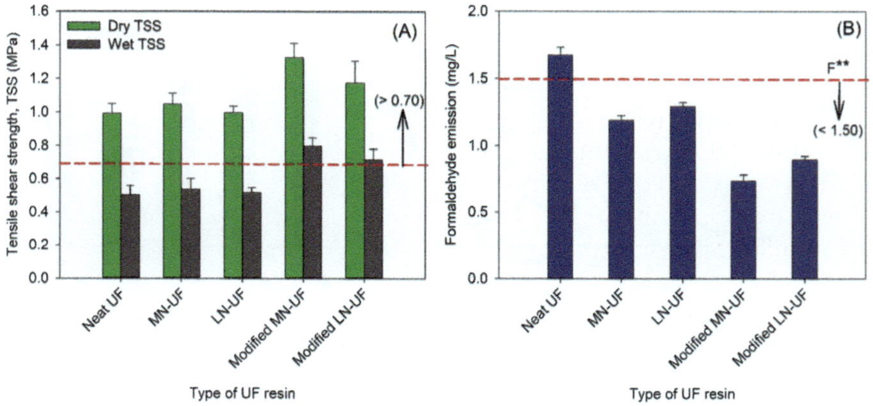

Fig. 7 Properties of plywood bonded with UF resin modified using nanoclay. **a** tensile shear strength, **b** formaldehyde emission

Formaldehyde emission of plywood panel was examined using desiccator method according to JAS No. 233 for plywood [19]. The result showed that plywood bonded with neat UF resin emitted around 1.67 mg/L of FE, indicating the plywood was classified to the F* class according to JAS No. 233 [19]. Addition of either pristine and modified nanoclay during the synthesis of UF resin reduced the FE from plywood panel remarkably. All plywood panel bonded with modified UF resin were classified to the F** class according to JAS No. 233 [19]. Addition of pristine MN during the synthesis of UF resin reduced the FE by 29.2% to 1.19 mg/L and around 1.29 mg/L with addition of pristine LN. By contrast, incorporation of modified nanoclay during the synthesis of UF resin further reduced the FE values. Addition of modified MN during the synthesis of UF resin reduced the FE by 56.5% to 0.73 mg/L and around 0.89 mg/L with addition of modified LN. This indicates that the intercaltion of nanoclay plays an important role in reducing the FE from plywood panel. It has been reported that a uniform dispersion of nanoclay could absorb and reduce the FE from the panel [21].

4 Conclusion

This paper attempted to enhance the adhesion strength and reduce formaldehyde emission (FE) of urea–formaldehyde (UF) resin bonded plywood by intercalating nanoclay during the synthesis of the resin. The crystal formation in UF resin of low formaldehyde/urea (F/U) mole ratio generates a poor adhesion at the expense of low FE. A way to prevent the crystallization in cured UF resin is by mixing nanoclay types during the addition reaction of resin's synthesis. Montmorillonite (MN) and laponite (LN) were used as the nanoclay in the form of pristine (original) and modified (TMI). XRD and FTIR analysis showed that the chemical structure of pristine nanoclay was crystal, which was similar to neat UF resin. The structure of modified MN-UF resin was partially amorphous, while modified LN-UF resin was still crystal. The neat UF resin had a crystallinity of about 54.1% after curing. But, after the addition of nanoclay during resin's synthesis, the crystallinity of cured UF resin remarkably decreased to 29.3% using modified MN and 47.5% using modified LN. The adhesion strength was enhanced and formaldehyde emission was further reduced. These results showed that the prevention of crystallization in UF resin could be done by mixing 5% modified MN during the synthesis, which subsequently improved the adhesion strength by 33.6% (1.37 MPa) and lowered the formaldehyde emission by 56.1% (0.73 mg/L). It is believed that the modified MN had been intercalated in UF resin polymers during synthesis and eventually partially blocked the crystal formation in cured UF resins. This could facilitate the formation of three-dimensional cross-linking structures in cured UF resin and led to a better adhesion strength and reduced formaldehyde emission of plywood bonded modified UF resin.

Acknowledgements This work was supported by integrated laboratory of bio-products (iLaB), Research Center for Biomaterials, National Research and Innovation Agency. Authors are also grateful for the publication support from Japan Society for the Promotion of Science (JSPS) and JSPS Alumnae Association Indonesia (JAAI).

References

1. UNECE, FAO (2020) Annual market review 2019–2020—forest products
2. Dunky M (1998) Urea–formaldehyde (UF) adhesive resins for wood. Int J Adhes Adhes 18:95–107
3. Myers G (1984) How mole ratio of UF resin affects formaldehyde emission and other properties: a literature critique. For Prod J 34:34–41
4. Dutkiewicz J (1983) Hydrolytic degradation of cured urea-formaldehyde. J Appl Polym Sci 28:3313–3320
5. Lubis MAR, Manohar SY, Permana R, Laksana B, Ismayati M, Falah F, Solihat NN, Sari FP (2021) The removal of cured urea-formaldehyde adhesive towards sustainable medium density fiberboard production: a review. J Sylva Lestari 9:23–44
6. Stuligross J, Koutsky JA (1985) A morphological study of urea-formaldehyde resins. J Adhes 18:281–299
7. Jada SS (1988) The structure of urea—formaldehyde resins. J Appl Polym Sci 35:1573–1592
8. Nuryawan A, Singh AP, Zanetti M, Park B, Causin V (2017) Insights into the development of crystallinity in liquid urea-formaldehyde resins. Int J Adhes Adhes 72:1–24
9. Pratt TJ, Johns WE, Rammon RM, Plagemann WL (1985) A novel concept on the structure of cured urea-formaldehyde resin. J Adhes 17:275–295
10. Wibowo ES, Park BD, Causin V (2020) Hydrogen-bond-induced crystallization in low-molar-ratio urea-formaldehyde resins during synthesis. Ind Eng Chem Res. https://doi.org/10.1021/acs.iecr.0c02268
11. Wibowo ES, Lubis MAR, Park BD (2021) In-situ modification of low molar ratio urea–formaldehyde resins with cellulose nanofibrils for plywood. J Adhes Sci Technol 1–14
12. Lubis MAR, Yadav SM, Park B-D (2021) Modification of oxidized starch polymer with nanoclay for enhanced adhesion and free formaldehyde emission of plywood. J Polym Environ. https://doi.org/10.1007/s10924-021-02101-w
13. Yadav SM, Lubis MAR, Sihag K (2021) A comprehensive review on process and technological aspects of wood-plastic composites. J Sylva Lestari 9:329
14. Wibowo ES, Lubis MAR, Park B-D, Kim J-S, Causin V (2020) Converting crystalline thermosetting urea—formaldehyde resins to amorphous polymer using modified nanoclay. J Ind Eng Chem. https://doi.org/10.1016/j.jiec.2020.03.014
15. Lubis MAR, Park B-D (2020) Enhancing the performance of low molar ratio urea–formaldehyde resin adhesives via in-situ modification with intercalated nanoclay. J Adhes 00:1–20
16. Delavernhe L, Pilavtepe M, Emmerich K (2018) Cation exchange capacity of natural and synthetic hectorite. Appl Clay Sci 151:175–180
17. Ma CHI, Eggleton RA (1999) Cation exchange capacity of kaolinite. Clays Clay Miner 47:174–180
18. Yadav SM, Lubis MAR, Park B-D (2021) Modification of nanoclay with different methods and its application in urea-formaldehyde bonded plywood panels. Wood Mater Sci Eng 1–10
19. JAS (2003) JAS 233: plywood. Japanese agric. Stand
20. Muñoz F, Moya R (2018) Effect of nanoclay-treated UF resin on the physical and mechanical properties of plywood manufactured with wood from tropical fast growth plantations. Maderas Cienc y Tecnol 20
21. Yadav SM, Lubis MAR, Wibowo ES, Park B (2020) Effects of nanoclay modification with transition metal ion on the performance of urea–formaldehyde resin adhesives. Polym Bull. https://doi.org/10.1007/s00289-020-03214-3

22. Younesi-Kordkheili H, Kazemi Najafi S, Behrooz R (2017) Influence of nanoclay on urea–glyoxalated lignin–formaldehyde resins for wood adhesive. J Adhes 93:431–443

Determination of Antifungal Activity from Hydrophilic Extracts of *Swietenia Macrophylla* King Bark

Masendra and Ganis Lukmandaru

Abstract This study aimed to investigate hydrophilic compounds and their antifungal activity from the inner and outer bark of *Swietenia macrophylla*. The hydrophilic compound of *S. macrophylla* in the inner and outer bark parts was detected by Gas Chromatography Mass Spectra (GC–MS) and antifungal activity was determined through the inhibition of sample to growth of *Phanerochaete chrysosporium*. The detection of hydrophilics observed several sugar and phenolic compounds in inner and outer bark of *S. macrophylla*. The antifungal activity in outer bark sample was greater (IC_{50} of 1536.27 $\pm$ 18.02 μg/ml) than that of inner bark (2433.89 $\pm$ 103.62 μg/ml). This suggests that the presence of (+)-catechin, (−)-epicatechin, and pyrocatechol in the outer bark layer tends to inhibit the white-rot fungal activity. Therefore, the phenolic compounds of *S. macrophylla* bark is potent for natural preservative.

Keywords *Swietenia macrophylla* · Flavonoids · Bark extracts · White-rot · Hydrophilics

1 Introduction

Swietenia macrophylla is woody species commonly used for furnitures. The utilization of mahogany wood as above mentioned discards the bark as residue. However, bark contains more polyphenol and lignin. Consequently, the utilization of bark is increased due to the important value of the extractives through natural preservative

Masendra
Department of Forest Products Technology, Faculty of Forestry, Universitas Gadjah Mada, Yogyakarta, Indonesia

Masendra · G. Lukmandaru (✉)
Department of Bioresources and Veterinary Technology, Vocational College, Universitas Gadjah Mada, Yogyakarta, Indonesia
e-mail: glukmandaru@ugm.ac.id

M. A. Rahandi Lubis (ed.), *Proceedings of the 10th International Symposium for Sustainable Humanosphere*, Lecture Notes in Civil Engineering 255,
https://doi.org/10.1007/978-981-19-1422-5_3

compounds. Previously, it was reported that hydrophilc compound of cinnamaldehyde [1], latifolin and its derivatives [2], stilbenes, pinosylvin, and pinosylvin monomethyl ether [3] were potential for antifungal activities. Furthermore, the lipophilic compounds of β-sitosterol and serrate-14-en-3β,21β–diol were reported to have antifungal activity [4, 5].

In the previous work, we investigated the lipophilic and hydrophilic compounds from the inner and outer bark of *S. macrophylla* such as triterpenoids, steroids, fatty acids, fatty alcohols [6], and monophenolic compounds, namely, 4-hydroxybenzyl alcohol, catechol, benzoic acid, resorcinol, 2,3,4-trimethoxycinnamic acid, 4-methylcatechol, syringol, antiarol, and syringic acid [7]. However, the presence of flavonoid such as (+)-cathecin and (−)-epicatechin, swietemacrophyllanin [8], quercitin, kaempferol, as polyphenolic compound that found in *S. macrophylla* bark [9] was not detected in the inner and outer bark of *S. macrophylla*. Thus, this study aim was to investigate polyphenolic compounds of inner and outer bark of *S. macrophylla* together with their antifungal activity.

Recently, the use of synthetic preservatives is under question concerning to the environmental condition. Thus, the research of natural preservatives from the plant is necessary. The several studies on the natural preservative for antifungal such as white-rot were reported [2, 10]. The ability of hydrophilic and lipophilic in inhibiting the growth of white-rot promotes those extractives acting as natural preservative.

2 Materials and Method

2.1 Extraction

The bark of *S.macrophylla* (±20 year-old) from Jepara (Indonesia) was manually separated into inner and outer bark. The 500 g of inner and outer bark sample was successively refluxed for 6 h with *n*-hexane, methanol (MeOH), and water. Dried extract was weighed as the result after sample evaporation with a rotary evaporator. The extractive content of *n*-hexane, MeOH, and hot water extract was reported in previous work [6]. In this study, the MeOH soluble extract of the inner and outer bark was subjected to antifungal test and GC–MS analysis.

2.2 Antifungal Activity

The *Phanerochaete chrysosporium* (white-rot) was used for antifungal activity test. The white-rot (code: IPBCC 930,259) was purchased from Institut Pertanian Bogor-Culture Center, IPB University. The growth inhibition of MeOH soluble extract of inner and outer bark against *P. chrysosporium* was conducted according Lukmandaru [11]. The sample of MeOH soluble extract from the inner and outer bark (300 μl) was spread on the surface of potato dextrose agar medium. After the solvent drying, the *P. chrysosporium* was inoculated to a Petri dish. The blank was made by spreading sample without extract. In comparison, a biocide was used as positive control. The growth inhibition of sample was calculated by Eq. 1 as follows:

$$\text{Growth inhibition } (\%) = A1/A0 \times 100 \quad (1)$$

where $A = \pi \times (d/2)^2$, d = diameter of sample growth, A0 is growth inhibition of blank and A1 is growth inhibition of sample. In this experiment, the IC_{50} value was made with concentration of 500, 1000, and 2000 μg/ml and each sample was conducted in three replications.

2.3 Gc–ms

The gas chromatography mass spectra (GC–MS) analysis was conducted through the method by Preciado et al. [12]. The sylilation process involved the dilution of 1 mg sample (MeOH soluble extract of inner and outer bark of *S. macrophylla*) with 50 μl of pyridine and mixture of 100 μl of BSA/ N,O-bis (trimethylsilyl) acetamide and TMCS/trimethylchlorosilane (99/1, v/v). The sample was then heated for 30 min at 100 °C. After heating with oven, the sample was cooled at room temperature and 1 ml of MeOH was added to the sample solution. For analysis, 1 μl of sample was injected to the GC–MS machine and data collected with a GCMS-QP 2010 (Shimadzu, Japan). The GC condition: Rtx-5MS capillary column (30 m × 0.25 mm I.D. and 0.25 μm; GL Sciences, Tokyo, Japan); detection temperature of 285 °C; column temperature from 70 °C (2 min) to 290 °C at 5 °C/min; injection temperature of 200 °C; acquisition mass range of 50–500 amu using helium as the gas carrier. Finally, the mass spectra of the sample were compared with the NIST11 (2002) library and the literatures [10, 11], while the quantification of the compound was calculated by peak relative method [7].

2.4 Statistical Analysis

The results from inner and outer bark antifungal activity was subjected to independent t-test analysis with 95% confidence interval using SPSS 20 (IBM, USA) with significant result is shown by the $p < 0.05$.

2.5 Chemicals

The TMCS, BSA, pyridine, were purchased from Sigma-Aldrich (Germany), while biocide (fungicide product) was obtained from Bioindustries, Yogyakarta, Indonesia. The active compound of biocide is methylene bis thiocyanate and 2-thyocyanomethyl thiobenzothiazole.

3 Results and Discussion

3.1 Fungal Inhibition

The antifungal of MeOH soluble extract from *S. macrophylla* bark is shown in Fig. 1. Based on t-test, there was a significant difference in the antifungal activity of MeOH soluble extract between the inner bark and outer bark ($p < 0.01$). The outer bark showed greater activity (IC_{50} of 1536.27 ± 18.02 μg/ml) than inner bark (2433.89 ± 103.62 μg/ml). However, the activity of both inner and outer bark was still in low activity in comparison with biocide (260.25 μg/ml). Compare with other literatures [2, 3, 10], the antifungal activity of the hydrophilic from *S. macrophylla* inner and outer bark was also in low activity. Related to hydrophilic extractives from the inner and outer bark of *S. macrophyla*, the sugars compound (Table 1) was detected in greater amount. The abundance of sugars in the bark samples might not supportthe inhibition of white-rot growth rather than phenolic of neoflavonoid [2] and stilbenes [3].

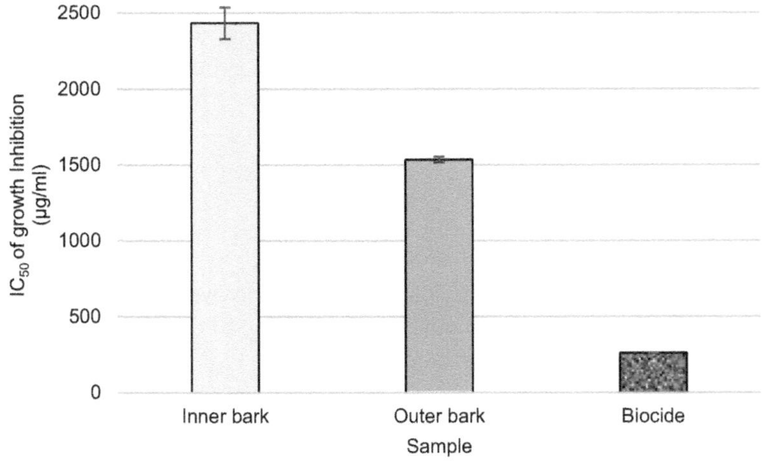

Fig. 1 Antifungal inhibition of MeOH soluble extract of *S. macrophylla* inner and outer bark

Table 1 Hydrophilics detected from *S. macrophylla* bark (identified as TMS derivative)

No	Ret. time (min)	Components	Concentration (%)		Similarity index (%)
			Inner bark	Outer bark	
1	15.3	Glycerol[a]	tr	3.23	65
2	16.2	Pyrocatechol[a]	–	12.6	87
3	28.8	D-(−)-Tagatofuranose[a]	6.63	–	
4	29.0	D-Psicofuranose[a]	15.42	–	84
5	30.2	(+)-Catechin[bc]	–	4.56	67
6	30.7	Glucopyranose[a]	9.24	tr	74
7	31.4	Mannitol[a]	tr	75.15	94
8	31.9	Myo-Inositol[a]	53.9	tr	88
9	32.6	β-D-(+)-Talopyranose[a]	14.7	tr	88
10	47.3	(−)-Epicatechin[c]	–	4.43	70

Compared with NIST library (a), Proestos and Komaitis [16] (b), and Preciado et al., [12] (c)
(-): not detected; (tr): trace

In comparison, the outer bark sample in this study exhibited higher inhibition than inner bark. It is assumed that the greater activity of *S. macrophylla* outer bark is due to this layer as protector for inner bark and other living tissues. Most of the outer bark part contain dead and suberinised cells. The suberins or lipophilics in the outer bark control the water loss from the inner tissues and to protect from tree pathogens and extreme weather. Previously, it was reported that the presence of lipophilics in the bark has an ecologically role by protecting the tree from outer conditions [13–15].

3.2 *Hydrophilic Compounds*

The soluble MeOH extract compound from *S. macrophylla* bark detected by GC–MS is displayed in Table 1. Both inner and outer bark sample dominated by sugars compound (peak no. 1, 3, 4, 6,7, 8, and 9). However, the presence of polyphenolic compounds ((+)-catechin and (−)-epicatechin) was only found in outer bark. In the inner bark, the dominant compound was myo-inositol (53.9%) while in the outer bark was mannitol (75.15%). The polyphenolic compounds confirmed in the outer bark sample were pyrocatechol (12.6%), (+)-catechin (4.56%), and (−)-epicatechin (4.43%).

The identification of (+)-catechin, (−)-epicatechin, pyrocatechol, and other hydrophilic compounds in this study was confirmed by NIST library and comparison with the literatures [12, 16]. In the outer bark, the presence of phenolic compound might be due to the function of this layer to protectthe inner layer. Furthermore, the presence of sugars especially in the inner bark is due to this layer as storage nutrients of the tree. The presence of sugars in the inner bark layer also was reported in the six species of *Pinus* [17].

In the outer bark, the greatest amount of sugar observed was only for mannitol while in the in the inner bark were myo-inositol with other sugars (glycerol, d-(−)-tagato-furanosea, d-psicofuranose, glucopyranose, and β-d-(+)-talopyranose). The dominant sugars like myo-inositol and mannitol are grouped into non-structural carbohydrates (NSC) that essential for the growth of a tree. The presence of NSC in woody plants is contributed by the inner bark [18]. While the presence of NSC such as starch is limited in the outer bark. The limitation of starch in plants was reported from outer bark of *Tectona grandis* in a previous work [19]. However, the present study showed that NSC of mannitol was in high abundance in the outer bark of *S. macrophylla* (Table 1). This suggests that the percentage of NSC is different due to different of layer tissues and species.

3.3 Correlation Between Antifungal Activity and Related Compounds

The higher antifungal inhibition of inner bark than outer bark was supported by the presence of phenolics such as pyrocatechol, (+)-catechin, (−)-epicatechin, and pyrocatechol. Previously, the presence of phenolic compounds, namely, lignan in species of *Pinus merkusii* wood [3] and latifolin in species of *Dalbergia latifolia* wood [2] showed to have antifungal activity.

The antifungal activities of those both inner and outer bark of *S. macrophylla* were categorized as under mild activity. This due to the involvement of sugars as major compounds in the bark of *S. macrophylla* that might not enhance to the antifungal activity or their role unlike phenolic compounds. The phenolic compounds were detected in outer bark sample of *S. macrophylla* in Fig. 2. They have a catechol structure which posseses two hydroxyls. The presence of hydroxyl groups and their hydrophobicity may inhibit the growth of *P. chrysosporium* white-rot. The attribution of hydroxyl structure and its hydrophobicity to activity of antifungal also was reported in a previous work [2].

4 Conclusions

The antifungal observation on the inner and outer bark of *S. macrophylla* was elucidated. The antifungal activity level in outer bark was higher than that inner bark. This finding was suggested by the presence of polyphenolic compounds in the outer bark, namely, pyrocatechol, (+)-catechin, and (−)-epicatechin. The presence of concentrated sugar compounds rather than polyphenolics in both inner and outer bark of *S. macrophylla* gave under mild activity of antifungal for those samples.

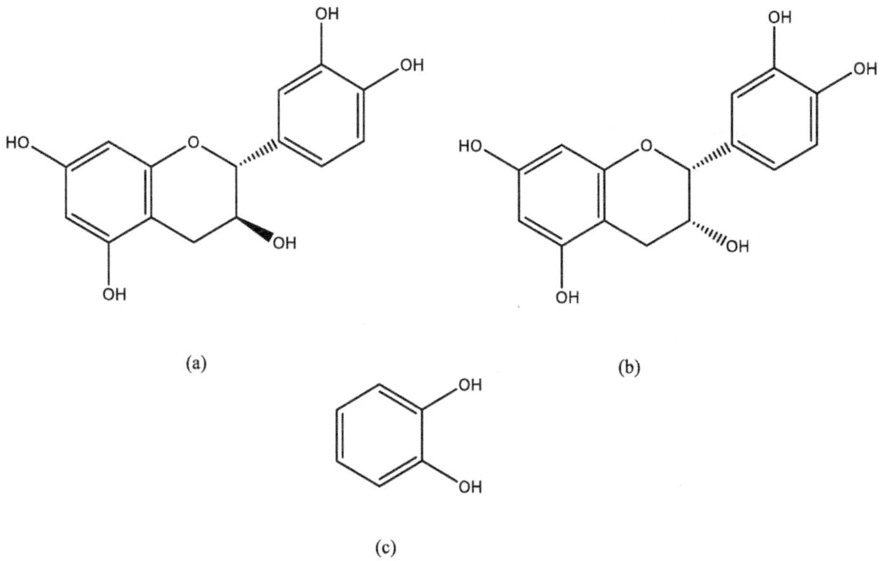

Fig. 2 Chemical structure (+)-catechin (**a**), (−)-epicatechin (**b**), and pyrocatechol (**c**)

Acknowledgements We thank to Srikandiratu Furniture for providing the inner and outer bark samples.

References

1. Lin CY, Wu CL, Chang ST (2007) Evaluating the potency of cinnamaldehyde as natural wood preservative. International research group wood preservation, SwedenIRG/WP07–30444
2. Sekine N, Ashitani T, Murayama T, Shibutani S, Hattori S, Takahashi K (2009) Bioactivity of Latifolin and its derivatives against termites and fungi. J Agric Food Chem 57(13):5707–12
3. Wijayanto A, Dumarçay S, Gérardin-Charbonnier C, Rita KS, Syafii W, Gérardin P (2015) Phenolic and lipophilic extractives in *Pinus merkusii* Jungh. et de Vries Knots and Stemwood. Indus Crops Prod 69:466–471
4. Masendra, Purba BAV, Lukmandaru G (2020) Antifungal activity of triterpenoids and steroids isolated from *Pinus merkusii* bark against *Phanerochaete chrysosporium*. Wood Res J 11(2):65–71
5. Masendra, Purba BAV, Lukmandaru G (2021) An evaluation of the antifungal and antioxidant activity of *Pinus merkusii* bark ethyl acetate extract. Jurnal Ilmu Kehutanan 15(1):102–111
6. Arisandi R, Masendra, Purba BAV, Wati FZ, Ihda FV, Sumantri F, Lukmandaru G (2019) Lipophilic extractives of Mahogany (*Swietenia macrophylla* King) barks. In: Proceeding of 9th international symposium of Indonesian wood research society, pp 92–102
7. Masendra, Purba BAV, Lukmandaru G (2021) Antioxidant activity of *Swietenia macrophylla* King bark extracts. Wood Res 66(1):57–70
8. Falah S, Suzuki T, Katayama T (2008) Chemical constituents *from Swietenia macrophylla* bark and their antioxidant activity. Pak J Biol Sci 11(16):2007–2012

9. El-Zalabani SME, El-Askary HI, Mousa OM, Issa MY, Zaitoun AA, Abdel-Sattar E (2012) Acaricidal activity of *Swietenia mahogany* and *Swietenia macrophylla* ethanolic extracts against Varroa destructor in honeybee colonies. Exp Parasitol 130:166–177
10. Xie Y, Wang Z, Huang Q, Zhang D (2017) Antifungal activity of several essential oils and major components against wood-rot fungi. Ind Crops Prod 108:278–285
11. Lukmandaru G (2013) Antifungal activities of certain components of teak wood extractives. Jurnal Ilmu Teknologi Kayu Tropis 11(1):11–8
12. Preciado LM, Lobo-Echeverri T, PereaÃ±ez JA, Zopata K, Rojano BA (2016) Antioxidants from three *Swietenia* Species (Meliaceae). J Medicinal Plants Res 10(2):8–17
13. Seki K, Orihashi K, Sato M (2012) Accumulation of constitutive diterpenoids in the rhytidome and secondary phloem of the branch bark of *Larix gmelinii* var. japonica. J Wood Sci 58:437–445
14. Morris H, Brodersen C, Schwarze FWMR, Jansen S (2016) The parenchyma of secondary xylem and its critical role in tree defense against fungal decay in relation to the CODIT model. Front Plant Sci 7(1665):1–18
15. Masendra, Ashitani T, Takahashi K, Lukmandaru G (2018) Triterpenoids and steroids from the bark of *Pinus merkusii* (Pinaceae). BioResources13(3):6160–6170
16. Proestos C, Komaitis M (2013) Analysis of naturally occurring phenolic compounds in aromatic plants by RP-HPLC coupled to diode array detector (DAD) and GC-MS after silylation. Foods 2:90–99
17. Masendra, Ashitani T, Takahashi K, Lukmandaru G (2018) Lipophilic extractives of the inner and outer barks from six different *Pinus* species grown in Indonesia. J Forest Res 29(5):1329–36
18. Rossel JA, Piper FI, Jimenez-Vera C, Virgilio PCB, Marcati CR, Castorena M, Olson ME (2021) Inner bark as a crucial tissue for non-structural carbohydrate storage across three tropical woody plant communities. Plant, Cell Environ 44:156–170
19. Islam MA, Begum S (2011) Distribution of starch, lipid, and nuclei in xylem and phloem of *Tectona grandis* Linn. J Bio-Sci 19:29–35

The Effect of Harvesting Time on the Fiber Quality of Kenaf and Roselle

Muhammad Rasyidur Ridho, Elvara Windra Madyaratri, Ismadi, Danang Sudarwoko Adi, Apri Heri Iswanto, Mala Murianingrum, Taufiq Hidayat, Nurindah, Riana Anggraini, and Widya Fatriasari

Abstract The Kenaf and rosella plants can grow in various types of land, and the fibers have been used as a reinforcement material for polymer composites, the automotive industry, rope, and paper. Some factors influenced the fiber properties, including the plant species, the growth site type, the plant's axial position, and harvesting time. Still, variety in the fiber properties before the harvesting period and axial position of plants have not been reported before. Thus, this study investigated the effect of the harvesting period and axial position of kenaf and roselle (top, middle, and bottom) on fiber dimension properties and their derived value. The harvesting period tends to increase the all-fiber dimension properties of kenaf. However, this tendency only occurs in the cell wall thickness, and fiber length of roselle fibers, while the fiber diameter and lumen cell before harvesting period are higher than during the harvesting period. The axial position of plants results in a variety of fiber dimensions. Except for the kenaf after harvesting, the fiber length of the kenaf and roselle stems increases from the base to the top of the stem in three axial positions. Based on fiber derivative values and fiber dimensions, roselle plants

M. R. Ridho · Ismadi · D. S. Adi · W. Fatriasari (✉)
Research Center for Biomass and Bioproducts, National Research and Innovation Agency (BRIN), Bogor, Indonesia
e-mail: widya.fatriasari@biomaterial.lipi.go.id

E. W. Madyaratri
Departement of Forest Product Technology, Bogor Agricultural University, Bogor, Indonesia

A. H. Iswanto
Departement of Forest Product, Faculty of Forestry, Universitas Sumatera Utara, Kota Medan, Indonesia

M. Murianingrum · T. Hidayat · Nurindah
Indonesian Sweetener and Fiber Crops Research Institute (ISFCRI), Ministry of Agriculture. Malang, East Java, Indonesia

W. Fatriasari
Department of Agricultural, Faculty of Forestry, Universitas Sumatera Utara, Kota Medan, Indonesia

R. Anggraini
Faculty of Agriculture Forestry, Universitas Jambi, Jambi, Indonesia

M. A. Rahandi Lubis (ed.), *Proceedings of the 10th International Symposium for Sustainable Humanosphere*, Lecture Notes in Civil Engineering 255,
https://doi.org/10.1007/978-981-19-1422-5_4

are more recommended as pulp and paper raw materials before harvesting time and during harvesting time than kenaf plants, with the proper stem position for pulp and paper raw materials being in the middle position of the stems before harvesting time.

Keywords Fiber dimension · Flowering time · Natural fiber · Fiber derived value · Axial position

1 Introduction

In tropical and subtropical climes, plants such as kenaf and roselle are frequently planted. Kenaf plants reach a height of 3–5 m and a diameter of 25–51 mm, producing between 4 and 8 tons of dry fiber per hectare, depending on growing circumstances [1]. According to global kenaf production data, production decreased by 19 percent between 2012/2013 and 2018/2019 [2]. Kenaf and roselle fibers can be used to reinforce polymer composites, as well as automobile parts and accessories [3–7]. Kenaf fiber can be used as a starting material for manufacturing currency paper [8], whilst rosella fiber can be used in the nautical, agricultural, and shipping industries as a rope and yarn [9].

As natural bioresources, natural fibers vary in size and quality due to complex interactions with environmental factors, transportation methods, storage duration and conditions, and fiber extraction. Understanding fundamental aspects of natural fibers, such as their morphological characteristics, is vital for choosing the optimal applications, such as high-quality bio-composite raw material [6]. Kenaf and roselle are annual plants that mature in 3–5 months (kenaf) and 7–8 months (roselle) and generate 1.5–3.0 tons of fiber per hectare [6, 8]. Kenaf fibers are 1.5–12 mm in length, 7–41 mm in diameter, 7–13 mm in lumen width, and 4–9 mm in cell wall thickness, whereas roselle fibers are 1.2–6.3 mm in length, 10–44 mm in diameter, 3–15 m in lumen width, and 4–15 m in cell wall thickness [9]. Fiber size variations might also be a result of changes in plant species and soil conditions [10, 11]. Due to the fact that fibers are produced from bast, the plant's height and diameter have an effect on the fiber quality [12]. Also, several factors, including plant type, age, harvest period, and axial position on the plant stem, influence the fiber size in plants [13, 14]. The age difference of one year in teak wood has no effect on the anatomical structure. When the age gap is wider, this structure is expected to have a stronger impact [15]. There is currently no information on fiber diameter or plant axial position prior to harvest. Photoperiodicity has an effect on kenaf and roselle growth. If the plants are planted on short days (less than 12 h of sunlight per day), they will blossom rapidly and their height will be lowered. On the other side, if they are planted on a long day, they will flower later. This condition will have an effect on the fiber's quality. The purpose of this study is to ascertain the difference in fiber diameters in the axial position of kenaf and roselle fibers due to harvestin time differences. It focuses on the analysis of fiber diameters and fiber derived value derivatives.

2 Material Methods

2.1 *Material*

The materials used in this study were kenaf variety kenafindo 2 (KI2) and one of the roselle accession (R) plants taken from the Indonesian Sweetener and Fiber Crops Research Institute (ISFCRI), Malang, Indonesia. The roselle and kenaf samples were taken before the harvesting or flowering period (3 and 5 months, respectively) and the harvest or flowering period of the plant (7 and 4 months, respectively), when 50% of the population had bloomed. The stem of the plant was divided into three axial positions, namely the base (a), middle (b) and top (c) parts. Before measuring its dimension, the analytical grade chemicals used for fiber maceration were hydrogen peroxide (H_2O_2) 30% and acetic acid glacial 100%.

2.2 *Methods*

The fiber dimensions, such as fiber length, diameter, lumen diameter, and cell wall thickness, were determined by macerating each sample using Franklin's solution [16]. The sample was placed in a test tube with hydrogen peroxide (H_2O_2) and glacial acetic acid (1:1), then heated in a water bath until the sample turned pale. In addition, the sample was washed with aquades 4–5 times, then stained with the safranin solution overnight. The macerated samples were observed and captured with the Olympus BX-63 equipped with a digital camera Dp73. The fiber length was taken at 4× of the objective lens, while the other dimensions were 10× of the objective lens in the middle of the fiber. ImageJ software was used to measure fiber dimensions in each sample, which included up to 30 fibers. The fiber derivative value was subsequently calculated to determine the potential fiber quality of pulp and paper raw materials, i.e. runkel ratio, muhlsteph ratio, felting power, flexibility ratio, and the coefficient of rigidity. The Runkle ratio was (2× cell wall thickness) divided by the lumen diameter, while the felting power was a comparison of fiber length and diameter. The flexibility ratio was determined by lumen diameter per fiber diameter; coefficient of rigidity was calculated by cell wall thickness divided by the fiber diameter; and the muhlsteph ratio was calculated as ((fiber diameter^2-diameter lumen^2)/fiber diameter^2) × 100% [17].

3 Results and Discussions

3.1 *Fiber Dimensions of Kenaf and Roselle*

The fiber length, lumen cell, diameter, and cell wall thickness of kenaf and roselle plants are listed in Table 1, whereas the fiber derivative value is listed in Table 2. The fiber dimensions measured on the stems of kenaf and roselle plants throughout the harvesting time revealed a range of values. The fiber dimensions of kenaf (fiber length, fiber diameter, lumen cell, and cell wall thickness) are greater during harvesting time than before harvesting time, while in roselle, fiber length and cell wall thickness were greater during harvesting time than before harvesting time, and yet fiber diameter and lumen cell were higher before harvesting i.e.,. Before and after flowering, the length of kenaf and roselle fibers is comparable to that of bamboo, banana fiber, jute, and sisal fibers [18]. At 3–5 and 5–7 months after planting, kenaf and roselle both exhibit higher fiber diameters than bamboo (14.0–17.8 m). Kenaf and roselle have thicker cell walls (1.2–1.5 m) than banana fibers [18].

Jahan and his colleagues reported that the fiber length increases with plant life. The fiber length of kenaf and roselle is longer during the harvesting period [19]. The fiber diameter of kenaf is wider during the harvesting time, and it will increase along

Table 1 Fiber dimensions of kenaf and roselle

Sample code		Fiber length (mm)	Score	Fiber diameter (μm)	Score	Lumen diameter (μm)	Score	Cell wall thickness (μm)	Score	Total score
Before harvesting time	KI_2a	2.710	3	15.088	1	7.988	2	3.550	9	15
	KI_2b	2.703	2	15.137	2	7.924	1	3.606	8	13
	KI_2c	2.970	12	16.483	3	9.681	3	3.401	11	29
	Average	2.795		15.569		8.531		3.519		
	Ra	2.743	4	19.618	7	12.617	8	3.501	10	29
	Rb	2.525	1	22.846	9	16.299	12	3.274	12	**34**
	Rc	2.945	11	18.404	6	10.729	6	3.837	5	28
	Average	2.738		20.289		13.215		3.537		
Harvesting time	KI_2a_2	2.882	9	23.282	10	13.034	9	5.124	2	**30**
	KI_2b_2	2.861	8	24.606	12	14.353	10	5.126	1	31
	KI_2c_2	2.786	7	20.166	8	12.177	7	3.994	4	26
	Average	2.843		22.685		13.188		4.748		
	Ra_2	2.742	5	17.435	4	10.036	4	3.700	6	19
	Rb_2	2.757	6	17.651	5	10.382	5	3.634	7	23
	Rc_2	2.902	10	23.724	11	14.748	11	4.488	3	35
	Average	2.800		19.603		11.722		3.941		

Note (KI_2): Kenaf, (R): Roselle, (a: base, b: middle, c: top (before harvesting time)), (a_2: base, b_2: middle, c_2: top (harvesting time))

Table 2 Fiber derivative value of kenaf and roselle

Sample code		Runkel ratio			Felting power			Flexibility ratio			Coefficient of rigidity			Muhlsteph ratio			
		Values	Class	Score	Values	Class	Score	Values	Class	Score	Values	Class	Score	Values	Class	Score	Total
Before harvesting time	KI_2a	1.192	IV	1	202.108	I	11	0.501	III	1	0.250	IV	1	72.876	III	1	15
	KI_2b	1.159	IV	2	191.833	I	10	0.503	III	2	0.248	IV	2	72.835	III	2	18
	KI_2c	1.002	IV	3	205.619	I	12	0.542	III	4	0.229	IV	4	68.428	III	4	27
	Ra	0.718	III	11	153.415	I	6	0.616	II	11	0.192	III	11	60.310	II	11	**50**
	Rb	0.474	II	12	120.307	I	1	0.694	II	12	0.153	II	12	50.785	II	12	49
	Rc	0.963	III	6	173.279	I	8	0.545	III	6	0.227	III	6	68.190	III	7	33
Harvesting time	KI_2a_2	0.997	III	4	131.649	I	2	0.537	III	3	0.232	IV	3	69.483	III	3	15
	KI_2b_2	0.908	III	7	135.561	I	4	0.558	III	8	0.221	IV	8	67.239	III	8	35
	KI_2c_2	0.781	III	9	150.895	I	5	0.586	III	9	0.207	IV	9	64.544	III	9	41
	Ra_2	0.902	III	8	172.521	I	7	0.552	III	7	0.224	IV	7	68.248	III	6	35
	Rb_2	0.994	III	5	178.253	I	9	0.543	III	5	0.228	IV	5	68.312	III	5	29
	Rc_2	0.752	III	10	133.300	I	3	0.599	III	10	0.200	IV	10	62.736	III	10	**43**

Note (KI_2): Kenaf, (R): Roselle, (a: base, b: middle, c: top (before harvesting time)), (a_2: base, b_2: middle, c_2: top (harvesting time)) with the increase in the harvesting period. However, fiber diameter and lumen width decrease at harvesting time in the roselle plant, which is similar to the findings of Shakhes et al. [13]. They reported that the kenaf fiber's fiber diameter and lumen decrease as the plant ages, related to environmental factors. The thickness of cell walls of kenaf and roselle plants increases from samples before the flowering phase to samples after the flowering phase due to the thickening of cell walls with the rising age of plants [20, 21]. The fiber dimension results of kenaf are in the range of the fiber dimension of the reported fiber [18]. They found that kenaf fiber lengths range from 1.4 to 11.0 mm, with fiber diameters ranging from 4.0 to 36.0 mm, lumens ranging from 5.4 to 11.1 mm, and cell wall thickness ranging from 1.6 to 12.6 mm. The fiber length of the roselle before and after the blooming phase is 2.10–2.8 mm, as reported [22]. While the fiber diameter of the roselle reported [23] is 13.1 μm, this result is less. With a cell wall thickness of 3.0 μm, the lumen width of 16.1 μm is larger than that of the roselle. Except for cell wall thickness, [24] evaluated the fiber dimensions of roselle, which were 17.54 μm, 6.88 μm, and 5.33 μm, respectively, higher than the fiber dimensions of our work. The hardwood fiber length is 1.16 mm, with a fiber diameter of 19.72 μm and a cell wall thickness of 4.84 μm, respectively. Softwood fibers have 2.72 mm, 25.52 μm, and 6.99 μm fiber lengths, diameters, and cell-wall thicknesses, respectively [25]. Natural fibers like kenaf and roselle have longer fibers than hardwood fibers.

Fiber measurement of kenaf and roselle at axial positions (base, middle, and top) of the plant has a difference in size that can be seen in Table 1. On the stem (base, middle, and top), each part has a difference in size-the measurement results of fiber kenaf and roselle at three stem positions. The measure of three-position fiber length (base, middle, and top) stem of kenaf and roselle plants obtained fiber length increased from the base to the end of the stem. This is the same as the results of Voulgaridis et al. [14] except at kenaf at harvesting time, where the fiber length increases from end to base. A similar report is also found in gmelina wood fiber [26]. The fiber dimensions of kenaf and roselle at harvesting time have almost the same fiber dimensions so that the fiber can be used as a rope [7]. Kenaf plant is usually used as paper making material, fiber carpet, and currency paper [6, 27], while roselle fiber, on the other hand, has been used in automotive parts and accessories [4, 5]. Roselle also has fiber strength that can be used for textiles and pulp and paper [21]. In the axial position of the plant, kenaf and roselle fibers before harvesting and during the harvesting period have a similar fiber form, as illustrated in Fig. 1.

3.2 Fiber Derivative Value of Fiber

Table 2 presents the fiber derivative value of kenaf and roselle that is calculated based on the fiber dimension of each fiber. While Table 3 depicts summary of scoring value assessment of fiber dimension and fiber derived value.

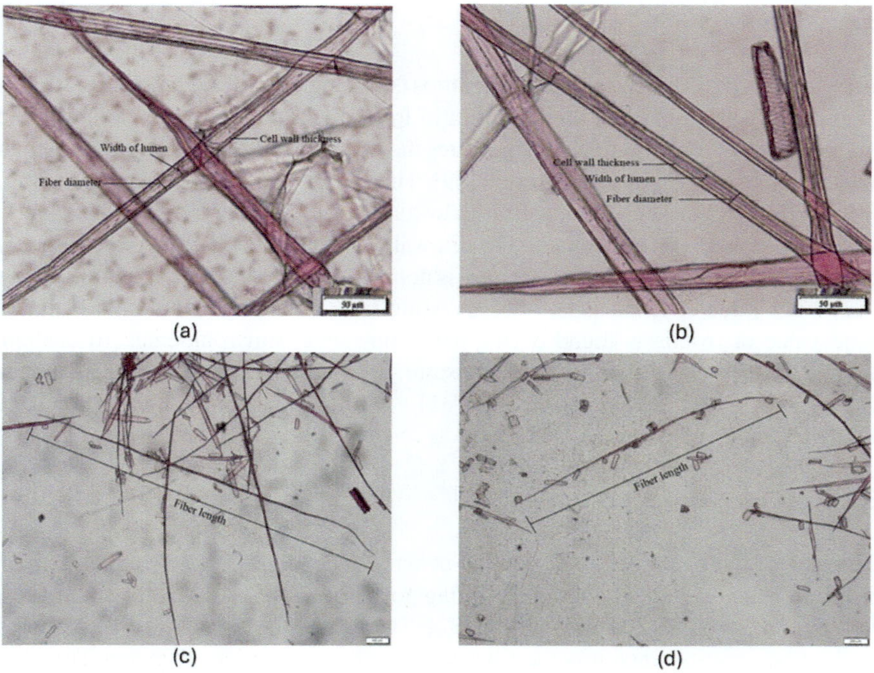

Fig. 1 Observations with BF microscope **a** fiber dimension of kenaf, **b** fiber dimension of roselle, **c** fiber length of kenaf, **d** fiber length of roselle

Table 3 Summary of the assessment of the fiber dimension and fiber derived value of kenaf and roselle fibers

Sample code		Score of fiber dimension	Score of fiber derived value	Total
Before harvesting time	KI_2a	15	15	30
	KI_2b	13	18	31
	KI_2c	29	27	56
	Ra	29	50	79
	Rb	34	49	**83**
	Rc	28	33	61
Harvesting time	KI_2a_2	30	15	45
	KI_2b_2	31	35	66
	KI_2c_2	26	41	67
	Ra_2	19	35	54
	Rb_2	23	29	52
	Rc_2	35	43	**78**

Note Roselle, (a: base, b: middle, c: top (before harvesting time)), (a_2: base, b_2: middle, c_2: top (harvesting time)) (KI_2): Kenaf, (R)

3.2.1 Runkel Ratio

In kenaf and roselle, the value of runkel ratios is still higher than the results of Shakhes et al. [28], who found that the runkel ratio for the entire kenaf stalk is 0.518. The higher or lower runkel ratio value indicates the fiber's strength. Fiber stiffens at high values and breaks easily at low levels [29]. The higher runkel ratio fibers are stiffer, less flexible, and form bulkier paper with lower bonded areas [17]. Lower runkel ratio values are directly proportional to the cell wall thickness and inversely proportional to lumen diameter [30]. The runkel ratio is not related to the harvesting season, as the maximum value of kenaf is before harvesting time. In contrast, the highest runkel ratio value of roselle is found during harvesting time. Environmental parameters, such as moisture content, pH, soil moisture, and rainfall intensity, influence fiber dimensions, according to Syafitri et al. [31].

3.2.2 Felting Power

The ratio of felting power to slenderness in kenaf and roselle is more valuable than in plants in general. Hardwoods have a felting power of 55–75, while softwoods have a felting power of 95–120, according to [32]. The value of felting power is connected to the fiber quality since it is the fiber length ratio to the fiber diameter. The fiber becomes more flexible as the felting power increases, while the fiber becomes stiffer as the felting power decreases [30]. The tear strength of this fiber is influenced by its felting power, which means that a high felting power value results in the high tear strength of the paper. Fiber length is a more important factor in creating links between fibers, as fiber length affects the tear strength of paper [33]. This is demonstrated by the fact that kenaf has the highest felting power at the base stem (KI_2c), implying that the highest-fiber length is also at KI_2c. On the other hand, roselle has the lowest felting power at the middle stem (Rb). Consequently, the shortest fiber length is also at Rb.

3.2.3 Flexibility Ratio

The flexibility ratio based on Jayusman and Hakim [34], high elastic fibers (>75%), elastic fibers (50–75%), rigid fibers (30–50%), and very stiff fibers (30%) are the four classes of fibers based on their flexibility ratio. Thus, the flexibility ratio of kenaf and roselle fibers is included in the elastic fiber. In kenaf, the value of the flexibility ratio is almost the same as the results of Kardiansyah and Sugesty [35], who found that the flexibility ratio for the kenaf stalk is 0.59. The flexibility ratio is the ratio of lumen diameter to fiber diameter, with the result having a link to tensile strength. High-flexibility fiber has a thin cell wall thickness and is easy to form. This deformability makes contact between the fiber surfaces more flexible, resulting in greater fiber bonding and stronger pulp sheets if utilized in papermaking [36].

3.2.4 Rigidity Coefficient

The highest rigidity coefficients were attained in the middle steam of the roselle before harvesting time, according to Table 2. The assertion of Ekhuemelo and Udo [37] that the younger the plant age, the greater the stiffness coefficient value backs up these findings. Thin cell walls and a large lumen diameter are linked to a high stiffness coefficient [29]. According to this study, roselle in the middle stem (Rb) had the lowest cell wall value, whereas Rb had the maximum lumen diameter value. In this study, the coefficient of rigidity has the lowest value obtained by roselle on the middle stem (Rb) of 0.153. This fiber allows it to be used as a pulp raw material. According to Fatriani and Banjarbaru [38], it must have a low coefficient of rigidity in the pulping process. It has a negative correlation with tensile strength, which means that the higher coefficient of rigidity causes lower tensile strength. On the other hand, the lower rigidity coefficient causes higher tensile strength.

3.2.5 Muhlsteph Ratio

Kenaf and roselle can be separated into two groups by mulhsteph ratio, according to Smook [32]. The muhlsteph ratio values for class II are 60.310 and 50.785, respectively, with a proportion of 60–80% of roselle species attained before harvesting time on the base and middle stems. Kenaf species were acquired both before and during harvesting time in complete stalks for class III, which had a percentage of up to 80%. The magnitude of the mulhsteph ratio influenced the density of the pulp sheet. The lower the mulhsteph ratio, the higher the strength and density of the pulp sheet and vice versa [30]. The high tensile strength and tear strength properties of the pulp are also affected by the muhlsteph ratio. Conversely, if the fiber has thick cell walls and small diameters, it tends to maintain its shape during sheet formation so that the contact area between cells is small, resulting in low tensile and tear strength [35].

Rb (roselle in the middle position of the stem before harvesting time) has the greatest score value, as shown in Table 3. The similar approach assessment method to determine the most proper kind for pulp and paper has been reported by Fatriasari and Hermiati [39]. In comparison to the other samples, Rb appears to be the most suitable as a pulp and paper raw material. In other hand, Rc_2 (roselle on the top position of the stem after harvesting time) is also more suggested as a raw material for pulp and paper than others during harvesting time.

4 Conclusions

Based on the fiber properties and value of fiber derivatives from plants kenaf and roselle before and after harvesting time, it can be concluded that the length of fiber in both plants is longer at harvesting time. The fiber diameter dimensions of fiber will increase as harvesting time increases. Except for kenaf in the harvesting period, fiber length grows from the base to the top. The difference in harvesting time of kenaf and roselle plants did not affect the fiber derivative value substantially, while the three axial positions resulted in value variation. According to scoring assessment on fiber derivative values and fiber dimensions, roselle plants before harvesting time and after harvesting time are more recommended as pulp and paper raw materials compared to kenaf plants. While the appropriate stem position as pulp and the paper raw material are in the middle position before harvesting time and in the top position at harvesting time.

Acknowledgements This work was supported by Productive-Innovative Research Program (RISPRO) by Indonesia Endowment Fund for Education (LPDP), Ministry of Finance in a collaboration with Ministry of Research and Technology/National Research and Innovation Agency (Kemenristek-BRIN), under the National Research Priority (PRN) program (grant number 273/E1/PRN/2020 and B-4528/IPH/KS.02.04/VII/2020). The authors would also like to thank the Japan-ASEAN Science, Technology and Innovation Platform (JASTIP) for the publication fee. And the authors acknowledge facilities, scientific and technical support from Advanced Characterization Laboratories Cibinong - Integrated Laboratory of Bioproducts, Indonesian Institute of Sciences through E-Layanan Sains Lembaga Ilmu Pengetahuan Indonesia

References

1. Webber III CL, Bhardwaj HL, Bledsoe VK (2002) Kenaf production: fiber, feed, and seed, Trends new Crop. New uses, pp 327–339
2. Rome (2019)FAO, "Jute, kenaf, coir and allied fibres" in Statistical Bulletin 2018
3. Thiruchitrambalam M, Athijayamani A, Sathiyamurthy S, Thaheer ASA (2010) A review on the natural fiber-reinforced polymer composites for the development of roselle fiber-reinforced polyester composite. J Nat Fibers 7(4):307–323
4. Akil HM, Omar MF, Mazuki AAM, Safiee S, Ishak ZAM, Abu Bakar A (2011) Kenaf fiber reinforced composites: a review. Mater Des 32(8–9):4107–4121
5. Chandramohan D, Bharanichandar J (2013) Natural fiber reinforced polymer composites for automobile accessories. Am J Environ Sci 9(6):494–504
6. Sudjindro (2011) Prospek serat alam untuk bahan baku kertas uang. Perspektif 10(2):92–104
7. Navaneethakrishnan S, Athijayamani A (2016) Mechanical properties and absorption behavior of CSP filled roselle fiber reinforced hybrid composites. J Mater Environ Sci 7(5):1674–1680
8. Karimah A, Ridho MR, Munawar SS, Adi DS, Ismadi, Damayanti R, Subiyanto B, Fatriasari W, Fudholi A (2021) A review on natural fibers for development of eco-friendly bio-composite: characteristics, and utilizations. J Mater Res Technol 13:2442–2458
9. Shamsuddin A, Van der Vossen HAM (2003) *Hibiscus cannabinus* L. In: Brink M, Escobin RP (eds) Plant resources of South-East Asia, No.17, Fiber Plants. Backhuys Publishers, Leiden
10. Supiani UD (2012) Pengaruh jenis tanah terhadap dimensi serat dan nilai turunan serat kayu akasia daun lebar (*Acacia mangium* Willd). Jurnal Hutan Tropis 13(1):1–10
11. Krisdianto (2006) Anatomi dan kualitas serat lima jenis kayu kurang dikenal dari Lengkong, Sukabumi. Jurnal Penelitian Hasil Hutan 24(3):201–218
12. Hamida R, Nugraheni SD (2011) Fotoperiodisitas dan hubungannya terhadap pertumbuhna dan produksi kenaf (*Hibiscus cannabinus* L.) dan roselle (*Hibiscus sabdariffa* L.), pp 181–186
13. Shakhes J, Zeinaly F, Marandi MAB, Saghafi T (2012) Effects of harvest time and cultivar on yield and physical properties fibers of kenaf (*Hibiscus cannabinus* L.). Afr J Biochem Res 6(6):69–74

14. Voulgaridis E, Passialis C, Grigoriou A (2002) Anatomical characteristics and properties of Kenaf stems (*Hibiscus cannabinus*). IAWA J 21(4):435–442
15. Wahyudi I, Priadi T, Rahayu IS (2014) Karakteristik dan sifat-sifat dasar kayu jati unggul umur 4 dan 5 tahun asal Jawa Barat. Jurnal Ilmu Pertanian Indonesia (JIPI) 19(1):50–56
16. Dodd RS (1986) Fiber length measurment systems: a review and modification of an exsisting method. Wood And Fiber Sci 18(2):276–287
17. Ververis C, Georghiou K, Christodoulakis N, Santas P, Santas R (2004) Fiber dimensions, lignin and cellulose content of various plant materials and their suitability for paper production. Ind Crops Prod 19(3):245–254
18. Al-Oqla FM, Salit MS (2017) Materials selection for natural fiber composites. In: Jones G (ed) Woodhead Publishing. Matthew Deans
19. Jahan MS, Tasmin RSB, Chowdhury DAN, Noori A, Al-Maruf A (2009) Effect of harvesting age on the chemical and morphological properties of Dhaincha (*Sesbania aculeata*) and its pulpability and bleachability. BioResources 4(2):471–481
20. Ayre BG, Stevens K, Chapman KD, Webber CL, Dagnon KL, D'souza N (2009) Viscoelastic properties of kenaf bast fiber in relation to stem age. Textt Res J 79(11):973–980
21. Razali N, Salit MS, Jawaid M, Ishak MR, Lazim Y (2015) A study on chemical composition, physical, tensile, morphological, and thermal properties of roselle fibre: Effect of fibre maturity. BioResources 10(1):1803–1823
22. Akubueze EU, Ezeanyanaso CS, Muniru OS, Nwaeche FC, Tumbi MI, Igwe CC, Elemo GN (2019) Extraction and characterization of bast fibres from roselle (*Hibiscus sabdariffa*) stem for industrial application. J Mater Sci Res Rev 2(4):1–7
23. Kalita BB, Jose S, Baruah S, Kalita S, Saikia SR (2017) *Hibiscus sabdariffa* (Roselle): a potential source of bast fiber. J Nat Fibers 16(1):49–57
24. Nadlene R, Sapuan SM, Jawaid M, Ishak MR, Yusriah L (2016) A review on roselle fiber and its composites. J Nat Fibers 13(1):10–41
25. Pulkkinen I, Ala-Kaila K, Aittamaa J (2006) Characterization of wood fibers using fiber property distributions. Chem Eng Process 45(7):546–554
26. Prabawa SB (2017) Pengaruh posisi pada batang terhadap sifat fisik dan dimensi serat kayu gmelina hasil penjarangan (The Influence of Stem Position on Physical Properties and Fiber Dimension of Gmelina from Thinning Activity). Jurnal Ilmu Teknologi Kayu Tropis 15(1):19–28
27. Ansell MP, Mwaikambo LY (2009) Handbook of textile fibre structure volume 2: natural, regenerated, inorganic and specialist fibres. In: Eichhorn SJ, Hearle JW, Jaffe M, Kikutani T (eds) 2nd ed)
28. Shakhes J, Zeinaly F, Marandi M, Saghafi T (2011) The effects of processing variables on the soda and soda-aq pulping of kenaf bast fiber. BioResources 6(4):4626–4639
29. Istikowati WT, Aiso H, Sutiya B, Ishiguri F, Ohshima J, Iizuka K, Yokota S (2016) Wood, chemical, and pulp properties of woods from less-utilized fast-growing tree species found in naturally regenerated secondary forest in South Kalimantan, Indonesia. J Wood Chem Technol 1–9
30. Aprianis Y, Rahmayanti S (2008) Fiber dimensions and their derived values of seven wood species from Jambi Province. J Forest Product Res 27(1):11–20
31. Syafitri A, Vauzia DM (2019) Dimensions of Fiber and Jabon Wood Fiber Derivative Value (*Anthocephalus cadamba* [Roxb] Miq.) in Sialang Dharmasraya and Tabing Padang Regions. Biosciences, pp 40–49
32. Smook GA (1997) Handbook for pulp & paper technologists. Angus Wilde Publications, Vancouver
33. Syafii W, Siregar IZ (2006) Chemical properties and fiber dimension of *Acacia mangium* Willd, from Three Provenances. J Tropical Sci Technol 4(1):28–32
34. Jayusman HL (2021) Comparison of the wood anatomy and fibers derived from Indonesian *Toona sinesis* Roem. and *Toona sureni* Merr. Bioresources 16(3):4769–4779
35. Kardiansyah T, Sugesty S (2014) Characteristic of chemimechanical pulp from Kenaf (*Hibiscus cannabinus* L.) for linier paper. J Cellulose 4(1):37–46

36. Haryadi F, Susanti C, Gunawan E (2017) Fiber dimension of pandanaus tectorius park, leaf for natural fiber product raw material. J Forest Papuasia 3(1):33–44
37. Ekhuemelo DO, Udo AM (2016) Investigation of variations in the fibre characteristics of *Moringa oleifera* (Lam) steam for pulp and paper production. Int J Sci Technol 5(1):19–25
38. Fatriani B (2017) Fiber dimension and derivatives value of sugar palm frond in various positions. Agric Forestry Fisheries 6(2):66–68
39. Fatriasari W, Hermiati E (2008) Analisis morfologi serat dan sifat fisis-kimia pada enam jenis bambu sebagai bahan baku pulp dan kertas. Jurnal Ilmu dan Teknologi Hasil Hutan 1(2):67–72

Response of *Rhizophora apiculata* Seed Growth to Application of Various Types of Fertilizer in Lubuk Kertang Village, Pangkalan Brandan Barat, North Sumatra

Yunasfi, H. L. Gultom, and M. Basyuni

Abstract Mangrove forests have soil characteristics that are muddy, wet, and rich in organic matter and if organic and inorganic fertilization activities are carried out, it is expected to affect both the soil and plants. This study aims to determine whether there is an effect on the growth of *Rhizophora apiculata* seedlings fed with NPK fertilizer, chicken manure, cow manure, quail manure or not. This research was conducted from October 2020 to May 2021 in the Mangrove Forest of Lubuk Kertang Village, Pangkalan Brandan Barat, North Sumatra. This study used a non-factorial completely randomized design consisting of 5 treatments and 4 replications. Variables observed were increase in seedling height, increase in seedling diameter, increase in number of leaves, temperature, plant wet weight, plant dry weight, leaf area. The results showed that the treatment with P3 (cow manure 200 g/plant) gave the best effect and results when compared to P0, P1, P2, and P4 because P3 fertilizer had the largest dose.

Keywords Fertilization · Manure · NPK fertilizer · Rhizophora apiculata

1 Introduction

The condition of mangrove forest areas, especially in North Sumatra, is currently quite alarming, it is known that the area of mangrove forests is decreasing with the change from mangrove forests to industrial areas, residential areas, illegal logging carried out arbitrarily, conversion of non-mangrove land uses such as aquaculture, oil palm plantations, agriculture and urban development and land in mangrove forests is used as new land by irresponsible parties [1].

Mangrove forests have many roles and functions that are very important for living things, it is necessary to maintain the preservation of mangrove forest vegetation,

Yunasfi (✉) · H. L. Gultom · M. Basyuni
Faculty of Forestry, Universitas Sumatera Utara, Medan, Indonesia
e-mail: yunasfijamhar@yahoo.co.id

M. A. Rahandi Lubis (ed.), *Proceedings of the 10th International Symposium for Sustainable Humanosphere*, Lecture Notes in Civil Engineering 255,
https://doi.org/10.1007/978-981-19-1422-5_5

one of which is by planting and providing fertilizer for mangrove species, namely *Rhizophora apiculata* or mangrove trees, according to information from people living in mangrove forest areas. In Lubuk Kertang Village, Pangkalan Brandan Barat, very few species of *R. apiculata* were found even though this species is a type of mangrove with a dominance level of 90%, often growing on muddy, smooth, deep soil and waterlogged at normal tides. *R. apiculata* does not like a harder substrate mixed with sand, likes tidal waters and has a strong influence on fresh water input, growth length and flowering throughout the year [2].

There is still very little research on direct fertilization of land or soil in mangrove forests with actual field conditions because there are several factors that are considered by someone to carry out this activity, namely the soil in mangrove forests is different from soil in general, the difference is the characteristics of the soil. In muddy mangrove forests which are always wet, humid and field conditions and mangrove plants are influenced by tides, the process of inserting fertilizer into the soil is quite difficult, and the soil in this mangrove forest is rich in organic matter where fertilization is simply defined as an activity in adding materials. or material in planting media and plants to meet the nutrient needs needed by these plants with the aim of increasing the productivity of a plant and improving and increasing soil fertility. What happens if the soil, which is already rich in organic matter, is still given fertilizer, whether it has a good impact on plants or does not have an effect at all, even as a result, the condition of plants and soil becomes worse [3].

2 Materials and Method

1. *Study Site*

The research was conducted in the Mangrove Forest, Lubuk Kertang Village, Brandan Barat District, Langkat Regency, as well as at the Forest Cultivation Laboratory, Faculty of Forestry, University of North Sumatra. The research was carried out from October 2020 to May 2021.

2. *Materials*

The materials used in this study were *R. apiculata* seeds, NPK fertilizer, chicken manure, cow manure, and quail fertilizer, name tags, plastic rope, sample envelopes, millimeter paper.

The tools used in this research are analytical balance, ruler, caliper, stationery, oven, camera, laptop, spss application, and calculator.

3. *Research procedure*

This research is an experimental method using a Completely Randomized Design (CRD) consisting of 5 treatments and 4 replications. Each replication consisted of 1 plant, so there were 20 experimental units, as follows:

P0 = No fertilizer application (control)
P1 = NPK fertilizer (12 g/plant)
P2 = Chicken Fertilizer (100 g/plant)
P3 = Cow Manure (200 g/plant)
P4 = Quail Fertilizer (5 g/plant)

Completely Randomized Design Linear Model (CRD) is as follows:

$$Yij = \mu + Ʈi + \sum ij$$

where

i = P0, P1, P2, P3, P4 (Treatment)
j = 1, 2, 3, 4 (Repeat)
Yij = Response or observation value in the i-th treatment and j repetition
μ = General average
Ʈi = Effect of the i-th practice
$\sum$ij = Error of the i-th treatment experiment and the j-th replication

4. *Land preparation*

First of all taken before carrying out the research is to determine the area to be used by sterilizing or cleaning the research site from weeds, garbage, and other impurities that can interfere with plants, so that the research area is clean. Then 2 plots were made for laying the seeds where 1 plot was a backup plot.

5. *Planting*

Before planting, R. apiculata seedlings were selected with the same number of leaves, namely 2 strands. The seeds used are healthy seedlings with characteristics of strong stems and fresh green leaves and no pest bite holes and uniform growth which are both 4 months old before being transferred to the land to be ready for planting. Planting activities were carried out manually by placing R. apiculata seeds in each planting hole. The selected seeds were then planted in 2 plots that had been prepared with a spacing of 0.5 m × 0.5 m. Then also prepared some seeds that have been planted and used for backup seeds for embroidery. After that it is labeled as an indicator if the seeds are for research purposes.

6. *Fertilizer application*

The fertilizers given in this study were NPK fertilizer as much as 12 g/seed, chicken fertilizer 100 g/seed, cattle fertilizer 200 g/seed, and quail fertilizer 5 g/seed. Fertilization is only done once in 3 months of observation. Fertilizer is given by inserting it into the dug holes around the seeds that have been made and then backfilled with soil.

7. *Maintenance of Rhizophora apiculata seedlings*

Embroidery should be done immediately (within 14 days after planting) if there are seeds that die/do not grow again. For embroidery, reserve seeds that have been provided or prepared in advance are used.

Every plant is not protected from the risk of pest and disease attacks, therefore it is necessary to control the attack. Control of caterpillar pests is done manually, namely by quoting pests directly from the attacked *R. apiculata* seedlings.

8. *Harvesting*

Harvesting was carried out on *R. apiculata* seedlings after the 6th observation and then manually removed.

9. *Observation and data collection*

In this research activity, observations were carried out every two weeks for 3 months. Data collection was carried out 6 times by observing several parameters.

(a) *Seedling height (cm)*

Measurement of seedling height was carried out by measuring the height of the seedling starting from the base of the lower stem 3 cm above the soil surface which was then given a marker as a measurement point, measured to the highest growing point using a meter every two weeks.

(b) *Seedling diameter (cm)*

Measured using a caliper. Diameter measurements were carried out at a height of 3 cm above the ground every two weeks.

(c) *Number of leaves (strands)*

Measurement of the number of leaves is carried out only on seedlings whose leaves can be counted, have grown and are perfectly formed. Measurements are carried out every two weeks.

(d) *Temperature (°C)*

This is done by observing the temperature in the field every two weeks.

(e) *Wet weight of the plant (g)*

This is done by weighing the entire plant that has been divided into several parts of leaves, stems, and roots using a scale at harvest before being put in the oven at 75 °C for 48 h until it reaches a constant state and is expressed in grams (g)/plant.

(f) *Plant dry weight (g)*

This is done by weighing the entire plant using a scale after removing it from the oven.

(g) *Leaf Area (cm^2)*

The measurement of leaf area was carried out after the observation ended by using the millimeter block paper method, placing the leaves on millimeter block paper and then drawing the leaves following the leaf pattern. Count the number of squares in the leaf area to calculate the leaf surface area.

The data collected during the research is primary data. Primary data were taken directly from the field, in the form of the results of measurements made on *Rhizophora apiculata* seedlings.

10. *Leaf Area Index (ILD) Analysis*

Leaf area index is expressed as the ratio of the leaf surface area (only one surface) to the area of land covered by the canopy. The average ILD price is stated as follows:

$$ILD = \frac{LD}{A}$$

where:

LD = Leaf Area
A = Indicates the area of crown cover if the plants have not yet touched the area of the individual canopy cover and when the plants are in contact, the area of crown cover is based on the spacing.

11. *Plant Growth Rate*

The calculation of the growth rate value is done by subtracting the value in the last observation data from the value at the first observation, so that the value in the result of the reduction will be the growth rate value for R. apiculata seedlings.

12. *Data analysis*

The data obtained from the results of the study were analyzed using Analysis of Variance and if there were significantly different results ($P < 0.05$) between treatments, a Duncan Multiple Range Test (DMRT) was carried out with an error rate of 5%.

3 Results and Discussion

1. *Seedling Height*

The results of the variance of seedling height showed the treatment of giving P1 (NPK fertilizer 12 g/plant), P2 (chicken fertilizer 100 g/plant), P3 (cow fertilizer 200 g/plant), P4 (quail fertilizer 5 g/plant) and P0 (control/not given fertilizer) had no significant effect on the increase in the height of R. apiculata seedlings, but at each observation there was an increase in seedling height. The results of the growth of the average height of seedlings can be seen in Tables 1 and 2.

Table 1 Analysis of variance seed height of *R. apiculata*

	Sum of squares	Df	Mean square	F	Sig
Between groups	118.875	4	29.719	1.817	0.178
Within groups	245.382	15	16.359		
Total	364.257	19			

Table 2 Average height of *R. apiculata* seedlings for 12 weeks

Treatment	Repetition				Total	Average
	I	II	III	IV		
P0	56.1	48.6	47.2	51.7	203.6	50.9^{ab}
P1	42.1	45.7	47.1	50.4	185.3	46.3^{a}
P2	50.8	59.3	55.2	49.2	214.5	53.6^{b}
P3	55.5	48.5	50.8	48.9	203.7	51^{ab}
P4	50	57.9	46.7	53.8	208.4	52.1^{ab}
Total	254.5	260	247	254	1015.5	

Note The average followed by the same letter in the same column or row shows no significant difference in the DMRT test at 5% level

Table 2 shows that the treatment of giving P2 (chicken fertilizer 100 g/plant) resulted in the highest growth value of seedling height compared to other treatments, which was 214.5 cm with an average of 53.6 cm. And the treatment of giving P1 (NPK fertilizer 12 g/plant) resulted in the lowest growth value of seedling height compared to other treatments, which was 185.3 cm with an average of 46.3 cm. This can happen because NPK fertilizer is a fertilizer that is easily soluble in water, therefore fertilizer is quickly carried away by water before it is well absorbed in the soil. According to Hamzah et al. [4] statement which states that NPK fertilizer is a fertilizer containing elements of Nitrogen, Phosphate and Potassium, where these elements have content that is easily soluble in water and cannot withstand exposure to sunlight (heat) so that evaporation is easy. (volatilization). P0 (control/ no fertilizer applied) was not significantly different from P1 (NPK fertilizer 12 g/ plant), P2 (chicken fertilizer 100 g/plant), P3 (cow manure 200 g/plant), P4 (quail fertilizer 5 g/plant). This indicates that the addition of manure can give a good and positive effect on the height growth of R. apiculata seedlings although the growth is not significant.

The application of P2 (chicken manure 100 g/plant) resulted in a better increase in plant height compared to the application of other types of fertilizers, this was because chicken manure contained high levels of N, P, and K elements which were useful for plant growth. Dewanto et al. [5] stated that chicken manure has a N content of 3 times greater than other manures with N content: 1.70%, P2O5: 1.90% and K2O: 1.50%. Where the element N is needed for plant vegetative growth. According to Anata et al. [8] stated that N promotes vegetative growth and stimulates the development of

Table 3 Analysis of variance seed diameter of *R. apiculata*

	Sum of squares	Df	Mean square	F	Sig
Between groups	0.118	4	0.030	0.923	0.476
Within groups	0.480	15	0.032		
Total	0.598	19			

Table 4 Average diameter of *R. apiculata* seedlings for 12 weeks selama

Treatment	Repetition				Total	Average
	I	II	III	IV		
P0	14.5	9.5	10.7	10.5	45.2	11.3[a]
P1	13.7	11	10.2	11.3	46.2	11.5[a]
P2	7.5	10.8	12.5	7.7	38.5	9.6[a]
P3	10.5	10.8	12.5	10.5	44.3	11.07[a]
P4	9.5	12.2	11	12.5	45.2	11.3[a]
Total	55.7	54.3	56.9	52.5		

Note The average followed by the same letter in the same column or row shows no significant difference in the DMRT test at 5% level

stems and leaves. The development of stems and leaves can be seen from the height, number of branches and also the wet and dry weight of leaves and stems.

2. *Seedling Diameter*

The results of the variance of seedling diameter showed that the treatments were P1 (NPK fertilizer 12 g/plant), P2 (chicken fertilizer 100 g/plant), P3 (cow fertilizer 200 g/plant), P4 (quail fertilizer 5 g/plant) and P0 (control/no fertilizer applied) had no significant effect on the increase in diameter of R. apiculata seedlings. The results of the increase in the average diameter of seedlings can be seen in Tables 3 and 4.

Table 4 shows that the treatment with P1 (NPK fertilizer 2 g/plant) resulted in the highest increase in seed diameter compared to other treatments, which was 46.2 with an average of 11.5. And the treatment of giving P2 (chicken fertilizer 100 g/plant) resulted in the lowest seed diameter increase value compared to other treatments, which was 38.5 with an average of 9.6. P0 (control/not given fertilizer) was not significantly different from P1 (12 g NPK fertilizer/plant), P2 (chicken fertilizer 100 g/plant), P3 (cow manure 200 g/plant), P4 (quail fertilizer 5 g/plant).

3. *Number of Leaves (strands)*

The results of the variance in the number of leaves showed the treatment of giving P1 (NPK fertilizer 12 g/plant), P2 (chicken fertilizer 100 g/plant), P3 (cow fertilizer 200 g/plant), P4 (quail fertilizer 5 g/plant) and P0 (control/not given fertilizer) significantly affected the increase in the number of leaves on *R. apiculata* seedlings. Thus, Duncan's further test was carried out with an error rate of 5% at P0 (control/no fertilizer) which was significantly different from P1 (12 g NPK fertilizer/plant), P2 (100 g chicken manure/plant), P3 (cow manure) 200 g/plant), P4 (quail fertilizer 5 g/plant) and P2 (chicken fertilizer 100 g/plant) were not significantly different from P3 (cow fertilizer 200 g/plant), and P4 (quail fertilizer 5 g/plant). The results of the average number of leaves can be seen in Tables 5 and 6.

Table 5 Analysis of variance number of leaves *R. apiculata*

	Sum of squares	Df	Mean square	F	Sig
Between groups	16.558	4	4.140	47.489	0.000
Within groups	1.308	15	0.087		
Total	17.866	19			

Table 6 Average number of leaves of *R. apiculata* for 12 weeks

Treatment	Repetition				Total	Average
	I	II	III	IV		
P0	5.3	5.5	5.3	4.6	20.7	5.2^{b}
P1	3.3	4.1	3.3	3.3	14	3.5^{a}
P2	5.3	5.8	6	5.5	22.6	5.7^{c}
P3	5.8	6	5.8	6	23.6	5.9^{c}
P4	6	6	5.8	6	23.8	6^{c}
Total	25.7	27.4	26.2	25.4	104.7	

Note The average followed by the same letter in the same column or row shows no significant difference in the DMRT test at 5% level

Table 6 shows that the effect of giving P1 (NPK fertilizer 12 g/plant), P2 (chicken fertilizer 100 g/plant), P3 (cow manure 200 g/plant), P4 (quail fertilizer 5 g/plant) and P0 (control/not given fertilizer) can increase the number of leaves of *R. apiculata* seedlings. Treatment with P4 (quail fertilizer 5 g/plant) resulted in the highest number of leaves from other treatments, which was 23.8 strands with an average of 6 strands. And the treatment of giving P1 (NPK fertilizer 12 g/plant) resulted in the lowest number of leaves from other treatments, namely 14 strands with an average of 3.5 strands.

Based on the observations made, the application of P4 (quail fertilizer 5 g/plant) gave better leaf growth results when compared to other types of fertilizers, this was presumably because the quails, which are poultry, were given feed sourced from industry and usually the rations were rich. of minerals and protein. According to Adiningsih [6] Feed rations that are rich in minerals and protein if given to animals will make animals able to produce feces and urine which are rich in nitrogen and other minerals. The nutrient that plays a role in leaf growth is nutrient N. By providing quail manure it will be able to increase the total or total N in the soil, so that what is needed by plants can be fulfilled properly. While the application of P1 (NPK fertilizer 12 g/plant) resulted in lower leaf growth results, this happened because there were insect bite marks in the form of caterpillars on the leaves so that there were several holes. This factor greatly affected the number of leaves on R. apiculata plants, so that the decrease in leaf number was not entirely caused by the administration of P1 (NPK fertilizer 12 g/plant). This is in line with the statement of Roidah [7] who said that in general the problems that occur in the growth of mangrove species are caused by the presence of bagworms that consume the lower surface of the leaves excessively. The sign is if there is a hole in the leaf, then this is most likely caused by the bagworm and the next stage the hollow leaf will turn yellow and then fall. Thus the number of leaves is reduced during the observation.

For the next factor that affects the growth of the number of leaves on R. apiculata seedlings, namely temperature as in Table 4, it is known that the type of *Rhizophora* sp. grows optimally at a temperature of 26 °C–28 °C and production will decrease further if the temperature is higher (Table 7). According to Anata et al. [8] statement which states that fresh leaves in mangrove plants will fall if the temperature is below the optimum and mangrove plants will also delay the production of new leaves if the temperature in the mangrove forest environment is above the optimum temperature. Leaves are the most important part of plants, this is because leaves are a tool used by plants to obtain food in addition to utilizing their roots in absorbing nutrients from the soil. If there are a lot of leaves on a plant, the photosynthesis process that takes place will be more and more and the food produced will also be more and more.

4. *Plant Wet Weight*

The results of the wet weight variance on the stems, roots and leaves of *R. apiculata* seedlings indicated the treatment of P1 (NPK fertilizer 12 g/plant), P2 (chicken fertilizer 100 g/plant), P3 (cow manure 200 g/plant), P4 (quail fertilizer 5 g/plant) and P0 (control/no fertilizer applied) had no significant effect on stem and root wet weight, but significantly on leaf wet weight for *R. apiculata* seedlings. Thus,

Table 7 Average temperature in Lubuk Kertang Mangrove Forest

Observation time (Day)	Temperature (°C)
0	26
14	27
28	26
42	28
56	25
70	25

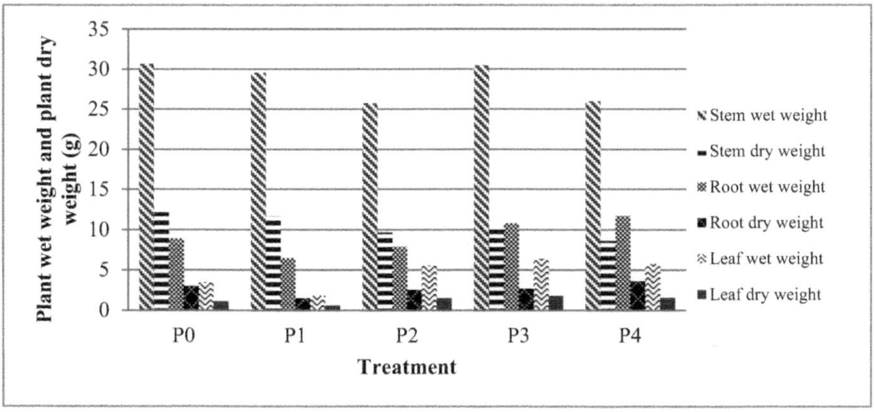

Fig. 1 Wet weight and average dry weight of *R. apiculata* seedlings for 12 weeks

Duncan's further test at P0 was not significantly different from P1 (NPK fertilizer 12 g/plant) but significantly different from P2 (chicken fertilizer 100 g/plant), P3 (cow manure 200 g/plant) and P4 (cow manure 200 g/plant) quail 5 g/plant). The average wet weight yield on stems, roots and leaves of *R. apiculata* seedlings can be seen in Fig. 1.

Figure 1 presents that the highest wet weight on the stems of *R. apiculata* seedlings was at P0 (control/not given fertilizer) of 30.649 g, the wet weight on the stems of R. apiculata seedlings was the lowest at the application of P2 (chicken fertilizer 100 g/ plants) of 25.709 g. Then the highest wet weight at the roots of R. apiculata seedlings was given P4 (quail fertilizer 5 g/plant) of 11.696 g, the lowest wet weight at the roots of R. apiculata seedlings was given P1 (NPK fertilizer 12 g/plant) of 6.436 g. And the highest wet weight on leaves of R. apiculata seedlings was given P3 (cow fertilizer 200 g/plant) of 6.402 g, wet weight on leaves of R. apiculata seedlings the lowest was on P1 (NPK fertilizer 12 g/plant) of 1.802 g.

The application of P4 (quail fertilizer 5 g/plant) wet weight to the roots of R. apiculata seedlings resulted in the greatest value among other treatments. This happens because the benefits of quail manure are able to correct the biological, chemical and physical properties of the soil and affect both soil fertility and roots in nutrient absorption [9]. For the results of weighing the wet weight of the plant will affect the result of weighing the dry weight of the plant. If the wet weight of the plant is high, it will produce a high dry weight of the plant.

5. *Plant Dry Weight*

The results of the dry weight variance on the stems, roots and leaves of *R. apiculata* seedlings showed the treatment of giving P1 (NPK fertilizer 12 g/plant), P2 (chicken fertilizer 100 g/plant), P3 (cow manure 200 g/plant).), P4 (quail fertilizer 5 g/plant) and P0 (control/no fertilizer applied) had no significant effect on stem and root dry weight, but had a significant effect on leaf dry weight on R. apiculata seedlings. Thus, Duncan's follow-up test at P0 (control/not given fertilizer) was not significantly different from the other treatments. However, P1 (NPK fertilizer 12 g/plant) was significantly different from P2 (chicken fertilizer 100 g/plant), P3 (cow manure 200 g/plant) and P4 (quail fertilizer 5 g/plant). The average dry weight yield on stems, roots and leaves of *R. apiculata* seedlings can be seen in Fig. 1.

From Fig. 1 it can be seen that the highest dry weight on the stems of *R. apiculata* seedlings was at P0 (control/no fertilizer) of 12.092 g, the dry weight on the stems of R. apiculata seedlings was the lowest at the application of P4 (5 g/quail fertilizer). plants) of 8.607 g. Then the highest dry weight at the roots of *R. apiculata* seedlings was at P4 (quail fertilizer 5 g/plant) of 3.575 g, the lowest dry weight at the roots of *R. apiculata* seedlings was at P1 (NPK fertilizer 12 g/plant) of 1.467 g. And the highest dry weight on leaves of R. apiculata seedlings was given P3 (cow manure 200 g/plant) of 1.797 g, the lowest dry weight on leaves of *R. apiculata* seedlings was given P1 (NPK fertilizer 12 g/plant) of 0.582 g.

The process of removing the moisture content or water content present in plant material is the definition of plant dry weight. Absorption of substances or nutrients by plants, irradiation from sun exposure, and CO_2 and water uptake affect the production of dry weight in plants [10] Dry weight shows the expertise of plants in taking components or nutrients from the growing media to support their growth. The increase in plant dry matter has something to do with plant metabolism or there is a more appropriate state of plant development in the occurrence of plant metabolic activities such as photosynthesis. Thus, continuing to be a large dry weight, will continue to be an effective process of photosynthesis. The higher the dry weight, the more efficient the photosynthesis activities that occur and the productivity and growth of tissue cells continues to be large and fast, so that plant development is better.

6. *Leaf Area*

The results of the variance in the number of leaves showed the treatment of giving P1 (NPK fertilizer 12 g/plant), P2 (chicken fertilizer 100 g/plant), P3 (cow fertilizer 200 g/plant), P4 (quail fertilizer 5 g/plant). plants) and P0 (control/no fertilizer applied) significantly affected leaf area of R. apiculata seedlings. Thus, Duncan's follow-up test at P0 (control/no fertilizer applied) was not significantly different from P1 (NPK fertilizer 12 g/plant), P2 (chicken manure 100 g/plant), P3 (cow manure 200 g/plant), and P3 (cow manure 200 g/plant). plants) and P4 (quail fertilizer 5 g/

plant). However, P1 (NPK fertilizer 12 g/plant) was significantly different from P2 (chicken fertilizer 100 g/plant) and P3 (cow manure 200 g/plant). The average leaf area yields can be seen in Tables 8 and 9.

Table 9 shows that the effect of giving P1 (NPK fertilizer 12 g/plant), P2 (chicken fertilizer 100 g/plant), P3 (cow manure 200 g/plant), P4 (quail fertilizer 5 g/plant) and P0 (control/not given fertilizer) can increase the leaf area of R. apiculata seedlings. The treatment of giving P3 (cow manure 200 g/plant) resulted in the highest leaf area value of the other treatments, which was 79.1 cm^2 with an average of 19.8 cm^2. And the treatment of giving P1 (NPK fertilizer 12 g/plant) resulted in the lowest leaf area value from other treatments, which was 52.3 cm^2 with an average of 13.1 cm^2.

Giving P3 (cow manure 200 g/plant) gave the best increase in leaf area where cow feces is rich in nutrients consisting of nitrogen (N), phosphorus (P) and potassium (K) and also has nutrients that are very useful for plants., one of which is in the process of fertilizing plants to make the development of these plants better and optimal. Nitrogen is the main nutrient that is absorbed and has an influence on the formation of leaf area. Nitrogen is an element that affects the development of leaves, because nitrogen is an element that has a role in the process of forming leaf green matter. Leaf green matter or chlorophyll is the green substance in leaves that is used for photosynthesis. If the chlorophyll formed in the leaves is getting bigger then the process of photosynthesis will be smoother and the photosynthate produced will be even greater. The results of photosynthesis will be used for the growth and development of plant organs, one of which is leaf formation which affects leaf area.

Table 8 Analysis of variance leaf area of *R. apiculata*

	Sum of squares	Df	Mean square	F	Sig
Between groups	126.193	4	31.548	3.996	0.021
Within groups	118.432	15	7.895		
Total	244.625	19			

Table 9 Average leaf area of *R. apiculata* for 12 weeks

Treatment	Repetition				Total	Average
	I	II	III	IV		
PO	15	17.2	16.4	13.8	62.4	15.6^{ab}
P1	10.5	14.5	12.5	14.8	52.3	13.1^{a}
P2	16.9	24.3	21.5	15.2	77.9	19.5^{b}
P3	19	19.1	25.2	15.8	79.1	19.8^{b}
P4	16.5	18.1	17.4	17.9	69.7	17.4^{ab}
Total	25.7	27.4	26.2	25.4	104.7	

Note The average followed by the same letter in the same column or row shows no significant difference in the DMRT test at 5% level

The growth of leaf area can be influenced by light intensity and environmental temperature, if the amount of light intensity needed is sufficient then leaf growth will be better and if the amount of light intensity is low then leaf growth will be bad. As stated by Roesmarkam and Yowono [11] which states that one of the factors of photosynthesis is light and photosynthesis is a process that is needed by green plants for growth. The process of respiration, transpiration and the physical structure of plants are influenced by light, the length of irradiation and the intensity of light.

Plants usually have the expertise to carry out photosynthetic activities and this skill is strongly influenced and determined by leaf area. The larger the leaf area, the more light rays can be captured by the plant. Because of this, the growth and yield of plants is strongly influenced by the leaf area. Where the leaf is a plant organ which is usually the place where photosynthesis takes place, by observing the leaf it can be seen its function as a light receiver and a means of photosynthesis. On this basis, leaf area is the main parameter. Because the rate of photosynthesis growth in most cases is strongly influenced by leaf area. If the leaf area increases, the resulting assimilation will also be greater. The large leaf area will cause the dry weight of the plant to increase as well.

7. *Leaf Area Index Analysis*

The results of analysis of variance in leaf area index showed that the treatment was P1 (NPK fertilizer 12 g/plant), P2 (chicken fertilizer 100 g/plant), P3 (cow manure 200 g/plant), P4 (quail fertilizer 5 g/plant). and P0 (control/not given fertilizer) significantly affected the analysis of leaf area index on R. apiculata seedlings. Thus, Duncan's follow-up test with an error rate of 5% At P0 (control/without fertilizer) was not significantly different from P1 (NPK fertilizer 12 g/plant), P2 (chicken fertilizer 100 g/plant), P3 (fertilizer cattle 200 g/plant) and P4 (quail fertilizer 5 g/plant). However, the treatment of giving P1 (NPK fertilizer 12 g/plant) was significantly different from P2 (chicken fertilizer 100 g/plant) and P3 (cow manure 200 g/plant). The results of the analysis of the average leaf area index on R. apiculata seedlings can be seen in Tables 10 and 11.

Table 5 presents that the treatment with P3 (200 g chicken fertilizer/plant) showed the highest total compared to other treatments, which was 158.2 with an average of 39.55 and P1 (NPK fertilizer 12 g/plant) showed the lowest total of the other treatments. that is equal to 104.4 with an average of 26.1. Leaf area in plants reflects the area of the part that carries out photosynthesis, whereas the leaf area index reflects the amount of sunlight absorbed by plants. Leaf area index is the ratio between leaf

Table 10 Analysis of variance of the leaf area index of *R. apiculata*

	Sum of squares	df	Mean square	F	Sig
Between groups	504.772	4	126.193	3.996	0.021
Within groups	473.730	15	31.582		
Total	978.502	19			

Table 11 Analysis of the average leaf area index of *R. Apiculata*

Treatment	Repetition				Total	Average
	I	II	III	IV		
P0	30	34.4	32.8	27.6	124.8	31.2^{ab}
P1	21	29	25	29.4	104.4	26.1^{a}
P2	33.8	48.6	43	30.4	155.8	38.95^{b}
P3	38	38.2	50.4	31.6	158.2	39.55^{b}
P4	33	36	34.6	35.8	139.4	34.85^{ab}
Total	155.8	186.2	185.8	154.8	682.6	

Note The average followed by the same letter in the same column or row shows no significant difference in the DMRT test at 5% level

surface area and plant spacing. Leaf area index is closely related to plant expertise in absorbing incoming solar radiation [12].

8. *Plant Growth Rate*

Based on the observations that have been made, it can be seen the value of the growth rate of R. apiculata seedlings with 6 observations for 12 weeks. The growth rate value is only calculated on the form of plant vegetative growth that is irreversible or unable to return to its original state, namely height, diameter, and number of leaves [13], [14].

The growth rate has been presented Table 12. Based on the growth rate observed in R. apiculata seedlings given the highest seedling height increase treatment, the highest growth rate was in the P3 treatment (cow manure 200 g/plant) and the highest seed diameter parameter was given P3 (cow manure 200 g/plant). And the P3 treatment (cow manure 200 g/plant) was able to increase the increase in stem diameter of R. apiculata seedlings by 1.2 cm/12 weeks. This is because cow feces has enormous benefits and uses in accordance with the statement [5] that cow feces can nourish plants so that plant growth will be better and optimal. In the parameter of the number of leaves, which had the highest number, the administration of P4 (5 g/plant quail fertilizer) was able to increase the number of leaves by 15 strands/12 weeks. Based on the calculation of the growth rate of R. apiculata seedlings treated with P1 (NPK fertilizer 12 g/plant) the lowest value was 11.2 cm/12 weeks.

Table 12 Growth rate of *R. apiculata* seedlings for 12 weeks

Observation parameters	Treatment					Unit (n/12 weeks)
	P0	P1	P2	P3	P4	
Average height	12.3	11.2	12.4	12.5	11.6	cm/12 week
Average diameter	0.4	1	0.8	1.2	0.7	mm/12 week
Number of Leaves	9	5	12	14	15	sheets/12 week

4 Conclusion

The most influential treatment on the growth of R. apiculata seedlings was given P3 (cow dung 200 g/plant) with the highest data values on 3 variables from 6 variables, namely the wet weight variable on the leaves with a value of 6.402 g, dry weight on leaves with a value of 1.797 g, and leaf area with a total value of 79.1 cm^2 and an average of 19.8 cm^2. Giving P3 (cow dung 200 g/plant) also affected the analysis of leaf area index and growth rate on the height and diameter of R. apiculata seedlings. It is better to do further research by giving a larger dose of organic fertilizer.

Acknowledgements The authors gratefully thank to the Research and Community Service Institute that has assisted in financing based on the NON-PNBP contract of the University of North Sumatera in accordance with the Letter of Assignment for the Implementation of Community Service Implementation of the Regular Year Mono Partnership Scheme Source of Non-PNBP Funds USU T.A. 2021 Number: 184/UN5.2.3.21/PPM/2021.

References

1. Basyuni M, Al Habib A, Slamet B, Sulistiyono N, Putri LAP, Bimantara Y, dan Lesmana I (2019) Assessment of one-year mangrove reforestation using Rhizophora apiculata seedlings in Lubuk Kertang village, North Sumatra. IOP Conf Series: Earth Environ Sci 260(1)
2. Noor, YR, Khazali M, & Suryadiputra INN (2006) Guide to the introduction of mangroves in Indonesia. Wetlands International Indonesia Programme, Bogor
3. Herison A, Romdania Y (2020) Mangrove for civil engineering. Mangrove Ecosyst for Development, 109–110
4. Hamzah AHP, Anggoro T, Puryono S (2020) The co-management concept in the mangrove environmental management approach in Lubuk Kertang Village, West Brandan, North Sumatra. J-MAS (J Manage Sci) 5(1):129–135
5. Dewanto FG, Londok JJ, Tuturoong RA, Kaunang WB (2017) Effect of inorganic and organic fertilization on corn crop production as a feed source. Zootec 32(5)
6. Adiningsih Y (2008) Changes in nitrogen, phosphate, and potassium content in NPK fertilizer storage. J Ind Technol Res 2(4)
7. Roidah IS (2018) Benefits of using organic fertilizers for soil fertility. Bonorowo J 1(1):30–43
8. Anata R, Sahiri N, Ete A (2014) Effect of various compositions of planting media and manure on growth and yield of Daun Dewa (Gynura Pseudochina (L.) Dc). Doctoral Dissertation, Tadulako University
9. Setyamidjaja D (1986) Fertilizer and fertilizer. CV Simplek Publisher, Jakarta
10. Utari V, Ekyastuti W, Orahami A (2017) Conditions of insect pest attacks on mangroves (*Rhizophora apiculata* BI) in PUP PT. Bina Oviviparian Universe West Kalimantan 5(4):999–1007
11. Roesmarkam A, Yowono NW (2002) Soil fertility science. Canisius, Yogyakarta
12. Sitompul SM, Guritno B (1995) Analysis of plant growth. Gadjah Mada University Press, Yogyakarta (ID), p 412
13. Robertson AI (1986) Leaf-burying crabs: their influence on energy flow and export from mixed mangrove forests (*Rhizophora spp.*) in northeastern Australia. J Experim Marine Biol Ecol 102(2–3):237–248
14. Trisnawati T, Wardati W, Yulia AE (2017) Growth of *Rhizophora sp.*mangrove seeds. In: Hydraquent medium given several doses of NPK, Doctoral Dissertation, Riau University

Chemical and Thermal Properties of Different Type of Indonesian Ramie Fibers

Maya Ismayati, Muhammad Adly Rahandi Lubis, Widya Fatriasari, Sudarmanto, Fahriya Puspita Sari, Faizatul Falah, Nissa Nurfajrin Solihat, Azizatul Karimah, Asri Peni Wulandari, Arif Rahman, Joko Kusmoro, Budi Irawan, and Suryana

Abstract Ramie (*Boehmeria nivea*) is an alternative Indonesian fiber and expected to replace imported cotton. An understanding of the chemical components and thermal properties of ramie will facilitate the application of ramie fibers. This study aims to determine the chemical component and thermal properties of bark ramie, decorticated ramie, Ina-grass ramie, degummed ramie 1, degummed ramie 2, and bleached ramie fiber. Chemical component was evaluated by Py-GCMS and wet-chemical analysis, meanwhile thermal properties was determined by TGA and DSC analysis. Py-GCMS results shown pyrolysis products originated from carbohydrates and lignin. The bark of ramie showed a lower carbohydrate, about 53.16%. Meanwhile, total carbohydrate of ramie and degummed fiber samples are above 90%. Similar result was shown by wet-chemical analysis, wherein the total carbohydrate for bark, de-corticated, Ina-grass, degummed ramie 1, degummed ramie 2 and bleached are 76.29%, 97.62%, 96.78%, 99.15%, 97.63%, and 98.92%, respectively. A slight higher Py-GCMS results than that of the wet-analysis may be due no-preparation needed on sample. Thermal properties shown that bleached ramie has the lowest thermal properties at 600 °C with the lowest lignin content. In the future, Indonesian ramie with high thermal properties can be used as an alternative fiber for composite and biomaterial products.

Keywords Ramie · Py-GCMS · TGA · DSC · Chemical component · Thermal properties

---

M. Ismayati (✉) · M. A. R. Lubis · W. Fatriasari · Sudarmanto · F. P. Sari · F. Falah · A. Karimah
Research Center for Biomass and Bioproducts, National Research and Innovation Agency (BRIN), Bogor, Indonesia
e-mail: maya_ismayati@biomaterial.lipi.go.id

N. N. Solihat · A. P. Wulandari · J. Kusmoro · B. Irawan · Suryana
Department of Biology, Universitas Padjajaran, Sumedang, Indonesia

A. Rahman
Program Study of Polymer Chemical Engineering, Ministry of Industry, Polytechnic STMI Jakarta, Jakarta, Indonesia

M. A. Rahandi Lubis (ed.), *Proceedings of the 10th International Symposium for Sustainable Humanosphere*, Lecture Notes in Civil Engineering 255,
https://doi.org/10.1007/978-981-19-1422-5_6

1 Introduction

Currently, cotton is the main raw material for the domestic textile industry. However, the local cotton varieties have not fulfilled the domestic demand of Indonesia. The International Trade Center (ITC) reported that Indonesia became the largest cotton importer in the world with an import value of USD 2.4 billion or 4.4% of the world's cotton needs in 2018. Whereas the domestic production of cotton fiber is less than 4% [1]. Reducing the level of dependence on cotton imports with natural fibers as an alternative fiber source is a strategic step to increase the competitiveness of the textile industry. Textile can be made from Indonesian local natural fibers. Ramie (*Boehmeria nivea*) as a strongest plant fiber with the highest strength and long fiber is a remarkable fiber for use in textile manufacture [2].

Ramie has been grown in Indonesia since 1985 after the era Japanese occupation [3]. Ramie or also known as "China grass" in accordance with its original name of ramie fiber that fulfill the 90% of total production in the world [4, 5]. The chemical component of raw ramie fiber are gums (20–30%), hemicellulose (8–16%), lignin (1–4%), pectin (26%), and wax (1–2%) [2, 6, 7]. Differences in the chemical components of ramie and its derivative products will affect its appropriate application. Textile industry and eco-friendly composites mainly light weight composites to substitute synthetic fiber-oriented composites become the primary application of natural fiber such as ramie recently including in Indonesia [8–11]. Previously, Aristri et al. [12] has been reported the use of bio-polyurethane resins derived from liquid fraction of lignin to modify the ramie fiber. As result this treatment can improve the thermal stability by 15%. A wide application of ramie is due to their properties as antibacterial and antiseptic that also shown a distinctive property, such as silky luster, high absorbency, high tensile strength, wrinkle resistance, and coolness [2, 13–16]. However, the flammability and thermal stability of ramie fibers tends to unstable [17].

In previous report, the characterization has been focused mainly in the China grass after degumming. While, a similar research on ramie from Indonesia (Ina Grass) has not been carried out. During Ina grass preparation, some by-products might be produced thus by this current investigation, possible utilization of them will be more opened after understanding these properties. In this study, the ramie sample consisting of bark of ramie, decorticated fiber, Ina-grass, degummed fiber (A), degummed fiber (B), and bleached fiber were comprehensively characterized for chemical component (Py-GCMS and wet-chemical analysis) and thermal properties (Thermo Gravimetry Analysis (TGA) and Differential Scanning Calorimetry (DSC)). A comprehensive information about this ramie is expected to facilitate the application of it in the future.

2 Materials and Methods

2.1 Preparation of Ramie Fiber Sample

The following is a sample of ramie fiber preparation were used in this study:

Bark. The bark (or bart) of ramie were harvested from the farm of CV RABERSA, Wonosobo, Central Java, Indonesia was cleaned from any other parts such as its leaves and flowers. The bast (without leaves) are separated from its outer layers (ramie-waste chip) and the fiber parts by a decorticator machine. The ramie-waste chip (measured around 1–3 cm in length) were the by-product biomass from ramie fiber process.

Decorticated. Decortication process of ramie involving with the help of water, subsequently the process called as wet decortication process. This process occurs when fresh-harvested ramie bast was decorticated along with water added to the decorticating machine. The decorticating process of ramie bast happens in repetitive manners resulting decorticated fiber product with brownish color and stiff fiber characteristics.

Ina-grass. Pre-treatment process: The decorticated fibers were soaked in water for approximately 2 nights/days. This process allows the excessive gum in the fiber dissolve in the water. Afterwards, the hand-combing process occurs to decrease the disarrangement degree/index of the fiber.

Degummed 1 (chemical). Pre-treatment process: For around 2 nights/days, the decorticated fibers were immersed in water. The extra gum in the fiber dissolves in the water as a result of this process. Following that, the disarrangement degree/index of the fiber was reduced by hand-combing.

Degummed 2 (biological). Ina-grass fibers were biologically degummed with pectinolytic fungus. The bio-degumming process occurs approximately for 72 h. The Ina-grass fiber will be fermented along with the fungus, then washed under running water to clean the residual materials from fermentation process. Then, the biological degummed fiber was mechanically brushed.

Bleached. Bleached fiber resulted from treated degummed fiber (1) with bleaching agent. The bleaching process happens for around 2 h. After the bleaching process the bleached fiber will undergo mechanical brushing to enhance the fiber quality (Fig. 1).

2.2 Py-GCMS Analysis

Chemical component of ramie fiber was analyzed by Py-GCMS. A 500 μg of ramie samples were analyzed by pyrolysis–gas chromatography-mass spectrometry (Py-GCMS). It was put in eco-cup SF PY1-EC50F and covered by glass wool. Furthermore, the eco-cup was pyrolyzed at 500 °C for 0.1 min using a multi-shot pyrolyzer (EGA/PY-3030D) which was interfaced (interface temperature 280 °C) with a GC/MS system QP-2020 NX (Shimadzu, Japan) equipped with an SH-Rxi-5Sil MS column (30 m × 0.25 mm i.d. film thickness. 0.25 μm), with electron impact of

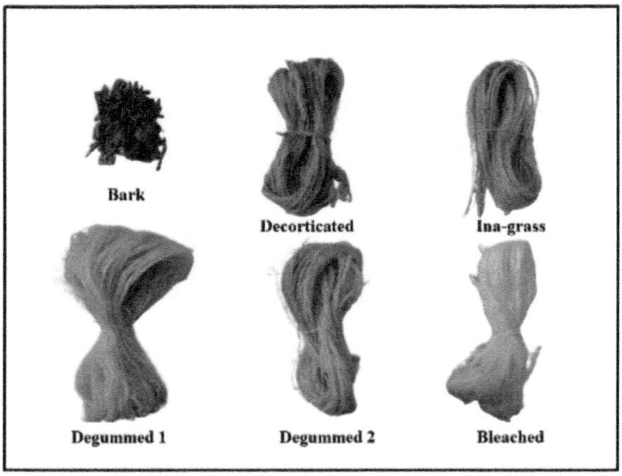

Fig. 1 Ramie samples were used in this experiment

70 eV and helium as a carrier gas. The pressure was 20.0 kPa (15.9 mL/min, column flow 0.61 mL/min). The temperature profile for GC was as follows: 50 °C hold for 1 min, then the temperature increases until 280 °C (5 °C/min), and 13 min at 280 °C. The pyrolysis products resulting were identified by comparing their retention times and mass spectra data with NIST LIBRARY 2017.

2.3 Chemical Analysis for Extractive, Lignin, Holocellulose, and Alpha-Cellulose

The sample determining the chemical composition of Ina-grass were collected by grinding. The chemical component of the ramie fibers was determined following several appropriate standards. Ethanol–benzene dissolved extractives, holocellulose, α-cellulose, acid insoluble lignin (AIL), and acid soluble lignin (ASL) were determined in accordance with the NREL/TP-510-42,619, [18, 19], NREL LAP 2012, and NREL LAP 2012).

2.4 Thermal Analysis

The thermal properties and stability of ramie fiber samples were monitored using Thermogravimetric Analyzer (TGA 4000, Perkin Elmer Inc., USA). About 10 mg ± 0.1 mg of sample was placed in a crucible pan and was heated from 25 °C to 750 °C at a heating rate of 10 °C/min. Nitrogen gas was purged with a flow of 20 mL/min during the analysis to prevent oxidation. The weight loss and derivative thermogravimetric (DTG) of each sample were calculated using Pyris software (Ver. 11.1.1, Perkin Elmer Inc., USA).

3 Results and Discussion

3.1 Chemical Component of Ramie by Py-GCMS and Wet-Chemical Analysis

Chemical component of ramie fiber samples was evaluated by Py-GCMS and also wet-chemical analysis. Figure 2 shown the Py-GCMS result for all ramie samples. The pyrogram generated the pyrolysis product peaks corresponding to carbohydrate and lignin, in more detail lignin is a combination of *p*-coumaryl unit (Lignin-H), coniferyl alcohol or guaiacol (Lignin-G) and synapyl alcohol or syringol (Lignin-S) [20, 21]. By Py-GCMS analysis, a macromolecule (carbohydrate, lignin) was fragmented into macromolecules as called pyrolysis products [22]. Pyrolysis products originated from carbohydrate were detected in the beginning of pyrogram, as readily fragmented followed by lignin pyrolysis products. While, an iconic peak of levoglucosane detected as a board with retention time, 25.93 min, respectively. A total relative abundance percentage was calculated as the total yield of carbohydrate and lignin. The bark of ramie showed a lower carbohydrate, 53.16%. Meanwhile, total carbohydrate of Ina-grass and degummed ramie samples are >90%. Similar result shown by wet-chemical analysis, wherein the total carbohydrate for bark, decorticated, Ina-grass, degumming (1), degumming fiber (2) and bleached are 76.29%, 97.62%, 96.78%, 99.15%, 97.63%, and 98.92%, respectively. In both analysis, degumming process by chemical and biological successfully removed lignin. A rapid screening of chemical component might play a role in biomass application, by using Py-GCMS is useful with small amount sample without any pre-treatment required [22, 23] and more environmentally friendly. But it has not determined hemicellulose and cellulose directly [22].

On lignin, this ramie fiber dominated by guaiacol unit with S/G ratio is 0.61, respectively. This S/G ratio is higher than banana fiber (0.55), jute fiber and bamboo [23, 24]. The higher lignin content presented by bark of ramie was followed by decorticated ramie fiber, about 46.85% and 12.12%, respectively. The result is similar with decorticated ramie fiber from India, about 9.25% [25]. The lignin-H unit was detected on program may come from extractives or a second fraction of benzene compounds. So, the lignin content may still be detected in lower than 5%. These chemical components of natural fiber will affect the properties of physical, mechanical and thermal [26].

3.2 Thermal Properties

TGA and DSC were used to acquire thermal characteristics of ramie fiber variations, as represent in Fig. 4. Figure 4a shows that all ramie fiber variations have similar pattern with three steps of thermal degradation. Region I was particularly attributed of the evaporation of fiber moisture with a slight weight loss at temperature 25 °C until

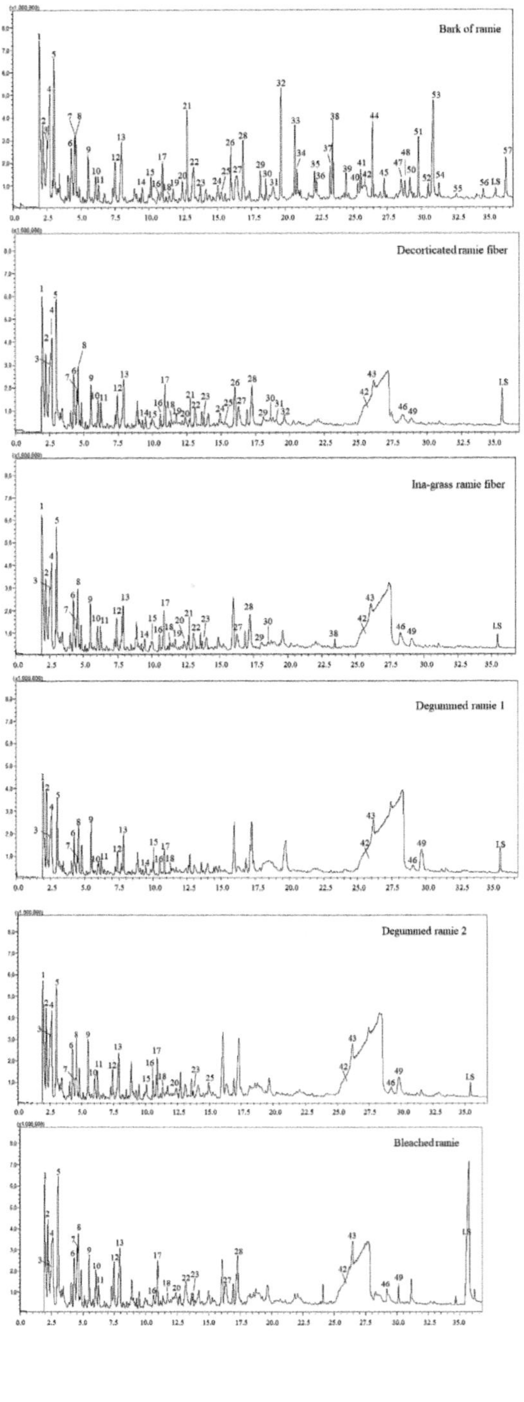

Fig. 2 Pyrogram Py-GCMS of ramie. Pyrolysis products were described in Table 1 with *n*-eicosane used as an internal standard (IS)

Table 1 The percentage of relative abundance of ramie samples were analyzed by Py-GCMS

No	Rt (min)	Pyrolysis products	Origin	Bark of ramie	Decorticated	Ina grass	Degumming (1)	Degumming (2)	Bleached
1	1.93	Ethyne, fluoro-/carbamic acid	C	8.73	10.99	8.81	7.75	9.47	9.19
2	2.18	1-Propen-2-ol, acetate	C	2.72	3.7	4.91	5.57	6.24	6.78
3	2.50	2,3-Butanedione	C	1.74	3.07	3.93	2.06	4.66	5.04
4	2.73	Acetic acid	C	6.73	6.09	5.25	5.3	6.32	4.01
5	3.03	2-Propanone, 1-hydroxy-	C	5.11	6.6	7.16	3.98	6.64	9.41
6	4.29	2-Butanone, 1-(acetyloxy)-	C	3.14	3.64	3.04	1.99	2.9	3.89
7	4.55	Propanal	C	1.94	1.6	1.02	0.67	0.84	3.42
8	4.62	Propanoic acid, 2-oxo-, methyl ester	C	1.53	2.3	2.44	2.04	2.85	3.08
9	5.52	Furfural	C	1.6	2.16	2.2	2.33	2.85	2.63
10	6.07	2-Hexanone, 3-methyl-	C	0.95	2.2	1.38	0.73	1.31	2.55
11	6.28	1,2-Ethanediol, diacetate	C	0.66	1.7	1.12	0.6	1.45	1.44
12	7.51	2(5H)-Furanone	C	1.9	2.3	1.6	1.16	1.67	2.87
13	7.98	1,2-Cyclopentanedione	C	3.32	4.5	2.85	2.36	3.07	5.25
14	9.47	Phenol	LH	0.76	0.73	0.55	0.37	N.d	N.d
15	10.13	Oxazolidine, 2,2-diethyl-3-methyl-	C	0.88	0.35	0.43	0.69	0.6	N.d
16	10.66	2-Cyclopenten-1-one, 2-hydroxy-3-methyl-	C	0.66	0.79	0.74	0.69	0.96	0.52
17	10.98	2-Cyclopenten-1-one, 2-hydroxy-3-methyl-	C	2	2.71	1.99	1.17	2.25	2.84
18	11.48	4-Methyl-5H-furan-2-one	C	0.36	0.55	0.16	N.d	0.29	0.48
19	11.74	Phenol, 2-methyl-	LH	0.55	0.49	0.47	N.d	N.d	N.d
20	12.42	Phenol, 3-methyl-	LH	0.72	1.32	0.61	N.d	0.64	1.01

(continued)

Table 1 (continued)

No	Rt (min)	Pyrolysis products	Origin	Bark of ramie	Decorticated	Ina grass	Degumming (1)	Degumming (2)	Bleached
21	12.80	Phenol, 2-methoxy-	LG	3.41	0.46	0.59	N.d	N.d	N.d
22	13.27	Pentanal	C	2.59	1.82	1.3	0.21	N.d	2.77
23	13.78	2-Cyclopenten-1-one, 3-ethyl-2-hydroxy-	C	0.47	0.59	0.25	0.5	0.26	0.52
24	15.06	2H-Pyran-2-one, 6-ethyltetrahydro-	C	0.9	0.85	N.d	0.49	5.85	N.d
25	15.31	2,3-Dihydroxybenzaldehyde	C	0.49	0.17	N.d	N.d	N.d	N.d
26	16.01	Creosol (4-methyl Guaiacol)	LG	2.06	3.51	N.d	N.d	N.d	N.d
27	16.45	Catechol	LH	2.67	2.5	1.36	N.d	N.d	3.03
28	16.89	Benzofuran, 2,3-dihydro-	C	2.84	3.96	3.21	5.8	N.d	5.1
29	18.15	1,2-Benzenediol, 3-methoxy- (2-methoxy catechol)	LH	1.18	1.21	0.53	N.d	N.d	N.d
30	18.56	Phenol, 4-ethyl-2-methoxy- (4-ethyl G)	LG	0.72	0.24	0.25	N.d	N.d	N.d
31	19.11	1,2-Benzenediol, 3-methyl- (3-methyl catechol	LH	0.97	0.87	N.d	5.06	N.d	N.d
32	19.69	2-Methoxy-4-vinylphenol (4-vinyl G)	LG	4.41	0.79	N.d	N.d	N.d	N.d
33	20.71	Phenol, 2,6-dimethoxy- Syringol	LS	2.57	N.d	N.d	N.d	N.d	N.d
34	20.85	Eugenol	LG	0.92	N.d	N.d	N.d	N.d	N.d
35	22.15	Vanillin	LG	1	N.d	N.d	N.d	N.d	N.d
36	22.28	trans-Isoeugenol	LG	0.48	N.d	N.d	N.d	N.d	N.d
37	23.31	3,5-Dimethoxy-4-hydroxytoluene (4-methyl S)	LS	1.13	N.d	N.d	N.d	N.d	N.d
38	23.51	trans-Isoeugenol	LG	2.87	N.d	0.32	N.d	N.d	N.d
39	24.47	Apocynin /Acetoguaiacone	LG	0.91	N.d	N.d	N.d	N.d	N.d
40	25.34	Benzene, 1,2,3-trimethoxy-5-methyl- (ethylS)	LS	0.4	N.d	N.d	N.d	N.d	N.d

(continued)

Table 1 (continued)

No	Rt (min)	Pyrolysis products	Origin	Bark of ramie	Decorticated	Ina grass	Degumming (1)	Degumming (2)	Bleached
41	25.55	2-Propanone, 1-(4-hydroxy-3-methoxyphenyl)- (Guacylacetone)	LG	0.74	N.d	N.d	N.d	N.d	N.d
42	25.93	Levoglucosane	C	1.9	21.52	36.88	38.64	33.14	19.48
	26.15	1,3-Di-O-acetyl-.alpha.-.beta.-d-ribopyranose	C	N.d	1.12	2.34	3.35	2.05	N.d
43	26.43	Phenol, 4-ethenyl-2,6-dimethoxy- (4-vinyl S)	LS	2.86	N.d	N.d	N.d	N.d	N.d
44	27.28	Phenol, 2,6-dimethoxy-4-(2-propenyl)- (propenylS cis)	LS	0.61	N.d	N.d	N.d	N.d	N.d
	28.24	1,6-Anhydro-.alpha.-d-galactofuranose	C	N.d	1.92	1.36	N.d	0.82	N.d
45	28.51	(E)-2,6-Dimethoxy-4-(prop-1-en-1-yl)phenol (Allyl S)	LS	0.82	N.d	N.d	N.d	N.d	N.d
46	28.79	Benzaldehyde, 4-hydroxy-3,5-dimethoxy- (syringaldehyde)	LS	0.62	N.d	N.d	N.d	N.d	N.d
	28.85	1,6-Anhydro-.beta.-D-glucofuranose	C	N.d	0.69	0.94	3.68	2.88	N.d
47	29.15	(E)-4-(3-Hydroxyprop-1-en-1-yl)-2-methoxyphenol (coniferyl alcohol cis	LG	0.85	N.d	N.d	N.d	N.d	N.d
48	29.79	Phenol, 2,6-dimethoxy-4-(2-propenyl)- (propenyl trans)	LS	2.18	N.d	N.d	N.d	N.d	N.d
49	30.49	Ethanone, 1-(4-hydroxy-3,5-dimethoxyphenyl)- (Acetosyringone)	LS	0.56	N.d	N.d	N.d	N.d	N.d
50	30.87	(E)-4-(3-Hydroxyprop-1-en-1-yl)-2-methoxyphenol (Coniferyl alcohol trans)	LG	6.5	N.d	N.d	N.d	N.d	N.d
51	31.28	Syringylacetone	LS	0.7	N.d	N.d	N.d	N.d	N.d

(continued)

Table 1 (continued)

No	Rt (min)	Pyrolysis products	Origin	Bark of ramie	Decorticated	Ina grass	Degumming (1)	Degumming (2)	Bleached
52	32.56	Propiosyringone	LS	0.37	N.d	N.d	N.d	N.d	N.d
53	34.55	Sinapyl alcohol (cis)	LS	0.52	N.d	N.d	N.d	N.d	N.d
54	36.23	trans-Sinapyl alcohol	LS	1.79	N.d	N.d	N.d	N.d	N.d
		Total Carbohydrate (%)		**53.16**	**87.89**	**95.31**	**91.76**	**99.37**	**91.27**
		Total LH (%)		6.85	7.12	3.52	5.43	0.64	4.04
		Total LG (%)		24.87	5	1.16	0	0	0
		Total LS (%)		15.13	0	0	0	0	0
		Total Lignin (%)		**46.85**	**12.12**	**4.68**	**5.43**	**0.64**	**4.04**
		S/G ratio		**0.61**	**0.00**	**0.00**	**0.00**	**0.00**	**0.00**
		L/C ratio		**0.88**	**0.14**	**0.05**	**0.06**	**0.01**	**0.04**

Note The values were means from three replicates; N.d means not detected; Rt (min) = Retention time (minutes); LH = Lignin H-unit; LG = Lignin G-unit; LS = Lignin S-unit; C = Carbohydates; L = Lignin; H = p-hydroxyphenyl; G = guaiacyl; S = syringyl

~100 °C. In this region, moisture content and low molecular weight materials such as extractives was removed and decomposed with a maximum evaporation temperature around ~50 °C which is depict in Fig. 4b. Region II was an primary stage of degradation of lignin-carbohydrate in ramie fibers with the range of temperature ~100 °C until ~400 °C. Lignin as one of main component in cell wall structure started to degrade from a temperature of 200 °C [12]. Decomposition of carbohydrates proceeded first from 120 to 260 °C, followed by lignin degradation at 200 to 480 °C [27]. On the primary stage, the weight loss was the highest, indicating that the cellulose and others major component of lignocellulose degraded and produced volatile molecules such as carbon monoxide, methane, and carbon dioxide with a maximum degradation temperature around ~350 °C (Fig. 4b). At temperatures ranging from ~400 °C to 600 °C, region III revealed the further breakdwon of thermal degradation intermediates such as levoglucosan and sugar to produce a volatile molecules such as hydrogen, ethylene, ethane, and tar [28, 29]. Also the aromatic ring of lignin structure was further decomposed. According to the results, bark has the highest thermal stabilization otherwise bleached ramie presents the lowest thermal stabilization, as evidenced by the residual weight value (the value of weight loss at final temperature or 600 °C on Fig. 3a), due to bark sample has the highest lignin content and bleached ramie has lower lignin content (as shown in Fig. 3). As well known, lignin has a complex structure that can interact with cellulose and leading a compact structure of biomass which increases the thermal stability [30].

According to DSC analysis (Fig. 4b), ramie fibers have two peaks, Tp1 as endothermic reaction peak and Tp2 as exothermic reaction peak. During the endothermic reaction at onset temperature of around ~25 °C, end temperature of around ~90 °C and maximum temperature of around ~50 °C (Table 1), ramie fibers absorbed heat from the environment to evaporate the moisture content [12, 29]. Variation of ramie fiber have different value of Tp1 peak (maximum temperature) which is indicating the moisture content of sample. Bleached ramie and degummed ramie-1 had the highest and lowest Tp1 peak value, as well as the highest and lowest moisture content. The Tp2 peak as maximum exothermic peak observed around ~350 °C indicating the degradation of hemicellulose and decomposition of α-cellulose in the fiber

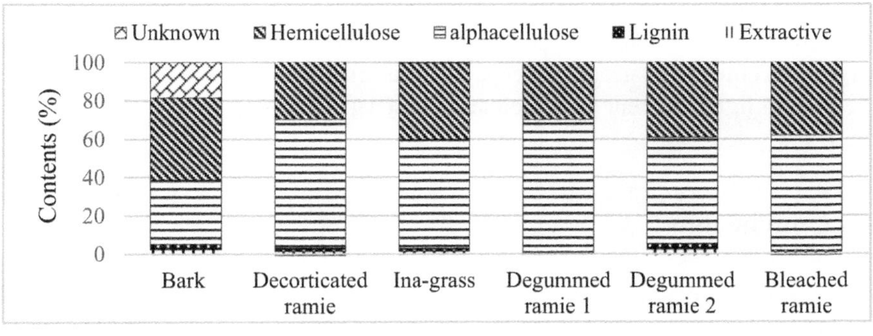

Fig. 3 Chemical composition on ramie fiber samples was evaluated by wet-chemical analysis

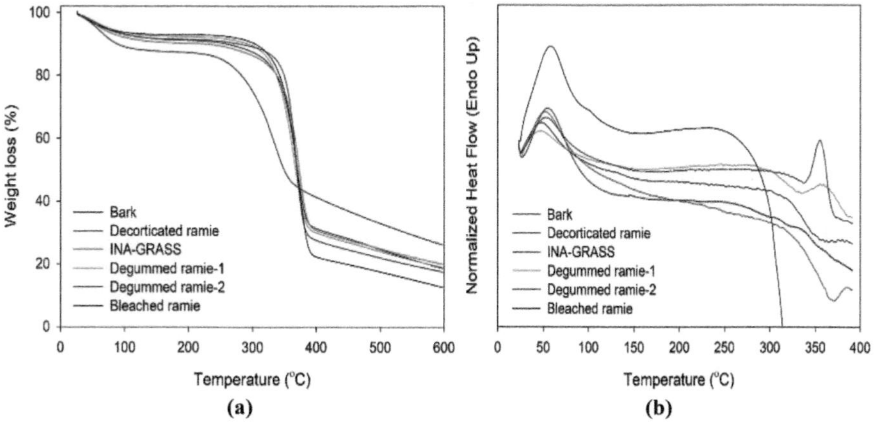

Fig. 4 TGA (**a**) and DSC (**b**) thermograms

Table 2 The Transition glass (*Tg*) value of ramie fiber

Ramie fibers	Tp_1 value (°C)			*Tg* value (°C)	
	Onset	Peak	End	Onset	End
Bark	29.20	56.53	91.54	224.67	247.18
Decorticated ramie	25.42	48.55	85.77	173.46	212.64
INA-GRASS	27.07	52.85	91.16	190.00	231.66
Degummed ramie 1	25.91	47.01	89.50	178.09	246.17
Degummed ramie 2	27.20	54.00	92.58	180.12	250.68
Bleached ramie	27.00	57.46	86.04	165.07	228.81

[27]. At this temperature, the fiber has been plasticized, which deformed the structure of the fiber and reduced its rigidity that is known as glass transition temperature (Tg) [17]. The Tg value of synthetic polymers is approximately 100–180 °C [17] but the Tg value (onset) of variation of ramie fiber (Table 2) was around 165–250 °C in this study. Bark and bleached ramie had the highest and lowest Tg values (onset), respectively, due to their high and low lignin concentrations. Because lignin have the largest -OH functional group that capable of forming hydrogen bonds with cellulose, the highest lignin content generate the highest Tg value [12].

4 Conclusions

The chemical component of ramie sample was successfully characterized by Py-GCMS and wet-chemical analysis. Total carbohydrate of Ina grass and degummed ramie samples are >90%. Similar result shown by wet-chemical analysis, wherein the total carbohydrate for bark, de-corticated, ina-grass, degummed 1, degummed 2 and bleached are 76.29%, 97.62%, 96.78%, 99.15%, 97.63%, and 98.92%. Bark of ramie have a higher lignin, followed by decorticated ramie, degummed 1, ina grass, bleached ramie and degummed 2. It correspondences to thermal properties, high lignin content can improve the thermal stability of bark of ramie, with a higher thermal stability.

Acknowledgements This work was supported by Productive-Innovative Research Program (RISPRO) by Indonesia Endowment Fund for Education (LPDP), Ministry of Finance in a collaboration with Ministry of Research and Technology/National Research and Innovation Agency (Kemenristek-BRIN), under the National Research Priority (PRN) program (grant number 273/E1/PRN/2020 and B-4528/IPH/KS.02.04/VII/2020). Author also would like to thank to JSPS/JAAI/JASTIP for publication fee on joining the 1st International Conference of Lignocellulose (ICON-LIG) in conjunction with The 10th International Symposium for Sustainable Humanosphere (ISSH).

References

1. https://www.bps.go.id/exim/masterhs.html. Accessed 28 July 2021
2. Qu Y, Zhao S, Shi Z, Zhang R, Liu L, Ji F, Yu J (2020) High-efficiency organosolv degumming of ramie fiber by autocatalysis of high-boiling alcohols: an evaluation study of solvents. Cellulose 27(8)
3. Subandi M (2012) The effect of fertilizers on the growth and the yield of ramie (Boehmeria nivea L. Gaud). Asian J Agric Rural Dev 2(393–2016–24012):126–135
4. Brühlmann F, Leupin M, Erismann KH, Fiechter A (2000) Enzymatic degumming of ramie bast fibers. J Biotechnol 76(1):43–50
5. Yuan J, Yu Y, Wang Q, Fan X, Chen S, Wang P (2013) Modification of ramie with 1-butyl-3-methylimidazolium chloride ionic liquid. Fibers and Polymers 14(8):1254–1260
6. Gu J, Catchmark JM (2012) Impact of hemicelluloses and pectin on sphere-like bacterial cellulose assembly. Carbohyd Polym 88(2):547–557
7. Li Z, Yu C (2014) Effect of peroxide and softness modification on properties of ramie fiber. Fibers and Polymers 15(10):2105–2111
8. Wan PP, Cheng L, Liu X (2016) Research of the intelligent splint for medicine on ramie fiber composite material. In: Key engineering materials (vol. 671). Trans Tech Publications Ltd, pp. 255–260
9. Bledzki AK, Gassan J (1999) Composites reinforced with cellulose based fibres. Prog Polym Sci 24(2):221–274
10. Meng C, Yang J, Zhang B, Yu C (2018) Rapid and energy-saving preparation of ramie fiber in TEMPO-mediated selective oxidation system. Ind Crops Prod 126:143–150
11. Karimah A, Ridho MR, Munawar SM, Adi DS, Ismadi, Damayanti R, Subiyanto B, Fatriasari W, Fudholi A (2021) A review on natural fibers for development of eco-friendly bio-composite: characteristics, and utilizations. J Mater Res Technol 13: 2442–245
12. Aristri MA, Lubis MAR, Laksana RBP, Falah F, Fatriasari W, Ismayati M, Wulandari AP, Nurindah, Ridho MR (2021) Bio-Polyurethane resins derived from liquid fractions of lignin for the modification of ramie fibers. Article in Press in Jurnal sylva Lestari
13. Choi HY, Lee JS (2012) Effects of surface treatment of ramie fibers in a ramie/poly (lactic acid) composite. Fibers and Polymers 13(2):217–223
14. Li Z, Chen J, Zhou J, Zheng L, Pradel KC, Fan X, Wang ZL (2016) High-efficiency ramie fiber degumming and self-powered degumming wastewater treatment using triboelectric nanogenerator. Nano Energy 22:548–557

15. Liu L, Xiang Y, Zhang R, Li B, Yu J (2019) Effect of NaClO dosage on the structure of degummed hemp fibers by 2, 2, 6, 6-tetramethyl-1-piperidinyloxy-laccase degumming. Text Res J 89(1):76–86
16. Meng C, Zhong N, Hu J, Yu C, Saddler JN (2019) The effects of metal elements on ramie fiber oxidation degumming and the potential of using spherical bacterial cellulose for metal removal. J Clean Prod 206:498–507
17. Buranov AU, Ross KA, Mazza G (2010) Isolation and characterization of Lignins extracted from flax shives using pressurized aqueous ethanol. Biores Technol 101(19):7446–7455
18. Wise LE (1949) Chlorite holocellulose, its fractionation and bearing on summative wood analysis and on studies on the hemicelluloses. Paper Trade 122:35–43
19. Rowell RM, Pettersen R, Han JS, Rowell JS, Tshabalala MA (2005) Cell wall chemistry. Handbook of wood chemistry and wood composites, 2
20. Lin SY, Dence CW (eds) (2012) Methods in lignin chemistry. Springer Science & Business Media
21. Ghalibaf M, Doddapaneni TRKC, Alén R (2019) Pyrolytic behavior of lignocellulosic-based polysaccharides. J Therm Anal Calorim 137(1):121–131
22. Ismayati M, Nakagawa-Izumi A, Kamaluddin NN, Ohi H (2016) Toxicity and feeding deterrent effect of 2-methylanthraquinone from the wood extractives of Tectona grandis on the subterranean termites Coptotermes formosanus and Reticulitermes speratus. Insects 7(4):63
23. Lourenço A, Gominho J, Pereira H (2018) Chemical characterization of lignocellulosic materials by analytical pyrolysis. In: Analytical pyrolysis. IntechOpen
24. Del Rio JC, Rencoret J, Marques G, Li J, Gellerstedt G, Jiménez-Barbero J, Matinez TA, Gutiérrez A (2009) Structural characterization of the lignin from Jute (Corchorus capsularis) Fibers. Agric Food Chem 57:10271–10281
25. Kalita BB, Gogoi N, Kalita S (2013) Properties of ramie and its blends. Int J Eng Res Gener Sci 1(2):1–6
26. Romagnoli M, Vinciguerra V, Silvestri A (2018) Heat treatment effect on lignin and carbohydrates in Corsican pine earlywood and latewood studied by PY–GC–MS technique. J Wood Chem Technol 38:57–70
27. Mimini V, Sykacek E, Hashim SNA, Holzweber J, Hettegger H, Fackler K, Potthast A, Mundigler N, Rosenau T (2019) Compatibility of kraft lignin, organosolv lignin and lignosulfonate with PLA in 3D Printing. J Wood Chem Technol 39(1):14–30
28. Listyanda RF, Wildan MW, Ilman MN (2020) Preparation and characterization of cellulose nanocrystal extracted from ramie fibers by sulfuric acid hydrolysis. Heliyon 6:1–10
29. Yani ZLÆ, Lili YÆ, Xiong ZLÆH (2007) Study on the cationic modification and dyeing of ramie fiber. Cellulose 14:337–345
30. Zhang N, Tao P, Lu Y, Nie S (2019) Effect of lignin on the thermal stability of cellulose nanofibrils produced from bagasse pulp. Cellulose 26:7823–7835

Decolorization Study of Acid Blue 25 Using Immobilized *Trametes Hirsuta* EDN082 in Light Expanded Clay Aggregate (LECA)

Oktan Dwi Nurhayat, Rayi Mishellia Chempaka, Sita Heris Anita, and Dede Heri Yuli Yanto

Abstract Acid blue 25 is one of the anthraquinone dyes that is frequently used in many industrial activities. This dye will harm the environment if this released without an efficient method. Therefore, the purpose of this research was to study the ability of the immobilized *Trametes hirsuta* EDN 082 in light-expanded clay (LECA) for decolorization acid blue 25 (AB25) dye. The comparative decolorization of AB25 dye was conducted using control-activated LECA, free and immobilized mycelia in LECA. The results showed that the fungal mycelia adsorbed and adhered into LECA, growing not only on the surface but also in the inside part of the LECA. The immobilization of fungal strain enhanced the decolorization up to 89.50% compared with free cell (61.1%) and only control LECA (9.25%) within 24 h of incubation time. Moreover, during all the contact times of decolorization test, treatment using immobilized mycelia exhibited higher laccase activities than that of free cell. The present study successfully demonstrated the use of LECA material to immobilize fungal mycelia of *T. hirsuta* EDN 082 for decolorization of AB25 dye and this technique could be a promising alternative for synthetic dye wastewater treatment in the future.

Keywords Decolorization · Immobilization · Laccase · LECA · *Trametes hirsuta*

O. D. Nurhayat (✉) · S. H. Anita · D. H. Y. Yanto (✉)
Research Center For Applied Microbiology, National Research and Innovation Agency, Cibinong, Bogor, Indonesia
e-mail: oktan.dwi.nurhayat@brin.go.id

D. H. Y. Yanto
e-mail: dede021@brin.go.id

R. M. Chempaka
Brawijaya University, Ketawanggede, Malang, Indonesia

M. A. Rahandi Lubis (ed.), *Proceedings of the 10th International Symposium for Sustainable Humanosphere*, Lecture Notes in Civil Engineering 255,
https://doi.org/10.1007/978-981-19-1422-5_7

1 Introduction

Today, providing clean water for all people around the world has become challenging for us as mankind. The increasing pollution from industrial and agricultural activities into water bodies has an impact on the reducing availability of clean water. Generally, the wastewater from industry activity that discharges into water bodies, contains distinct types of pollutants such as dyes, heavy metals, insecticides, and pesticides which will harm the environment [1]. One of the largest contributions to water pollution is wastewater from textile industries activities that consist of synthetic dyes. In the water bodies, synthetic dyes could remain stable due to their complex aromatic structure that resistant to light and moderate oxidizing agents and photodegradation [2]. In addition, synthetic dyes and their secondary compound have many toxic, mutagenic, and carcinogenic, impacts on the aquatic organism and human beings [3–5].

Several groups of dyes such as anthraquinone, azo, phthalocyanine, and triarylmethane dye, etc., are distinct based on their chemical structure. Among these dye groups, anthraquinone dyes, such as acid blue 25 (AB25), represent the second most frequently used and distributed in the market after azo dyes. Acid blue has wide application dyes that are used for wool, nylon, silk, paper, ink, aluminum, detergent, wood, fur, cosmetic, and biological stain [6, 7]. As an anthraquinone dye, acid blue 25 (AB25) is considered a toxic compound that may cause a problem in the respiratory system organ and irritates the skin and eyes [8]. Therefore, efficient methods and strategies should be devised to remove such compounds from wastewater.

In recent studies, physical and chemical methods are developed to remove synthetic dyes such as adsorption [9], membrane filtration [10], coagulation [11], advanced oxidation [12], and electrochemical process [13]. These methods have many limitations that require high energy, costly, time-consuming, and produce a large amount of sludge [7]. In contrast, many researchers prefer using biological methods especially using microbes for dye removal because these methods are cost-efficient and environmentally friendly. As a microbe, fungi play a role not only as lignin decomposers but also as bioremediation agents for wastewater treatment [14]. Many studies reported, basidiomycetes group, especially the white-rot fungi strain, capable to decolorize various synthetic dyes. White-rot fungi extracellularly produce the ligninolytic enzyme, such as lignin peroxidase (LiP), manganese peroxidase (MnP), and laccase (Lac). Those enzymes can degrade many aromatic and xenobiotic compounds such as dyes, due to their low specificity enzymatic activity [15].

The application of white-rot fungi to remove dyes in wastewater on large-scale land still has several limitations. The application of fungal biomass and mix-culture shows some drawbacks since the mycelia directly contacted with environmental stresses [16]. Therefore, developing an alternative technique that involving the immobilization of fungal cells has attracted some researchers to develop, due to its reusability, resists to environmental stress reduces contamination, and improves fungal activity [17]. Moreover, several studies have been reported the fungal immobilization can maintain the production of various enzymes on a high level compared to the free cell or pellet form [18, 19].

The attempt of immobilized white-rot fungi on various matrices has been successfully reported by some researchers such as activated carbon [20], polyurethane foam, stainless steel sponge, alginate [21], synthetic sponge, eco mat, coconut husk [22], silica [23], polypropylene washer, sorghum [24], brush, grid, and sawdust [19]. However, there are still few studies that conducted immobilization of white-rot fungi in light expanded clay aggregate (LECA). LECA is a material that was made from clay granule and heated at temperatures 1000–1300 °C. Generally, the compositions of LECA is dominated by 5–6 element, consist of 60–70% SiO_2, 15–18% AlO_3, 4–7% Fe_2O_3, 1–4% MgO, CaO, Na_2O, and less than 1% of the other elements [25]. Moreover, the characteristics of LECA are porous, light, non-toxic, inert, resistant to high temperature, insoluble in water, non-degradable against severe conditions [26]. Hence, LECA could be considered as a promising material for immobilization of fungi. The recent studies have demonstrated the immobilization process for fungal cell *Trametes hirsuta* D7 [27] and laccase enzyme from *T. hirsuta* EDN082 for RBBR dye removal in LECA [28]. The use of LECA for fungal immobilization application on RB4, RBBR, and AB129 dye degradation showed less toxicity on fibroblast cells of the human dermal [29]. However, there was no report about the application of the immobilized fungal cell *T. hirsuta* EDN082 on LECA to AB25 decolorization, which is one of the toxic dyes.

In this work, the fungal cell of *T. hirsuta* EDN082 immobilized on LECA was used to decolorized AB25 dyes. The characteristics of LECA-EDN082 after the decolorization process were also investigated. The comparison activity of laccase enzyme between free and immobilized mycelia *T. hirsuta* EDN082 also was investigated.

2 Material and Methods

All experiments were conducted in the Integrated laboratory (ILAB), National Research and Innovation Agency (BRIN). Hipolypeptone and acetate buffer (pH 4.5) were purchased from Wako Pure Chemical Industries, Ltd (Japan). Malt extract and glucose were provided by Merck. Acid Blue 25 (AB25) dye and 2,2-azino-bis-[3-ethyl benzothiazoline-6-sulphonic acid] (ABTS) were obtained from Sigma-Aldrich. Granular LECA was purchased from a local company in Ciamis, West Java, Indonesia.

2.1 Microorganism and Inoculum Preparation

In this experiment, the white-rot fungal strain *Trametes hirsuta* EDN 082 was isolated from Taman Eden 100 Toba, North Sumatera Indonesia. The fungal strain was precultured on malt extract agar (MEA) and incubated at 27 °C for 7 days. Next, ten plugs (Ø 5 mm) of 7 days old fungal mycelia were submerged into 300 mL Erlenmeyer flasks containing 100 mL of malt extract-peptone-glucose (MGP) medium. The medium consisted of 20 g/L malt extract, 20 g/L glucose, and 1 g/L hipolypeptone, pH 4.5. The fungal culture was then incubated statically at 27 °C in an incubator for 7 days. After that, the fungal mycelia were homogenized by using a waring blender and used as an inoculum for cell immobilization.

2.2 LECA Preparation

The granular LECAs used in this experiment were pretreated by steaming activation at 800–900 °C. After activation process, LECAs were rinsed with distilled water and dried at 60 °C for 24 h [30].

2.3 Fungal Cell Immobilization in LECA

The activated LECA was submerged into the Erlenmeyer flask containing MGP broth medium and sterilized at 121 °C for 15 min. The fungal cell inoculum was added to sterilized LECA with the proportion of 1:3 (w/v) and incubated in a reciprocal shaker (100 rpm) at 27 °C for 3 days. Next, the immobilized fungus in LECA was separated from the excess liquid inoculum and air-dried aseptically at room temperature in the biosafety cabinet [27]. The amount of immobilized fungal cell was determined following equation:

$$\begin{aligned}&\text{The immobilized fungal cell/g LECA}\\&\quad=\text{The dried fungal cell (g/mL)}\\&\quad\times\text{The adsorped media (mL)}\end{aligned}\tag{1}$$

The morphology of immobilized fungus in LECA (myco-LECA) and LECA was observed directly by digital microscope (KEYENCE VHX 6000, Japan).

2.4 Decolorization of Acid Blue 25 Experiment by Myco-LECA

Decolorization test by immobilized *T. hirsuta* EDN082 in LECA was conducted in a 100 mL Erlenmeyer flasks containing sterilized 100 mg/L of 40 mL Acid Blue 25 solution and 1% (w/v) glucose and incubated in reciprocal shaker 100 rpm, at 27 °C for 96 h. Decolorization of dye was monitored at 24, 48, 72, and 96 h using TECAN Infinite 200 pro multimode reader at 606 nm. The control LECA (without fungal inoculum) and free cell were also tested as a comparison. All decolorization experiment was done in triplicate and the reported results were shown as an average with a standard deviation. The percentage of dye decolorization was calculated according to the formula [2].

$$Decolorization(\%) = \frac{(Initial\ absorbance - final\ absorbance)}{Initial\ absorbance} \times 100 \quad (2)$$

Moreover, the decolorization process from each treatment was investigated by observing the morphology of control LECA, myco-LECA using digital microscope (KEYENCE VHX 6000, Japan) and free cells by microscope (Olympus BX-63, Japan) at 24 and 96 h.

2.5 Laccase Activity Assays

Laccase activity of all treatment samples was measured using TECAN Infinite 200 pro multimode reader by monitoring the oxidation of 0.5 mM 2,2-azino-bis (3-ethylbenzothiazoline-6-sulfonic acid) (ABTS) as substrate at 420 nm. The laccase will be oxidized non-phenolic dye ABTS to the cation radical. The intense blue-green color will be produced by cation radical and measured at 420 nm as enzyme activity. The laccase activity was determined at 24, 48, 72, and 96 h from all filtrate samples of decolorization assay. The assay mixture contained 50 μL of sample filtrate, 250 μL of 0.1 M acetate buffer, and 200 μL of 0.5 mM ABTS. Enzyme activity (U/L) was calculated by the following equation with molar absorptivity (ε) of 36,000 M^{-1} cm^{-1}:

$$Enzyme\ activity\left(\frac{U}{L}\right) = \frac{Abs.}{\varepsilon} \times \frac{Vmixture(\mu L)}{10^6} \times 10^6 \times \frac{60/t}{Venzyme(\mu L)/10^3} \times 1000 \quad (3)$$

3 Results and Discussion

3.1 Fungal Immobilization in LECA

The surface and inside observations were conducted to see the changes in the morphology of LECA (Fig. 1). On the surface of control LECA, the color was light brown and a rough texture with pores was observed (Fig. 1a1). On the other hand, the pores textures with black color were observed in the inside of control LECA. There were no mycelia or hyphae that grew in control LECA (Fig. 1a2). In contrast, *T. hirsuta* EDN082 mycelia were seen immobilized into LECA and grew inside to outside part of LECA. Moreover, the white fungal mycelia covered the surface (Fig. 1b1) and innard part of the LECA (Fig. 1b2). We assumed the fungal cells were adsorbed to the LECA by a physical adsorption mechanism. The adsorption method is one of the simple methods in immobilizing microbial to wastewater treatment technology, which involves the distribution of microbe cells to porous and inert support materials and followed by cell adhesion and colonization [31, 32]. In a previous study, the observation of immobilization of *T. hirsuta* D7 in LECA by using scanning electron microscopy (SEM), showed the growth and interaction between fungal mycelia and LECA as solid material [29].

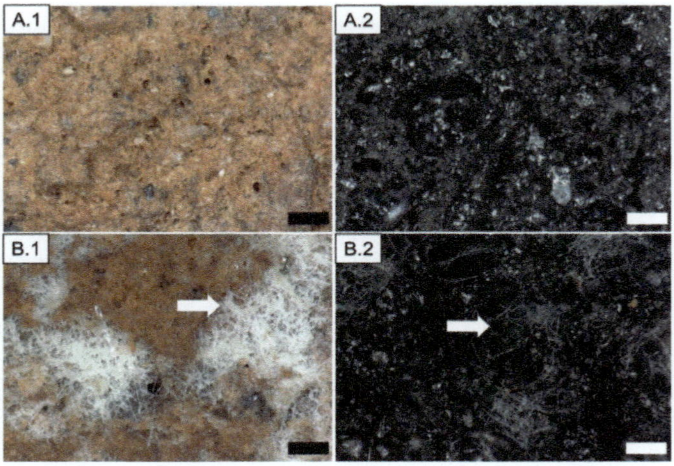

Fig. 1 A.1 Surface morphology of control LECA, A.2 innard morphology of control LECA. B.1 Surface morphology of myco-LECA. B.2 innard morphology myco-LECA. Arrow = Mycelia. Scale bar = 200 μm

3.2 Decolorization Assay

In this experiment, the immobilized *T. hirs*uta EDN082 is compared to control LECA, and the living free cell of *T. hirs*uta EDN082 to see the effectiveness of fungal immobilization to decolorized in 40 mL Acid blue 25 (100 ppm) solutions with 1% glucose. The differences in decolorization percentages were measured for all test treatments (Fig. 2). According to the results, the activated LECA (control LECA) could also give the decolorization of dye by increasing the contact time (Fig. 2). The percentage decolorization had a rapid increase in 48 h and slowly increases until the last time of assays by this treatment. The increase in contiguity between adsorbent and adsorbed correlated with the contact time, where the result of this experiment, is in line with results from a previous study [33]. Furthermore, the recent study reported that activated LECA could increase the removal RBBR dye up to 35% via adsorption mechanism [27]. The activation process by steaming would affect the enlargement of microporosity, meso- and macroporosity of material that could increase the dye decolorization by adsorption mechanism [30].

Overall, the immobilized *T. hirs*uta EDN082 gives a higher percentage of dyes decolorization than the two other treatments in each interval time of assay. We also investigate that the living free cell *T. hirs*uta EDN082 could decolorize AB25 dye but not as efficiently as the immobilized *T. hirs*uta EDN082. In the first 24 h of incubation time, the immobilized fungi had 89.50% a percentage of decolorization that is higher than the living free cell *T. hirs*uta EDN082 and control LECA which has 61.1% and 9.25%, respectively. The results indicated, the immobilized *T. hirsuta* EDN082 efficiently decolorized the AB25 dye within 1 day. The maximum percentage of decolorization by activated LECA, *T. hirs*uta EDN082 free cell, and the immobilized *T. hirs*uta EDN082 were reached after 96 h, respectively are 78.7%, 89.3%, and 96.1%. It was clear that the immobilized *T. hirs*uta EDN082 treatment had increased the percentage decolorization of AB25 dye. These results were in line with a recent

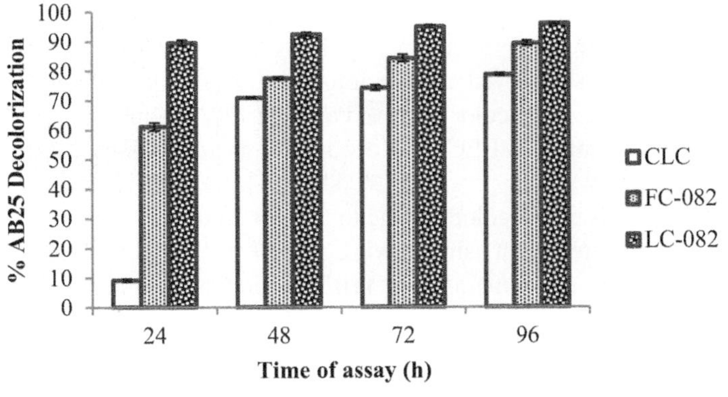

Fig. 2 Decolorization of AB25 dye by CLC (control LECA), FC-082 (free cell), and LC-082 (myco-LECA)

studies experiment which was immobilized *T. hirsuta* D7 in LECA, could increase anthraquinone dye decolorization [27, 29]. However, in their studies, almost complete decolorization of RBBR dye mixture was reached after 24 h by immobilizing fungus, while in this experiment it was observed after 96 h for AB25. The results showed that AB25 was more difficult to decolorized than RBBR dye. The differences in the chemical structure of dyes are expected to influence the decolorization rate of dyes [34, 35].

Decolorization of AB25 dye occurs through adsorption into LECA [36, 37], biosorption, and enzymatic degradation by *T. hirsuta* EDN 082 [38]. Thus, changes in the LECA in the decolorization process in this study were observed.

The data in Fig. 2 showed that the immobilized *T. hirsuta* EDN082 in LECA is effectively decolorized AB25 dye molecule compared to the other treatments. Besides, in this experiment, we have proved immobilization of *T. hirsuta* EDN 082 can enhance dye removal through biosorption and biodegradation besides adsorption mechanism via LECA material. To confirm that process, the comparison of changes that occur between the immobilized *T. hirsuta* EDN 082, free cell, and LECA treatment is observed (Fig. 3). The observation results of the free cell after 24 h under the microscope showed the hyphae color become blue (Fig. 3A.2) while the mycelia observed on the LECA surface become dark blue color. The color change in hyphae proves the primary decolorization process through biosorption in the first 24 h. Biosorption is a non-enzymatic process through adsorption and accumulation of dye through the cells [39]. However, after 96 h of treatment, the blue color is decreased from hyphae at free cell treatment as well as in *T. hirsuta* EDN. 082 immobilized in LECA. These indicate the decrease of color intensity is caused by increasing enzyme production or concentration of fungal cell mass that degradation of dye occurred both by enzymatic degradation and adsorption to the cell mass [40]. In previous studies has been reported that the biosorption mechanism and extracellular enzymes were involved in the biodegradation process of anthraquinone dyes [41, 42].

The observation of surface change of control LECA showed the difference in intensity of dyes that were adsorbed was affected by increasing the contact time (Fig. 3A.1, B.1). In the first 24 h (Fig. 3A.1), on the surface of activated LECA, the blue dye particle is observed and the amount of blue particle dye is increased at 96 h of treatments. The blue color observed on the LECA surface means, AB25 dye decolorization mechanism by LECA can occur through adsorption, and no enzymatic process involved.

All treatment tests give a color change to the AB 25 dye of each incubation time (Fig. 4). The control treatment using only LECA affects the reduction color intensity of AB25 but the blue color remains (Fig. 4a). This implies that the reduction color by AB25 only by adsorption mechanism into LECA material without changing chemical form AB25 dye. In contrast, the color intensity of Acid Blue 25 is reduced from blue into brown at total incubation time by free cell (Fig. 4b) and fungal cell immobilization treatment (Fig. 4c). It means that the decreasing blue color occurs through cleavage of the chromophore group [43]. However, the difference was observed at 24 h incubation. The appearance of AB25 liquid with free cell treatments the blue

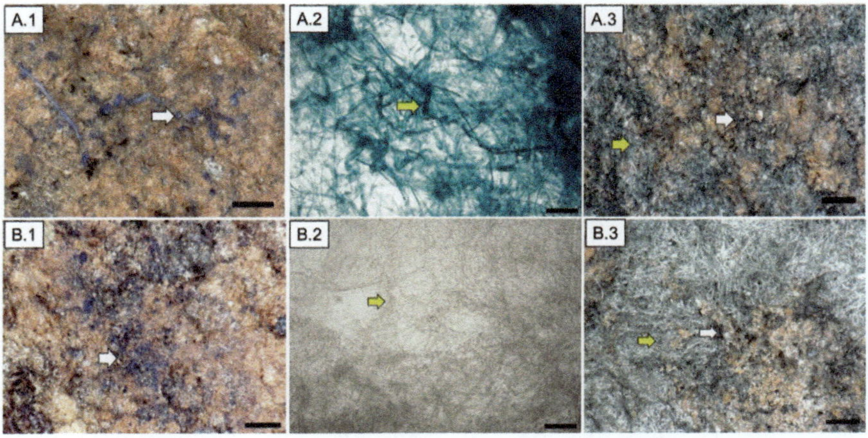

Fig. 3 Surface morphology of LECA and hyphae after AB25 dye decolorization process under the microscope. A.1. Control LECA, A.2. hypha appearances and A.3. myco-LECA after 24 h. B.1. Control LECA, B.2. hypha appearances, B.3. myco-LECA after 96 h. White arrow: dye. Yellow arrow: mycelia. A.1, A.3, B.1, B.3, Scale bar = 200 μm. A.2, B.2 Scale bar = 20 μm

color become purple (Fig. 4b) while in myco-LECA treatment was brown (Fig. 4c). These results were in line with the percentage decolorization that the myco-LECA increased the decolorization process of AB25 dye in 24 h (Fig. 2). The intensity of color change also confirmed that in free cell treatments dye removal was dominated with biosorption mechanism while in the myco-LECA treatment the biosorption, biodegradation, and adsorption occurred during dye removal in 24 h. Therefore the myco-leca treatment showed the most color change in AB25 liquid among the others.

In the previous study, the fungal biomass immobilization technique showed a positive impact on the effectiveness of synthetic dye removal. Immobilized process *Phanerochaete* sp. on a natural sponge loofa showed better results for dye removal than using biomass in media [44]. Decolorization of reactive azo dye 4 and reactive azo dye 2 was achieved 70% after 3 days of treatment using *Trametes versicolor* immobilized on a natural sponge[45]. Immobilized biomass of *T. hirsuta* on alginate was reported removed 69% of phenol red and 95% of indigotine after 24 h experiment [46].

3.3 Laccase Activity

The laccase activities were tested to monitor the association of enzyme activity to rate decolorization which is secreted extracellularly by fungus in free cell and myco-LECA treatments (Fig. 5). Based on the results, the activity of laccase enzyme involved in the dye removal. In this case, AB25 dye act as an enzymes substrate that can directly oxidize by laccase was utilized as a carbon source by this fungus.

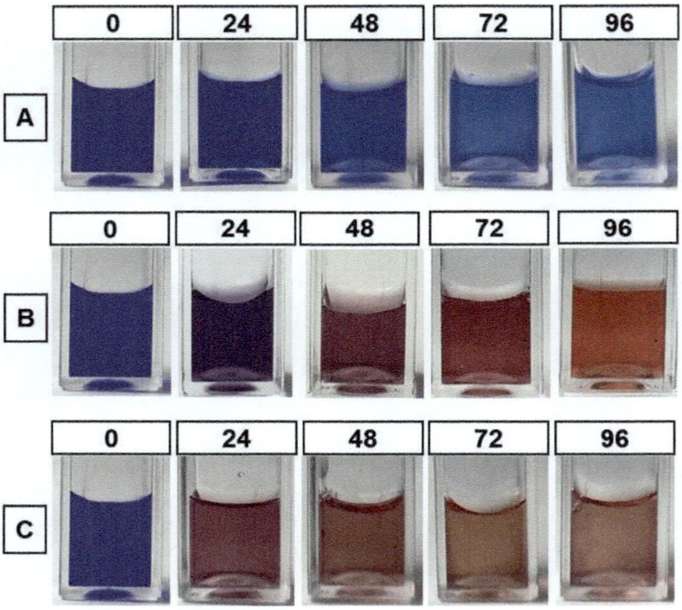

Fig. 4 The appearances of AB25 dye decolorization using three different treatments with 0–96 h incubation time. **a** Control LECA. **b** Free cell. **c** Myco-LECA

Fig. 5 Laccase activity by free (dash line) and immobilized (solid line) mycelia of *T. hirsuta* EDN082

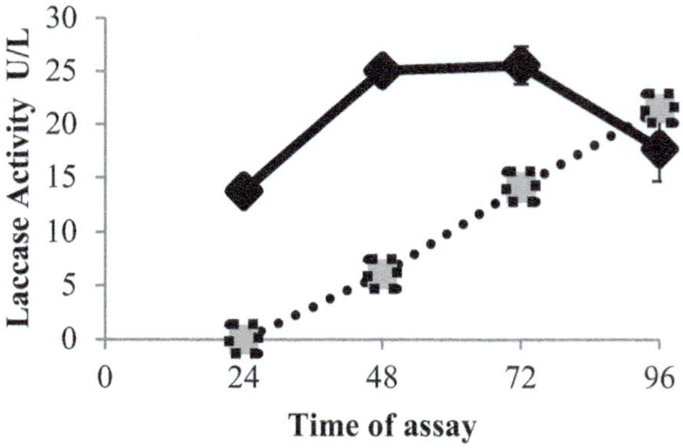

The ability of this enzyme to oxidized a broad range of inorganic and organic compounds such as aromatic amines, polyphenols, diamines, and diphenols by reducing oxygen to water is important in biodegradation process of synthetic dye [47]. The measurement of laccase activity showed myco-LECA treatment produced higher laccase activity than free cell treatments during all the incubation time. The immobilization of *T. hirsuta* EDN 082 in LECA clearly increase the production of laccase enzyme, which achieved a maximum activity of up to 25, 52 U/L, while the application of free cells of this fungus had lower activity (14.14 U/L) after 48 h incubation time. These results were in line with a previous study that demonstrated immobilized *Peniophora cinerea* on synthetic fiber, can increase 35-fold in laccase enzyme production when compared to the free cell [48].

4 Conclusion

Fungal mycelia of *T. hirsuta* EDN 082 was successfully immobilized in LECA. The Immobilized *T. hirsuta* EDN 082 enhanced the effectiveness of AB25 dye removal when compared to free cell and control LECA assay. Moreover, the fungal immobilized technique in LECA increased the production of laccase enzyme as well as improve decolorization of AB25 dye via biodegradation mechanism.

Acknowledgements This study was funded by the National Research and Innovation Agency (BRIN) through DIPA 2021 of the Deputy for Scientific Services, and by the Japan ASEAN Science, Technology and Innovation Platform (JASTIP). The acknowledgment also states by the author for the facilities and technical support of Integrated Laboratory Bioproducts (ILAB) at the BRIN. The authors also thank to Japan Society for the Promotion of Science (JSPS) for partial funding of the publication fee.

References

1. Hall J, Borgomeo E (2013) Risk-based principles for defining and managing water security
2. Lellis B, Fávaro-Polonio CZ, Pamphile JA, Polonio JC (2019) Effects of textile dyes on health and the environment and bioremediation potential of living organisms. Biotechnol Res Innov 3(2):275–290
3. Zhou Y, Lu J, Zhou Y, Liu Y (2019) Recent advances for dyes removal using novel adsorbents: a review. Environ Pollut 252:352–365
4. Köktürk M, Altindağ F, Ozhan G, Çalimli MH, Nas MS (2020) Textile dyes Maxilon blue 5G and Reactive blue 203 induce acute toxicity and DNA damage during embryonic development of Danio rerio. Comp Biochem Physiol Part—C Toxicol Pharmacol 242(November):2021
5. Gita S, Shukla SP, Deshmukhe G, Choudhury TG, Saharan N, Singh AK (2021) Toxicity evaluation of six textile dyes on growth, metabolism and elemental composition (C, H, N, S) of microalgae spirulina platensis: the environmental consequences. Bull Environ Contam Toxicol 106(2):302–309
6. Ghodbane H, Hamdaoui O (2010) Decolorization of antraquinonic dye, C. I. Acid Blue 25, in aqueous solution by direct UV irradiation, UV/H_2O_2 and UV/Fe(II) processes. Chem Eng J 160(1):226–231
7. Wang Y et al (2020) Resuscitation, isolation and immobilization of bacterial species for efficient textile wastewater treatment: a critical review and update. Sci Total Environ 730:139034
8. Al Kausor M, Chakrabortty D (2020) Optimization of system parameters and kinetic study of photocatalytic degradation of toxic acid blue 25 dye by Ag_3PO_4 @ RGO nanocomposite
9. Sharma A, Syed Z, Brighu U, Gupta AB, Ram C (2019) Adsorption of textile wastewater on alkali-activated sand. J Clean Prod 220:23–32
10. Ji D, Xiao C, An S, Zhao J, Hao J, Chen K (2019) Preparation of high-flux PSF/GO loose nanofiltration hollow fiber membranes with dense-loose structure for treating textile wastewater. Chem Eng J 363:33–42

11. Dotto J, Fagundes-Klen MR, Veit MT, Palácio SM, Bergamasco R (2019) Performance of different coagulants in the coagulation/flocculation process of textile wastewater. J Clean Prod 208:656–665
12. GilPavas E, Correa-Sánchez S, Acosta DA (2019) Using scrap zero valent iron to replace dissolved iron in the Fenton process for textile wastewater treatment: optimization and assessment of toxicity and biodegradability. Environ Pollut 252:1709–1718
13. GilPavas E, Arbeláez-Castaño P, Medina J, Acosta DA (2017) Combined electrocoagulation and electro-oxidation of industrial textile wastewater treatment in a continuous multi-stage reactor. Water Sci Technol 76(9):2515–2525
14. Silva R et al (2018) Fungi of biotechnological interest in the decolouration of textile effluents. Ann Biomed Technol Eng 1(1):1–4
15. Eichlerov I, Baldrian P (2020) Ligninolytic enzyme production and decolorization capacity of synthetic dyes by Saprotrophic White Rot , Brown Rot, and litter decomposing basidiomycetes 1–22
16. Bulai IM, Spina F, Varese GC, Venturino E (2018) Wastewater bioremediation using white rot fungi: validation of a dynamical system with real data obtained in laboratory. Math Methods Appl Sci 41(11):4195–4207
17. Mazmanci MA (2010) Decolorization of azo dyes by immobilized fungi. In: Handbook environmental chemistry. Springer-Verlag, pp 169–181
18. Elgueta S, Santos C, Lima N, Diez MC (2016) Immobilization of the white-rot fungus Anthracophyllum discolor to degrade the herbicide atrazine. AMB Expr 6(1)
19. Przystas W, Zabłocka-Godlewska E, Grabińska-Sota E (2018) Efficiency of decolorization of different dyes using fungal biomass immobilized on different solid supports. 9:285–295
20. Zhang F, Yu J (2000) Decolourisation of Acid Violet 7 with complex pellets of white rot fungus and activated carbon. Bioprocess Eng 23(3):295–301
21. Rodríguez Couto S, Sanromán MA, Hofer D, Gübitz GM (2004) Stainless steel sponge: a novel carrier for the immobilisation of the white-rot fungus Trametes hirsuta for decolourization of textile dyes. Bioresour Technol 95(1):67–72
22. Low JY, Abdullah N, Vikineswary S (2009) Evaluation of support materials for immobilization of Pycnoporus sanguniesus mycelia for laccase production and biodegradation of polycyclic aromatic hydrocarbon. Res J Environ Sci 3(3):357–366
23. Wan Mohd Zawawi WNI, Mansor AF, Othman NS, Mohidem NA, Nik Malek NAN, Mat H (2016) Synthesis and characterization of immobilized white-rot fungus Trametes versicolor in sol–gel ceramics. J Sol-Gel Sci Technol 77(1):28–38
24. Zahmatkesh M, Spanjers H, van Lier JB (2018) A novel approach for application of white rot fungi in wastewater treatment under non-sterile conditions: immobilization of fungi on sorghum. Environ Technol (United Kingdom) 39(16):2030–2040
25. Mlih R et al (2020) Light-expanded clay aggregate (LECA) as a substrate in constructed wetlands—a review. Ecol Eng 148(March):105783
26. Rashad AM (2018) Lightweight expanded clay aggregate as a building material—an overview. Constr Build Mater 170:757–775
27. Ardiati FC, Yanto DHY, Anita SHA, Watanabe T (2019) Immobilization of Trametes hirsuta D7 in light expanded clay aggregate for decolorization of synthetic dye
28. Anita SH et al (2020) Immobilization of laccase from Trametes hirsuta EDN 082 in light expanded clay aggregate for decolorization of Remazol Brilliant Blue R dye. Bioresour Technol Rep 12:100602
29. Alam R et al (2021) Biodegradation and metabolic pathway of anthraquinone dyes by Trametes hirsuta D7 immobilized in light expanded clay aggregate and cytotoxicity assessment. J Hazard Mater 405(July):124176
30. Rodríguez-Reinoso F, Molina-Sabio M, González MT (1995) The use of steam and CO_2 as activating agents in the preparation of activated carbons. Carbon NY 33(1):15–23
31. Kilonzo P (2012) Surface modifications for controlled and optimized cell immobilization by adsorption: applications in fibrous bed bioreactors containing recombinant cells. J Microb Biochem Technol 01(S8)

32. Zhang X, You S, Ma L, Li C (2015) The application of immobilized microorganism technology in wastewater treatment. In: Mmeceb, pp 103–106 (2016)
33. Shokoohi R, Taghi M, Reza M (2016) Comparing the performance of granular coral limestone and Leca in adsorbing Acid Cyanine 5R from aqueous solution. Saudi J Biol Sci 1–11
34. Anita SH, Sari FP, Heri D, Yanto Y (2019) Decolorization of synthetic dyes by ligninolytic enzymes from Trametes hirsuta D7. Makara 23(1):44–50
35. Yanto DHY, Auliana N, Anita SH, Watanabe T (2019) Decolorization of synthetic textile dyes by laccase from newly isolated Trametes hirsuta EDN084 mediated by violuric acid Decolorization of synthetic textile dyes by laccase from newly isolated Trametes hirsuta EDN084 mediated by violuric acid. IOP Conf Ser Earth Environ Sci 374
36. Bahmanpour H, Habashi R, Hosseini S (2017) Investigating the efficiency of lightweight expanded clay aggregate (LECA) in wastewater treatment of dairy industry. Anthropog Pollut J
37. Dordio A, Carvalho AJP (2013) Constructed wetlands with light expanded clay aggregates for agricultural wastewater treatment. Sci Total Environ 463–464:454–461
38. Wesenberg D, Kyriakides I, Agathos SN (2003) White-rot fungi and their enzymes for the treatment of industrial dye effluents. Biotechnol Adv 22(1–2):161–187
39. Ngo ACR, Devanadera MKP, Dedeles GR (2016) Decolorization of selected synthetic textile dyes by Yeasts from leaves and fruit peels. J Heal Pollut 6(10):42–55
40. Singh L (2017) Biodegradation of synthetic dyes: a mycoremediation approach for degradation/decolourization of textile dyes and effluents. J Appl Biotechnol Bioeng 3(5):430–435
41. Hadibarata T, Yusoff ARM, Kristanti RA (2012) Acceleration of anthraquinone-type dye removal by white-rot fungus under optimized environmental conditions. Water Air Soil Pollut 223(8):4669–4677
42. Hadibarata T, Yusoff ARM, Kristanti RA (2012) Decolorization and metabolism of anthraquionone-type dye by laccase of white-rot fungi polyporus sp. S133. Water Air Soil Pollut 223(2):933–941
43. Osma JF, Toca-Herrera JL, Rodríguez-Couto S (2010) Transformation pathway of Remazol Brilliant Blue R by immobilised laccase. Bioresour Technol 101(22):8509–8514
44. Iqbal M, Saeed A (2007) Biosorption of reactive dye by loofa sponge-immobilized fungal biomass of Phanerochaete chrysosporium. Process Biochem 42(7):1160–1164
45. Nilsson I, Möller A, Mattiasson B, Rubindamayugi MST, Welander U (2006) Decolorization of synthetic and real textile wastewater by the use of white-rot fungi. Enzyme Microb Technol 38(1–2):94–100
46. Domínguez A, Couto SR, Sanromán MÁ (2005) Dye decolorization by Trametes hirsuta immobilized into alginate beads. World J Microbiol Biotechnol 21(4):405–409
47. Kiiskinen LL, Viikari L, Kruus K (2002) Purification and characterisation of a novel laccase from the ascomycete Melanocarpus albomyces. Appl Microbiol Biotechnol 59(2–3):198–204
48. Silvério SC, Moreira S, Milagres AMF, Macedo EA, Teixeira JA, Mussatto SI (2013) Laccase production by free and immobilized mycelia of Peniophora cinerea and Trametes versicolor: a comparative study. Bioprocess Biosyst Eng 36(3):365–373

Characteristics and Quality of Pooti Damar (*Hopea Gregaria* V. Slooten) from Tahura Nipa-Nipa, Southeast Sulawesi

Niken Pujirahayu, Basrudin, Zakiah Uslinawaty, Nurhayati Hadjar, Abigael Kabe, Nurnaningsih Hamzah, and Widyastuti

Abstract Pooti (*Hopea gregaria* V. Slooten) is a species from the Dipterocarpaceae family with a limited distribution around the Tahura Nipa-Nipa area, Kendari, Southeast Sulawesi. This species can produce damar resin. However, the characteristics of *H. gregaria* resin are not widely known. This study aims to determine the characteristics and quality of *H. gregaria* sap resin. This research uses two methods, namely the visual test of SNI 2900 1.2012 and the laboratory quality test of SNI 2900 2.2013. Visual test results show clear yellow pooti resin and chunk size 3 × 3 cm. The impurity value of pooti damar is 0.44%, soft point value 75 °C, ash content value 0.01%, average value of turbidity in toluene 192.3 belonged to class II quality—the overall average was included in the first quality class. The visual and chemical-physical test of pooti resin shows that this damar meets the best quality requirements, class I based on SNI standards. These results also indicate that the resin quality of *H. gregaria* is equivalent to that of the best quality resins, such as resins from the *Shorea javanica* tree, and provides a great opportunity to utilize this pooti damar.

Keywords Pooti damar · *Hopea gregaria* · Damar characterisctic · Tahura Nipa-Nipa · Southeast Sulawesi

1 Introduction

Dipterocarpaceae is a large family with 506 species worldwide, including in 14 genera, most of which (76%) species grow in the Malesia region, especially in Indonesia [1]. This type of Dipterocarpaceae or known as the towering tree is a tree species whose growth is prolonged and is on the verge of extinction because this type of family is constantly being exploited [2].

N. Pujirahayu (✉) · Basrudin · Z. Uslinawaty · N. Hadjar · A. Kabe · N. Hamzah · Widyastuti
Department of Forestry, Faculty of Forestry and Environmental Science, Halu Oleo University, Kendari 93232, Indonesia
e-mail: rahayuken08@gmail.com

M. A. Rahandi Lubis (ed.), *Proceedings of the 10th International Symposium for Sustainable Humanosphere*, Lecture Notes in Civil Engineering 255,
https://doi.org/10.1007/978-981-19-1422-5_8

Damar from this family is one of the superior products from non-timber forest products in Indonesia. This sap is generally obtained from the genus Shorea, and Hopea such as *Shorea javanica, S. koordersii, Hopea dryobalanoides*, and *H. intermedia*. The resin commodity can make an immense contribution to economic growth because it has many benefits. This sap is used in various industries, including paints, inks, varnishes, frankincense, and food additives [3].

One species of this family Dipterocarpaceae is *Hopea gregaria* V. Slooten which is a species of the genus Hopea. This tree is medium in size and can reach a height of 35 m. This species has a narrow distribution and is only found in Southeast Sulawesi [4], and it has been reported as endangered by the IUCN since 1998 [5].

The *H. gregaria* tree, also known as the Pooti tree by local people, can produce white and yellow sap [6], known as cat's eye damar. However, the sap of the cat's eye from the pooti tree in Southeast Sulawesi has not been utilized because the surrounding community is not familiar with the types and benefits of pooti damar [7]. Pooti damar production research conducted by Munandar [8] at 30, 50, and 90 cm diameters on pooti trees (*H. gregaria*) with the tapping method obtained different damar production levels best production in the 45–50 cm diameter class. Futhermore, Uslinawaty et al. [7] conducted a resin quality test from a pooti tree with a diameter of 30 cm. They reported that *H. gregaria* resin had a clear yellow color and lump size > $3 \times 3\ \text{cm}^2$, which was included in the grade A quality based on the SNI 29001.1.2012 [9] standard. Meanwhile, based on the test results, the average ash content and acid number are included in class II quality. However, the quality of resin from trees with a diameter of 50 cm is not yet known.

Resin characteristics include physical and chemical properties [10], such as color and lump size, impurity content, ash content, turbidity, melting point, and acid number. These characteristics can be used as the basis for determining the quality and utilization. In resin processing, information about these properties is required, for example, to obtain a suitable method for resin purification. In addition, information about the characteristics and quality of this resin is also crucial for use in various industrial fields that use resin as a base material, where each form of utilization requires different special requirements.

Therefore, it is deemed necessary to research the characteristics and quality of pooti damar from various tree diameter classes. This study aims to determine the characteristics and quality of resin from pooti trees obtained from 50 cm diameter class from Tahura Nipa-Nipa region, Southeast Sulawesi.

Fig. 1 The size of incision was ± 10 cm wide, 10 cm high and ± 2 mm thick on the trunk of the pooti tree

2 Materials and Methods

2.1 Preparation of Pooti Resin

Pooti resin was obtained from three pooti trees with a diameter of 50 cm tapped following the method of Mampi et al. [11], namely by cleaning the part of the tree trunk where the bark will be tapped, then an initial incision was made with a tapping tool called a kadukul (a kind of knife specially used to cut tree bark). An incision was made with a 10 cm × 10 cm size and a depth of 2 cm (including skin thickness) Fig. 1. The resin that comes out was channeled through gutters and collected using coconut shells or small containers. Resin collection was done every seven days for one month.

2.2 Quality Testing Procedure

Visual Testing. Visual observation consists of color and size chunks. Visual test of sorting resin refered to SNI 2900-1-2012 [9].

Laboratory Testing. Resin testing in the laboratory referred to SNI 2900.2:2013 [12] used five parameters as measuring points or requirements for resin quality, as follows:

Acid Number. The test sample was weighed ± 2 g (M) and inserted into the in Erlenmeyer. Then the test sample was dissolved with 5 mL of a 1:1 toluene-ethanol mixture, then drop 2 mL of phenolphtalein (PP) indicator drops. Next, the solution titrated with 0.1 N Potassium hydroxide (KOH) until the color of the solution becomes pink; the volume of KOH solution required to the titration of the sample solution (V) is recorded. Lastly, solution titrated toluene-ethanol 1:1 with 0.1 N KOH to set a blank (V1). Acid numbers can be calculated using the following equation:

$$AN = \frac{(V_1 - V) \times N \times 56.1}{M} \quad (1)$$

With AN = Acid number, V = Volume of alkaline solution required to titrate a 1:1 toluene-ethanol solution containing the test sample (mL), N = Normality of alkaline solution (KOH), M = Weight of test sample (g), V1 = Volume of alkaline solution required to titrate the blank (mL), and 56.1 = Molecular weight of KOH.

Ash content. Ash content testing was carried out with reference to SNI 7636:2010; SNI 2900.2:2013 [12] as follows: The porcelain cup was heated at a temperature of 625 ± 5 °C for ± 20 min and put in a desiccator. Next, the porcelain dish was weighed (W_0), and the test sample made into a fine powder was carefully weighed ± 5 g of the sample in a porcelain cup (W_1). Then The cup and the test sample are put into the furnace at a temperature of 625 ± 5 °C until colored ash was obtained gray, or for ± 30 min. Next, cooled in a desiccator and weighed (W_2). Calculation of ash content can be written with the formula:

$$Ash\ content(\%) = \frac{W_2 - W}{W_1 - W} \times 100\% \quad (2)$$

where W is the weight of the empty cup (g), W_1 is the weight of the cup + test sample (g) and W_2 is the weight of the cup + ash (g).

Impurity Level. The amount of impurities was calculated by weighing 2.6 g of the test sample (W) and putting it into a 300 mL beaker. Adding 31.15 mL of toluene, heating (the temperature was below the boiling point of toluene), and shake until completely dissolved. Then filter the solution through a Gooch dish with hot toluene until the filtrate was clear. Then dry the Gooch dish containing the residue at a temperature of (105 ± 3)°C for 1 h until a constant weight is obtained. The level of impurities in the sample is calculated using the equation:

$$I = \frac{W_1 - W_2}{W} \times 100\,\% \quad (3)$$

With I = impurity content (%), W = initial sample weight (g), W_1 = weight of empty cup (g), W_2 = weight of the cup containing toluene insoluble residue (g).

Turbidity in toluene solution. The resin turbidity test was carried out by weighing 7 g of the test sample in 25 ml of toluene, then put it in a turbidimeter to calculate the turbidity.

Soft Points. Soft point testing is carried out based on the SNI 7636: 2011; SNI 2900.2:2013 [12] test with the following procedure: The resin is ground to a particular dimension in such a way that the resin changes from a lump to a fine powder. Then the powder was melted at a low temperature and put into a beaker evenly. Then insert the steel ball and thermometer into the beaker, then distilled water was filled into the beaker and heated slowly starting from 40 °C. Determining the soft point by calculating the average value of the initial temperature indicated by the steel ball starting to decrease and the final temperature indicated by the steel ball touching the plate. The mean value is then set as the soft point. Soft point can be determined by the following equation:

$$SP = \frac{T_1 + T_2}{\sum n(n_1 + n_2 \ldots)} \quad (4)$$

With SP = soft point (°C), T_1 = sum of initial temperature (°C), T_2 = sum of final temperature (°C), n = the number of data.

3 Results and Discussions

3.1 Colour and Size of Damar Pooti

The results of the observation of the quality class of damar pooti (*H. gregaria*) based on visually observed colour, according to the quality standard of SNI 2900.1:2012 are presented in Table 1.

In Table 1, the visual observations of the pooti damar (*H. gregaria*) based on the color and size of the pooti damar lumps are clear yellow, and the lump size is > 3 × 3 cm^2. The color and size of the lumps are included in the grade A quality based on the SNI 29001:2012. Previous research on visual observation of pooti resin from trees with a diameter of 30 cm showed similar results, namely pooti resin with a clear yellowish color with a lump size of > 3 × 3 cm^2 [7]. The quality of the damar will be higher if it has a more transparent yellow color and the size of the lump is getting bigger. On the other hand, the quality of the damar will be lower if it has a more brownish color and the size of the lump is small [13]. In the standard SNI 29001:2012, the highest quality standard of lumps is 3 × 3 cm^2 and the lowest quality standard is 0.5 × 0.5 cm^2. The visual appearance of the color and size of the pooti sap (*H. gregaria*) is shown in Fig. 2.

Table 1 Colour and size of damar pooti from Tahura Nipa Nipa

Resin type	Resin visual test			Quality based on SNI 2900.1:2012
	Sample	Colour (C)	Size of chunks (Z)	
H. gregaria	A1	Clear yellow	> (3 × 3) cm	A (C,Z)
	A2	Clear yellow	> (3 × 3) cm	A (C,Z)
	A3	Clear yellow	> (3 × 3) cm	A (C,Z)

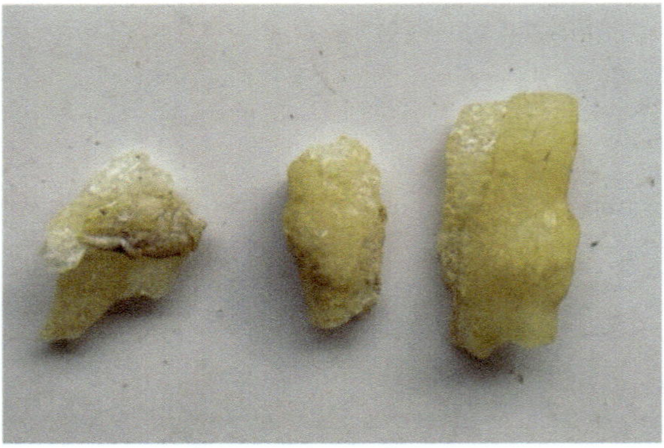

Fig. 2 Damar of *H. gregaria* from Tahura Nipa-Nipa

3.2 *The Physical and Chemical Characteristics of Pooti Damar*

Pooti resin physical and chemical properties refer to SNI. 2900.2:2013 presented in Table 2.

Ash content is the mineral content found in residue after resin burning. Ash is an inorganic substance leftover from the combustion of organic material. The ash content indicates that the material contains impurities and other minerals [14]. The higher the ash content, the worse the quality [13]. According to the quality class standard of SNI 2900.2.2013, the highest quality has an ash content percentage value of < 0.5%. In addition, the lowest quality standard is the one that has an ash content percentage > 1.0%.

The average indicating that the ash content of *H. gregaria* resin was 0.01%, including class 1 SNI 2900.2:2013. This value was better than the ash content of *H. gregaria* resin was tapped from trees with a diameter of 30 cm [7]. The ash content is closely related to the impurity content, the higher the impurity content, the higher the ash content, according to the results of this study, the impurity content is in class 1, and the ash content is also in class 1.

Table 2 Damar pooti (*H. gregaria*) quality is based on SNI 2900.2:13

Parametre	Sample				Quality based on SNI 2900.2:13
	A1	A2	A3	Average	
Impurity content (%)	0.56	0.52	0.24	0.44	I (< 1.5)
Softening point (°C)	78	76	71	75	I (75–85)
Turbidity in toluene	202	190	187	192.3	I (< 200)
Ash content (%)	0	0.04	0	0.01	I (< 0.5)
Acid number	26.7	28.22	29.1	28.0	I–II (20–30)

Toluene is an organic solvent that can dissolve resin so that the insoluble material in toluene shows the impurity level of resin. The higher the percentage of ingredients insoluble in toluene, the lower the quality of the resin [10]. The test results showed that the quality of the pooti resin based on the insoluble resin material in toluene was 192.3 (grade 1). The turbidity test in toluene was carried out to determine the level of impurities, namely the percentage of the amount of organic matter or inorganic material that was not soluble in toluene. Based on previous research conducted by Wijayanto [13], the turbidity value in toluene resin of cat's eye resin (*S. javanica*) ranged from 0.14 to 39.72 and was included in quality class II. This result means that the results obtained in this study have a better quality when compared to previous studies.

The Softening point is the temperature at which damar begins to change from a solid to a semisolid state. Based on the results of the analysis showed that the suitability of the average soft point value of damar pooti resin (*H. gregaria)* obtained a value of 75 °C with the quality class of SNI 2900.2.2013 included in the quality class I, namely 75–85 °C. According to Gusti and Zulnely [15], the presence of impurities that bind to compounds in resin gum affects the soft point value of resin gum. Damar with low impurities will produce a low softening point value as well. It is proven by the research results obtained, namely the resin of the cat's eye (*S. javanica*), where the soft point of the natural resin is 96–110 °C, while the soft point of the purified resin is 94–104 °C. This is in line with Larasati's [16] statement, which states that the lower the soft point, the better the quality class.

A high soft point of resin indicates resin that has been stored for a long time [17]. This is due to the reduction in the number of double bonds in the constituent compounds due to oxidation reactions. Therefore, the heat required to soften the resin will be greater. The longer the damar is stored, the higher the oxidation reaction rate and the higher the soft point value. The high soft point value is also influenced by the higher levels of impurities [10].

4 Conclusions

Based on the results of this study, it can be concluded that:

Visual test results of H. gregaria tree resin with a diameter of 50 cm were included in the quality class A based on the SNI 2900.1:2012 standard, with a clear yellow color and a lump size of $P > 3 \times 3\ cm^2$. Meanwhile, the results of testing the physical and chemical properties of the resin refer to SNI. 2900.2:2013 obtained the average value of impurity content, soft point, turbidity in toluene, and ash content were in class I, while the acid number is in class I–II. The damar quality of *H. gregaria* as a wholew was in grade I. Compared with the quality of the pooti resin from the 30 cm diameter class where the average value was in class II, the quality of pooti damar obtained from the 50 cm diameter tree was better. These results also indicate that the damar quality of *H. gregaria* is equivalent to that of the best quality resins, such as resins from the *Shorea javanica* tree, and provides a great opportunity to utilize this pooti damar.

References

1. Purwaningsih P (2004) Sebaran Ekologi Jenis Jenis Dipterocarpaceae Di Indonesia. Biodiversitas 5(2):89–95
2. Septria D, Fernando TM, Tavita GE (2018) Keanekaragaman Jenis Pohon Famili Dipterocarpaceae Di Hutan Adat Bukit Benuah Kecamatan Sungai Ambawang Kabupaten Kubu Raya. J Hutan Lestari 6(1):114–122
3. Mulyono N, Wijaya CH, Dedi F, Wuryaningsih SR (2012) Identifikasi Komponen Kimia Damar Mata Kucing (Shorea Javanica) dengan Metode Pirolisis-GC/MS. J Natiral Indonesia 14(2):155–159
4. Sahidin EHH, Syah YM, Juliawaty LD, Achmad SA, Latip J (2006) Tiga Oligomer Resveratrol dari Kulit Batang Hopea gregaria (Dipterocarpaceae) serta sifat toksik dan sitotoksiknya. Majalah Farmasi Indonesia 17(3):109–115
5. IUCN (Internasional Union for Conservation of Nature and Natural resources) (2019) IUCN red list of thereatened species. Version 2019.2
6. Soerianegara I, Lemmens RHMJ (1994) Timber tree: major commercial timbers. PROSEA 5(1):190–191
7. Uslinawaty Z, Hadjar N, Pujirahayu N, Hamzah N, Kabe A, Nurhafidzah A (2021) Quality of Pooti Dammar Resin (Hopea gregaria) based on visual test, acid number, and ash content. Perennial 17(1):1–4
8. Munandar YS (2020) Produksi Alami Getah Damar Pooti (*Hopea gregaria* V.Slooten) Pada Beberapa Ukuran Diameter. Undergraduate thesis] Universitas Halu Oleo, Kendari
9. SNI 29001:2012 (2012) Cat's eye Resin part 1: classification and quality requirements based on visual test. BSN, Jakarta
10. Sari RK, Wistara NJ, Wihayanto A, Waluyo TK (2013) Characterization of cat's eye dammar for revision of Indonesia National Standard. J Ilmu Teknologi Kayu Tropis 11(1):73–81
11. Mampi B, Hapid A, Mutmainnah (2018) Produksi getah Pinus (Pinus Merkusii Jung Et De Vriese) pada berbagai diameter batang menggunakan sistem koakan di Desa Namo Kecamatan Kulawi Kabupaten Sigi. J Warta Rimba 6(3):42–48
12. SNI 2900.2:2013 (2013) Cat's eye Resin part 2: classification, requirements and test methods based on laboratory tests. BSN, Jakarta
13. Wijayanto A (2012) Sifat fisiko-Kimia Damar Mata Kuicing (*Shorea javanica* K. Et V.) Hasil Klasifikasi Mutu di Pasar Domestik. Undergraduate thesis. IPB Bogor
14. Putra FY, Hermawan, Azhar (2021) Purification of Damar Mata Kucing (*Shorea Javanica*) with dissolving method. J Kelitbangan 9(1):27–40
15. Gusti REP, Zulnely (2014) Sifat Fisiko Kimia Damar Mata Kucing Hasil Pemurnian Tanpa Pelarut. J Penelitian Hasil Hutan 32(3):164–174
16. Larasati F (2007) Pemurnian Beberapa Mutu Damar Mata Kucing (Shorea javanica) dengan Sistem Pemanasan. Undergraduate thesis. Departemen Hasil Hutan, Fakultas Kehutanan, IPB Bogor, Bogor (2007)
17. Mulyono N (2009) Ekstrak damar untuk bahan pengeruh dan fosforilasi damar untuk bahan pemberat. Disertasion. Institut Pertanian Bogor, Bogor

Changes in the Characteristics of LIPI Fast-Grown Teak Wood at Different Ages and Growing Sites

Wahyu Dwianto, Danang S. Adi, Dwi A. Pramasari, Eka Lestari, Adik Bahanawan, Teguh Darmawan, Dimas Triwibowo, Yusup Amin, Prabu S. Sejati, Imran A. Sofianto, Sudarmanto, Mohamad Gopar, Fazhar Akbar, Betalini W. Hapsari, Witjaksono, Ratih Damayanti, Ganis Lukmandaru, Akihisa Kitamori, Takuro Mori, and Junji Sugiyama

Abstract This paper deals with the changes in characteristics of mutant fast grown teak wood developed by Indonesians Institute of Science (LIPI) at different ages and growing sites. Wood samples were taken from various ages at three different growing sites in Cibinong Science Center Area, Bogor. Anatomical observations of the wood samples followed the International Association of Wood Anatomists (IAWA) Committee. Fiber maceration were prepared by using modified Franklin

W. Dwianto (✉) · D. S. Adi · D. A. Pramasari · E. Lestari · A. Bahanawan · T. Darmawan · D. Triwibowo · Y. Amin · P. S. Sejati · I. A. Sofianto · Sudarmanto · M. Gopar · F. Akbar
Research Center for Biomass and Bioproducts, National Research and Innovation Agency, Bogor, Indonesia
e-mail: adik002@brin.go.id

B. W. Hapsari
Research Center for Biomass and Bioproducts, National Research and Innovation Agency, Bogor, Indonesia

Witjaksono
Research Center for Biomass and Bioproducts, National Research and Innovation Agency, Bogor, Indonesia

R. Damayanti
Forest Products Research and Development Center, Ministry of Environment and Forestry, Bogor, Indonesia

G. Lukmandaru
Faculty of Forestry, Gajah Mada University, Yogyakarta, Indonesia

A. Kitamori
Faculty of Engineering, Osaka Sangyo University, Osaka, Japan

T. Mori
Graduate School of Engineering, Hiroshima University, Hiroshima, Japan

J. Sugiyama
Division of Forest and Biomaterials Science, Graduate School of Agriculture, Kyoto University, Kyoto, Japan

M. A. Rahandi Lubis (ed.), *Proceedings of the 10th International Symposium for Sustainable Humanosphere*, Lecture Notes in Civil Engineering 255,
https://doi.org/10.1007/978-981-19-1422-5_9

method. Chemical components analyses were carried out using the Mokushitsu Kagaku Jikken Manual standard. Physical and mechanical property measurements were tested according to British Standard (BS) 373-1957. The results showed that anatomical characteristics of the 5 years-old (at Site 2) mutants teak wood were not different from community forests and conventional teak woods at the same ages. The fiber length increased from pith to the bark. Alcohol-benzene soluble extractive substances tended to increase, the lignin content did not vary, holocellulose content decreased, and the α-cellulose content were almost similar with increasing age of the wood. MOR and MOE of the 5 years-old (Site 2) wood were almost the same as 20–30 years-old conventional teak wood. However, MORs of 8–10 years-old (Site 3) wood were slightly lower than that of the 5 years-old (Site 2). Surprisingly, the MOR for 9 years-old wood which grown in Site 2 was even higher than that of 60–80 years-old conventional teak wood. Density, MOR, and MOE values depended on growing site, age of the trees, and parts of stem, however all of the 5, 8, 9, and 10 years-old wood which grown in Site 2 and 3 can be classified into Strength Class II according to Indonesian Forestry Vademecum. In addition, the compressive strength parallel to the grain values of 9 years-old teak wood which grown in Site 2 can be classified into Strength Class I. From these studies, it can be concluded that the mutant teak wood has very good prospects to be developed and cultivated as an alternative raw material for wood in the future and quite suitable material for timber frame structures of wooden construction with medium strength property.

Keywords Fast grown teak wood · Ages · Growing sites · Anatomical characteristics · Chemical · Physical · Mechanical properties

1 Introduction

Teak wood (*Tectona grandis* Linn. f) is a high quality major commercial timber that is well known and widely used for various purposes [1]. Besides having good strength, natural durability, dimensional stability, and easy to work with [2], this wood has a beautiful texture and gives the owner a luxurious impression due to its relatively expensive price. The demand for teak wood has increased along with the development of timber industry, however it cannot be met by the availability of its stands. The very long tree cycle is one of the main impediments causing the limited availability of teak wood from forest stands or plantations. The cutting period for teak trees was initially carried out at age class VIII equivalent to age of 80 years-old, but it has been shortened to class VI equivalent to age of 60 years-old.

To meet the needs for conventional teak wood, some fast grown teak clones had been introduced in mid 1990s with different brand names, such as *Jati Mas*, *Jati Super*, *Jati Unggul*, *Jati Biotropia* [3]. Some of those introductions which were advertised heavily on the tissue culture technique of propagation instead of the tree quality had failed the market, while some others have survived in the fields and can be seen standing distinctly in small patches across the production area in Centra and East Java.

Since 2006, the Research Center for Biology and Biotechnology, Indonesian Institute of Sciences (LIPI) has also been promoting and planting fast growing teak genotypes including the newly introduce clones and their mutants altogether have been referred as Platinum Teak, which was characterized with straight stem with minor secondary branching, but with wood quality yet to be determined. Therefore, Platinum Teak is a mixture original introduced clones and uncharacterized mutants induced by gamma irradiated tissue culture derived from one of the introduced teak genotypes. These mutants and the introduced clones were then propagated using tissue culture techniques to obtain clonal materials for planting in larger quantities. As fast-grown clonal trees, they are expected to have shorter harvesting period, but with characteristics comparable to conventional teak wood, especially in strength, so that they can substitute the need of major wood species for timber frame structures of wooden construction, as well.

Generally, the fast-grown wood with short harvest period negatively correlates with the quality of the wood. However, quantitative analysis on the wood quality would be needed to determine the appropriate harvest period and proper use of the wood. The characterization of the fast-grown teak planted in Cibinong Science Center has been conducted for monitoring purposes [4–9]. Our analysis on 5 years-old fast growing teak demonstrated comparable wood quality to 20–30-year-old conventional teak [9].

In this study, the latest data on the wood quality of the mutant fast grown-teak planted in Cibinong Science Center and Botanic Garden campus are reported. This information is important for determining the quality of the fast-grown teak and its appropriate use.

2 Materials and Method

2.1 Specimen Preparation

Wood samples were taken from teak tree stands with various ages at three different growing sites in Cibinong Science Center (CSC) area, Bogor (Fig. 1). The teak tree genotypes planted were consisted of mutants and control namely non-mutant genotypes with morphological resemblance to the currently commercially known as "Salomon teak". The saplings were derived from tissue culture propagation that were planted at different time and location around CSC. Site 1 was planted in 2011 with two selected clones arranged in three different planting distances (2 × 2, 3 × 3, and 4 × 4 m). Site 2 has a little bit of contour and was planted in 2011–2012 with also two selected clones and planting distance of 3 × 3 m. Moreover, Site 3 was planted in 2007 with saplings derived from an induced mutation experiment of one selected clone. The samples used for the wood characterization are summerized in Table 1.

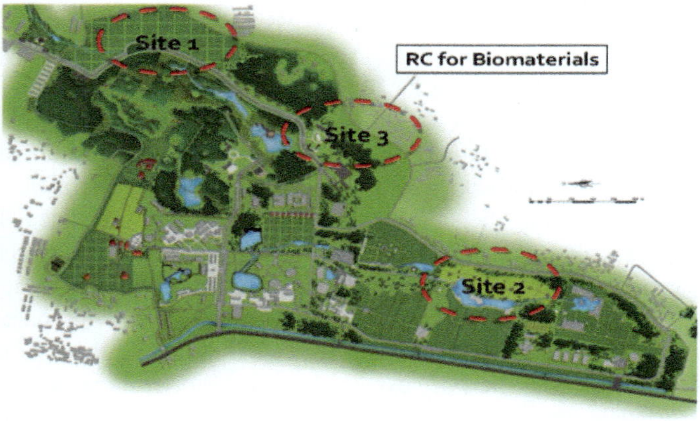

Fig. 1 Growing sites of the LIPI mutant teak trees in Cibinong Science Center Area

Table 1 Mutants' teak wood specimens for wood quality analysis collected from different tree ages at three different locations

No	Wood quality analysis	Tree age (years) at different planting locations at the time of sampling (years)		
		Site 1	Site 2	Site 3
1	Anatomical structure		5	
2	Chemical component	2, 3, 4	5	8, 9
3	Physical and mechanical properties (MOE, MOR)		5 and 9	8, 9, 10
4	Compressive strength parallel to the grain tests		5 and 9	

Tree sampling based on age was not representative of every growing location because in this study we focused more on the mechanical properties (MOE, MOR, compressive strength parallel to the grain) in order to determine the potential for the fast growing teak wood as an alternative raw material for timber frame structures of wooden construction. While observing the anatomical structure was just to compare with conventional teak woods. We did not compare the mechanical properties of the teak stands at Site 1 because they were immature.

Samplings of the wood were conducted as described. The sampled trees were cut at 130 cm above ground and sample of 30 cm long logs were taken aside. The 30 cm long log sample were also taken from the mid and the upper part of the trees. The sampled logs were then cut tangentially into plate of 3 cm thick (Fig. 2). The plates were then again cut into pieces of appropriate sizes for different analyses and measurements.

Fig. 2 Samples preparation of LIPI mutant teak wood

2.2 Anatomical Structure Observation

The microscopic structure of the cross section for anatomical and fiber observations were conducted on the incision of wood sample with a thickness of 40 μ-m using a sliding microtome, and then observed followed the International Association of Wood Anatomists (IAWA) [10]. The fiber maceration of wood samples were prepared by using modified Franklin method [11]. Matchsticks size samples were prepared and placed into a glass tube, then mixture of 50:50 (v/v) of glacial acetic acid and hydrogen peroxide was added until all the matchsticks were completely immersed. Test tubes were then placed in a water bath heated to 60 °C until the samples were softened. The tubes were shaken vigorously to separate the fibers and then stained with 2.5% safranin. In addition, the separated (macerated) wood cells (fibers, vessels and parenchyma) were placed on object-glass by using glycerin for temporary slide, and ready for observation. 30–50 fiber were observed and measured the dimensions, i.e. fiber length, cell wall thickness and lumen diameter, under a microscope.

2.3 Chemical Component Analysis

The analysis of chemical components were carried out using the Mokushitsu Kagaku Jikken Manual standard [12]. The testing processes are presented in Fig. 3.

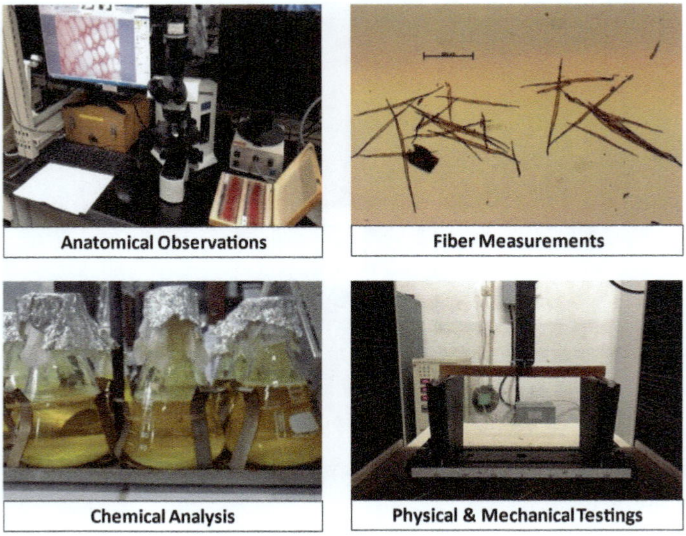

Fig. 3 Characterization of LIPI mutant teak wood

2.4 *Physical and Mechanical Property Measurements*

Small clear specimens were cut in radial direction from the bottom, middle and upper parts of the 10 m stem in length for physical and mechanical property measurements, and were tested according to British Standard (BS) 373-1957 [13]. MOR and MOE values were determined from samples with a size of 280 mm in length × 20 mm in width × 20 mm in thickness using the formula: MOR (N/mm^2) = $[(3PL)/(2bh^2)]$; MOE (N/mm^2) = $[(\Delta PL^3)/(4\Delta ybh^3)]$, where P = load (N), ΔP = load difference (N), L = span distance (mm), Δy = deflection (mm), b = sample width (mm), h = sample thickness (mm).

Wood samples with size of 40 mm in length × 20 mm in width × 20 mm in thickness were used for compressive strength parallel to the grain tests. The values were calculated using the formula: Compressive strength (N/mm^2) = $[P/(bh)]$, where P = load (N), b = sample width (mm), h = sample thickness (mm).

3 Results and Discussion

Anatomical observations of 5 years-old mutants' teak wood which was taken from Site 2 were similar to community forests and conventional teak woods, both in their macroscopic and microscopic features in tangential, radial, and cross sections [4]. Common features such as the growth ring boundaries were very clear. The early- and late-wood cells were clearly visible, and the heartwood and sapwood were clearly distinguished. The fiber length still increased from pith to the bark, showing that the wood consisted of 100% juvenile wood [14]. The result supported [1] reported that in 5-year-old fast grown teak, the fiber length tends to increase even though the pattern was slighter than conventional teak at the same age. Generally, mature wood is developed at 10–20 years or after 25 years [14, 15]. Study by Darmawan et al. [16], the trees of *Tectona grandis* produces mature wood at 10th–15th year. Determination of mature wood development in the mutant teak wood is needed.

Because many properties of wood are influenced by the age of the cells, then the mutant teak wood at that age probably has low dimensional stability. Another characteristics of the juvenile wood are short fibers, thin-walled cells, higher microfibril angle (MFA), low density and strength, low natural durability, and low drying properties [16–22].

Chemical component analysis of the mutant's teak wood based on tree ages, regardless of the growth location, showed that all variables responded linearly with age, even though with different R^2 values (Fig. 4). The alcohol-benzene soluble extractive substances increased linearly with age with relatively high slope ($R^2 = 0.6$). This indicates that prolonging the harvest time would still increase alcohol –benzene extractive content significantly, i.e., 0.29% per year. The lignin content tended to also increase linearly with age but with low slope and with low R^2 value, indicating that the lignin content was not significantly affected by the age of the trees up to 9 years. The increase may only be 1% per year. The holocellulose contents decrease with age of the trees with $R^2 = 0.6$, but the decrease per year was low and approximately 0.14% per year The α-cellulose content also decreased linearly with age with $R^2 = 0.8$, and per year the value decrease by 1.2%. The decrease in holocellulose content was due to decreased in hemicellulose and pectin contents. This was proven because the α-cellulose content of the wood did not change significantly with age.

Dimensional stability (T/R ratio) of 5 years-old (Site 2) mutants fast grown teak wood was lower (2.16) than that of 60–80 years-old conventional teak wood (1.86) [9] (Table 2). The age of the tree greatly affected the proportion of mature wood, where the older the tree, the more the proportion of mature cells increased [14].

Comparison of the MOR, MOE and wood density values of the mutant fast grown teak of 8–10 years (Site 3) with the published reports from the tree grown in the different site (Site 2) with earlier age (5 years-old), and the published report on the conventionally planted and harvested trees indicated several interesting phenomena (Table 3). The values of MOR and MOE of the 5 years-old of fast growing teak wood of Site 2 [9] were nearly the same as 20–30 years-old conventional teak wood [23], i.e. 91.97 N/mm^2 and 10,000.3 N/mm^2, respectively. It seems peculiar that the value of MORs and MOEs of the 8- to 10 years-old fast grown teak wood grown at Site 3 were slightly lower than that of the 5 years-old at Site 2. Although, the measurement of trees growing in the same area (Site 3) with little age different (8–10 years) showed no significant different in regard to the value of MOR and MOE (Fig. 5).

Compared to the older trees of 9- year-old at the same site (Site 2), these MOR and MOE values increased significantly by 30% and 20% respectively in 4 years. This MOR value of 121.20 N/mm^2 of 9-year-old fast growing teak was almost 20% higher

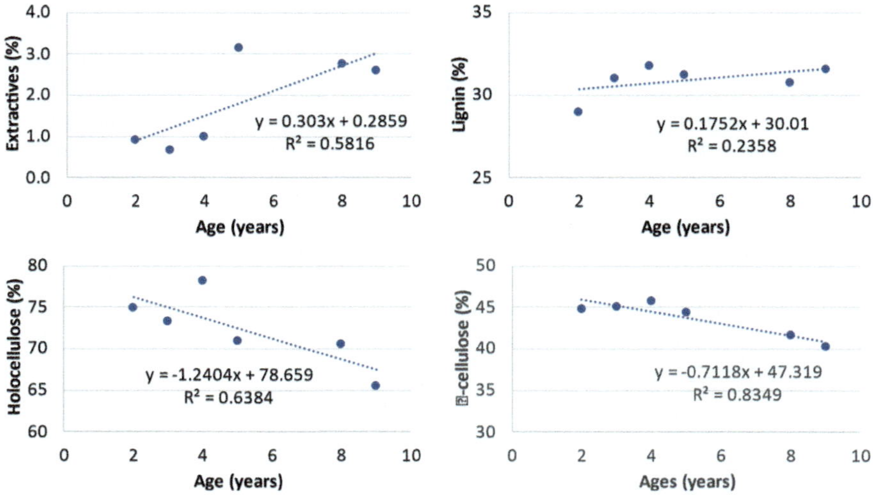

Fig. 4 Relationships between chemical component and tree ages of fast grown teak wood developed by LIPI

Table 2 Dimensional stability of 5 years-old (Site 2) mutants fast grown teak wood comparing to conventional teak

Teak woods	Oven-dried shrinkage		
	T	R	T/R ratio
5 years-old fast grown teak in Site 2 [9]	8.23	3.81	2.16
20–30 years-old conventional teak [23]	5.31	3.53	1.51
60–80 years-old conventional teak [24]	5.20	2.80	1.86

Table 3 Density, MOR and MOE values of mutant fast-grown teak wood at various ages and growing sites as compared to conventional teak woods

Teak woods	MOR	MOE	Density g/cm^3
	N/mm^2		
5 years-old platinum teak in Site 2 [9]	91.97	10,000.3	0.52
9 years-old platinum teak in Site 2	121.20	11,898.8	0.57–0.61
8 years-old platinum teak in Site 3	81.62	10,086.0	0.49
9 years-old platinum teak in Site 3	83.11	11,565.0	0.50
10 years-old platinum teak in Site 3	75.50	9221.0	0.46
20–30 years-old conventional teak [23]	93.43	10,338.0	0.63
60–80 years-old conventional teak [24]	103.10	12,770.0	0.67

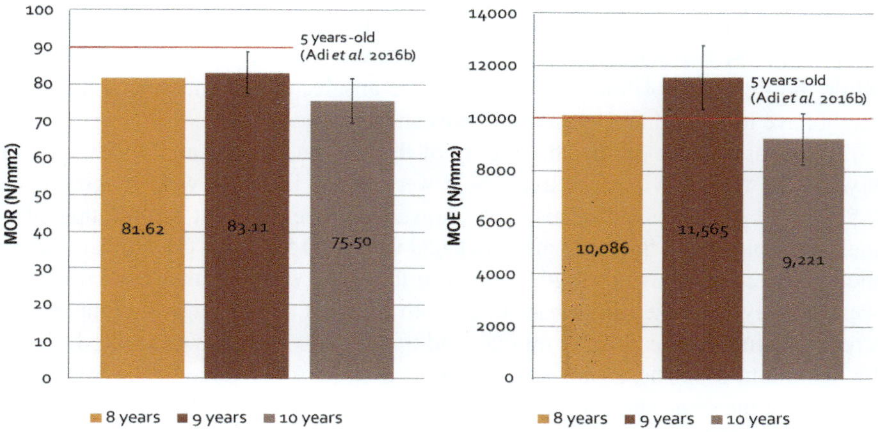

Fig. 5 MOR and MOE of 8–10 years-old (Site 3) fast grown teak wood developed by LIPI

than that of 60–80 years-old conventional teak wood, with value of 103.10 N/mm^2 (Table 3). However, this fast grown wood did not differ significantly compared to the 60–80-year-old tree in regard to MOE values. The wood density of the fast-grown teak at Site 2 showed notable increase with age, but was still lower and therefore lighter than the 60–80-year-old conventional teak wood.

In addition, the value of modulus of rupture (MOR), and modulus of elasticity (MOE) in the axial direction tended to decrease from bottom to the upper parts as indicated by wood sample from 9-years-old teak tree (Fig. 6), as well as the specific gravity (data not shown) which could be the predictor of the strength of the wood. However, in the radial direction the MOE and MOR values have an increasing tendency from the pith to the bark (data not published).

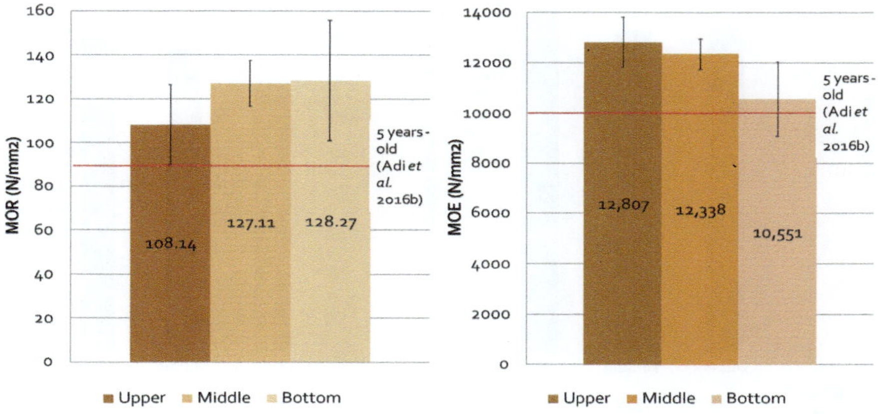

Fig. 6 MOR and MOE of 9 years-old (Site 2) fast grown teak wood developed by LIPI

Different in those values could possibly attributed to the genotypes of the trees grown on those two sites. The trees on Site 2 were composed of two selected clones, while the tree composition of the Site 3 were unselected mutants which therefore might have incurred more variability between trees.

The measurement of specific gravity of the bottom, middle and upper parts of 9 years-old fast growing teak tree which were grown in Site 2 were ranging from 0.57 to 0.61 g/cm^3 (Fig. 7). This value showed an increase about 10% compared to the 5-year-old tree of the same site which had value of 0.52 g/cm^3 [9]. However, even though the compressive strength parallel to the grain values in the radial direction were slightly decreased from bark to the pith of wood, the average of compression strength values of the bottom, middle, and upper parts of stem were only decrease by maximum 6%. These values were almost double of the value of the 5-year-old teak tree at the same site (Fig. 7).

Density, MOR, and MOE values depended on growing site, age of the trees, and parts of stem, however, all of the 5, 8, 9, and 10 years-old mutants fast grown teak developed by LIPI which grown in Site 2 and 3 can be classified into Strength Class II according to the Indonesian Forestry Vademecum (Table 4) [25]. In addition, the compressive strength parallel to the grain values of 9 years-old mutants fast grown teak wood which grown in Site 2 can be classified into Strength Class I, where the absolute compressive strength values were over 65 N/mm^2.

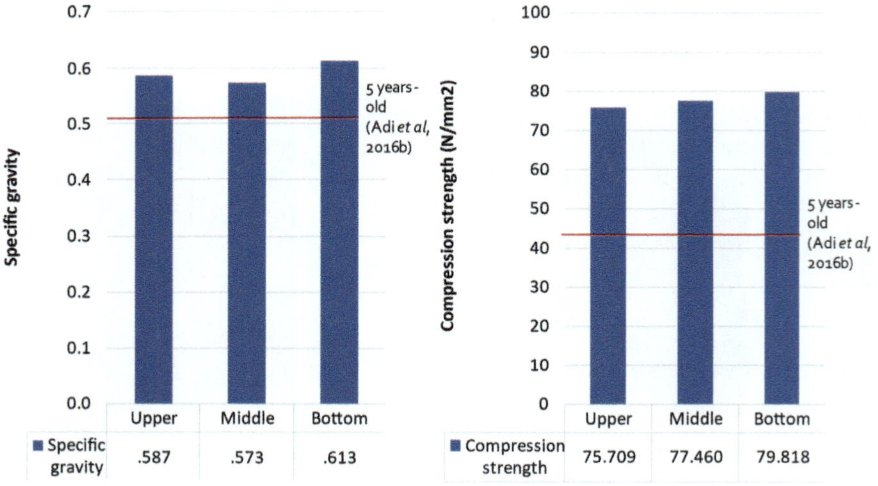

Fig. 7 Average specific gravity and compressive strength parallel to the grain of 9- years-old (Site 2) fast-grown teak developed by LIPI

Table 4 Strength class of Indonesian woods [25]

Strength class	Specific gravity	Absolute bending strength (N/mm^2)	Absolute compressive strength (N/mm^2)
I	Over 0.90	Over 110	Over 65
II	0.60–0.90	72.5–110	42.5–65
III	0.40–0.60	50–72.5	30–42.5
IV	0.30–0.40	36–50	21.5–30
V	Under 0.30	Under 36	Under 21.5

4 Conclusion

Anatomical observations of 5 years-old (Site 2) mutants fast grown teak wood developed by LIPI showed that there were not in a wide variation compared to community forests and conventional teak wood, both in their macroscopic and microscopic features. The fiber length still increased from pith to the bark, showing that the wood consisted of 100% juvenile wood.

There were changes in the chemical components with increasing age of mutants fast grown teak wood developed by LIPI, especially the extractive substances in alcohol-benzene and holocellulose/hemicellulose. Meanwhile, lignin and α-cellulose levels did not show a huge difference change with increasing wood age. This indicated that the two components were not affected by the age of the wood.

The mutants fast grown teak wood developed by LIPI up to 8 years of age could be categorized as Strength Class II. Meanwhile, the 9-year-old wood from Site 2 could be classified as Strength Class I.

The mutants fast grown teak wood developed by LIPI genotypes has very good prospects to be developed and cultivated as an alternative raw material for furniture, and medium to high quality for timber frame structures of wooden construction.

Acknowledgements Author would like to thank to the Japan—ASEAN Science and Technology Innovation Platform (JASTIP) Program, which has partly funded to conduct these researches for a period of 5 years from 2015 to 2019 in collaboration with Prof. Junji Sugiyama and Assoc. Prof. Akihisa Kitamori as the leaders of JASTIP projects, entitles Searching and Characterization of Economically Potential Utilizations of Tropical Wood Species and Development of ECO-house Concepts for A Traditional-Modern Combination Design of Wooden House Construction, respectively. Author would like to thank to Japan Society for the Promotion of Sciences (JSPS) for financial support, as well.

References

1. Damayanti R, Ozarska B, Pandit IKN, Febrianto F, Pari G (2018) Wood properties of 5-year-old fast grown teak. Wood Res J 9:29–34. https://doi.org/10.51850/wrj.2018.9.2.29-34

2. Wahyudi I, Priadi T, Rahayu IS (2014) Karakteristik dan sifat-sifat dasar kayu jati unggul umur 4 dan 5 tahun asal Jawa Barat (Characteristics and basic properties of 4 and 5 year-old of superior teakwoods from West Java). J Ilmu Pertan Indones 19:7
3. Ayobudidaya.com: Jenis-Jenis Pohon Jati di Indonesia yang Cepat Panen. https://ayobudidaya.com/perkebunan/jenis-jenis-pohon-jati-di-indonesia/
4. Adi D, Amin A, Darmawan T, Pramasari D, Dwianto W (2016) Anatomical features and fiber characteristics of 2 and 5 years-old platinum teak wood. In Proceedings of the 7th international symposium Indonesian wood research society, Bandung, Indonesia, pp 251–255
5. Pramasari D, Wahyuni I, Adi D, Amin A, Darmawan T, Dwianto W (2015) Effect of age on chemical component of platinum teak wood—a fast growing teak wood from LIPI. In: Proceedings of the 6th symposium of Indonesian wood research society, Medan, Indonesia, pp 211–216
6. Pramasari D, Lestari E, Adi D, Dwianto W, Lukmandaru G (2015) Analysis of extractive compounds and colour variation of five years-old platinum teak wood. In: Proceedings of the 5th international symposium of sustainable Humanosphere, Jakarta, Indonesia
7. Pramasari D, Lestari E, Bahanawan A, Adi D, Dwianto W (2015) Study of the chemical components of platinum teak based on tree age (II). In: Prosiding seminar Nasional XVIII MAPEKI, Bandung, Indonesia, pp 190–197
8. Lestari E, Pramasari D, Amin Y, Adi D, Bahanawan A, Dwianto W (2016) The chemical components changes of Platinum teak wood. In: Proceedings of the 6th international symposium for sustainable Humanosphere, Bogor, Indonesia, pp 165–171
9. Adi D, Sudarmanto S, Ismadi I, Gopar M, Darmawan T, Amin Y, Dwianto T (2016) Evaluation of the wood quality of Platinum Teak Wood. Teknol Indones 39:36–44
10. Committee IAWA (1989) IAWA list of microscopic features for hardwood identification. IAWA Bull 10:219–332
11. Damayanti R, Ilic J, Ozarska B, Pari G, Vinden P (2019) Effect of clone and sites for wood properties of fast grown teak and determination of the best clone for next cultivation in teak plantation industries. IOP Conf Ser: Earth Environ Sci 359:011001
12. Japan Wood Research Society (2000) Mokushitsu Kagaku Jikken Manual
13. BS (British Standard) (1957) British Standard 373-1957: methods of testing small clear specimens of timber. British Standards Institution, London
14. Panshin AJ, de Zeuw C (1980) Textbook of wood technology. McGraw-Hill Book Co., New York, Iowa
15. Shmulsky R, Jones P (2019) Forest products and wood science: an introduction. Wiley-Blackwell, West Sussex (GB)
16. Darmawan W, Nandika D, Kartikasari R, Sitompul A, Rahayu I, Gardner D (2015) Juvenile and mature wood characteristics of short and long rotation teak in Java. IAWA J 36:15
17. Gryc V, Vavrcik H, Horn K (2011) Density of juvenile and mature wood of selected coniferous species. J For Sci 57:123–130. https://doi.org/10.17221/18/2010-JFS
18. Perre´P, Keey RB (2014) 36 drying of wood: principles and practices. In: Handbook of industrial drying. CRC Press, Boca Raton, pp 822–872
19. Mahmud S, Hashim R, Saleh A, Sulaiman O, Saharudin N, Ngah M, Husain H (2017) Physical and mechanical properties of juvenile wood from *Neolamarckia cadamba* planted in West Malaysia. Maderas-Cienc Tecnol 19:225–238. https://doi.org/10.4067/S0718-221X2017005000020
20. Arsyad W, Basri E, Hendra D, Trisatya D (2019) Termite resistance of impregnated jabon wood (*Anthocephalus cadamba* Miq.) with combined impregnants. J Korean Wood Sci Technol 47:1–8. https://doi.org/10.5658/WOOD.2019.47.4.1
21. Cahyono T, Wahyudi I, Priadi T, Febrianto F, Darmawan W, Bahtiar ET, Ohorella S, Novriyanti E (2015) The quality of 8 and 10 years old samama wood (*Anthocephalus macrophyllus*). J Indian Acad Wood Sci 12:22–28. https://doi.org/10.1007/s13196-015-0140-8
22. Basri E, Wahyudi I (2013) Sifat dasar kayu Jati Plus Perhutani dari berbagai umur dan kaitannya dengan sifat dan kualitas pengeringan. J Penelit Has Hutan 31:93–102

23. Suprapto E, Marsoem N (2007) Axial and radial variations in the physical and mechanical properties of teak wood (*Tectona grandis* Linn. F) thin felling (KU III) from KPH Purwakarta
24. Martawijaya A, Kartasujana I, Kadir K, Prawira SA (2005) Atlas Kayu Indonesia Jilid I (Indonesian Wood Atlas). Forest Products Research and Development Centre, Indonesian Ministry of Forestry, Bogor
25. Abdurachman, Akbar A, Supriadi A, Santoso A, Ismanto A, Salim AG, Andianto, Wibowo A, Soka A, Hidayat A, Sukmana A, Widiarti A, Utari AD, Windyoningrum A, Harjadi B, Bintoro, Narendra BH, Leksono B, Lastiantoro CY, Martono D, Darwo, Indrawati DR, Yuniati D, Indrawan DA, Djarwanto, Hendra D, Octavia D, Wicaksono D, Dulsalam, Puspasari D, Basri E, Subiandono E, Widyati E, Syahadat E, Satiti ER, Pasaribu G, Pari G, Roliadi H, Nuroniah HS, Krisnawati H, Gunawan H, Khotimah H, Dharmawan IWS, Sulastiningsih IM, Nugroho IA, Anggraeni I, Rahmawati I, Pramono IB, Yeny I, Hakim I, Samsoedin I, Balfas J, Malik J, Jasni, Maharani KE, Kuntadi, Andadari L, Dewi LM, Wibowo LR, Abdullah L, Iskandar MI, Muslich M, Magdalena, Wardani M, Lugina M, Minarningsih, Yulianti M, Murniati, Haryanti N, Lelana NE, Sakuntaladewi N, Sari N, Mindawati N, Wahyuningrum N, Nugroho NP, Hadjib N, Pratiwi, Setio P, Purwanto, Djaenuddin RD, Garsetiasih R, Irianto RSB, Damayanti R, Mutiaradevi R, Agustarini R, Maryani R, Imanuddin R, Dewi R, Kwatrina RT, Wibowo S, Irawanti S, Soenarno, Suhartana S, Suharti S, Subarudi, Ekawati S, Kosasih S, Ratina S, Sylviani, Rostiwati T, Kalima T, Setyawati T, Waluyo TK, Yuwati TW, Herawati T, Sriyati T, Darmawan UW, Arsyad WOM, Adinugroho WC, Endom W, Darwiati W, Rochmayanto Y, Siswiyanti Y, Adalina Y, Nuraeni Y, Heryati Y, Yulianti, Yuniawati (2020) Vademecum Kehutanan Indonesia 2020. Badan Penelitian, Pengembangan, dan Inovasi, Kementerian Lingkungan Hidup dan Kehutanan, Jakarta, Indonesia

Hot Pressing Treatment on Physical and Mechanical Properties of Com-Ply Bonded with Citric Acid

Dimas Triwibowo, Narto, Sudarmanto, Fazhar Akbar, Yusup Amin, Sukma Surya Kusumah, and Yusuf Sudo Hadi

Abstract Comp-ply is one of composite panels. This study aims to determine the effect of the manufacturing method and hot-pressing time on the physical and mechanical properties of composites made of sengon veneer and bamboo betung particles using citric acid adhesive. Based on the calculation results, the com-ply manufacturing method in stages with a hot-pressing time of 10 min has better physical and mechanical properties than the others. However, the mechanical properties of the compound are still below JIS A 5908 2003 requirements.

Keywords Bamboo · Citric acid · Com-ply · Sengon · Veneer

1 Introduction

Bamboo is a non-timber forest product that has the potential to be used as an alternative material to replace wood. This is because bamboo in Indonesia has 157 types of bamboo [1]. Bamboo also has a fast-growing and cutting period of 3–5 years [2]. In 2018, the total amount of bamboo production reached more than 20.3 million sticks [3]. This number increased compared to the total production in 2017 which reached more than 14.8 million sticks [4]. The use of bamboo by the community is generally used as handicrafts such as weaving, tables, chairs, cabinets, and shelves. The development of technology, the use of bamboo as a building material is no longer used as a whole. The development of bamboo-based building materials a few years later made

D. Triwibowo · Narto · Sudarmanto · F. Akbar · Y. Amin · S. S. Kusumah
Research Center for Biomaterials, Indonesian Institute of Sciences, Cibinong Science Center—Botanical Garden (CSC-BG), Cibinong-Bogor, Indonesia

Y. S. Hadi
Departement of Forest Products, Faculty of Forestry and Environment, IPB University, Bogor, Indonesia

Y. S. Hadi (✉)
School of Forestry, Papua New Guinea University of Technology, Lae, Papua New Guinea
e-mail: yshadi@indo.net.id

M. A. Rahandi Lubis (ed.), *Proceedings of the 10th International Symposium for Sustainable Humanosphere*, Lecture Notes in Civil Engineering 255,
https://doi.org/10.1007/978-981-19-1422-5_10

it a composite panel. Some composite panel products that utilize bamboo include oriented strand board (OSB) [5–7], laminated bamboo wood [8, 9], cross-laminated bamboo [10], and particle board [11]. The manufacture of bamboo composite panels generally uses synthetic adhesives such as urea–formaldehyde (UF). However, UF as a synthetic adhesive has formaldehyde emissions that can harm the environment and human health. The impact of formaldehyde emission causes irritation, shortness of breath, and decreased lung function [12]. Based on this, natural adhesives are needed to replace synthetic adhesive emissions. One of the natural ingredients that can be used as an adhesive is citric acid.

Citric acid is a weak acid that comes from lemons, oranges, tomatoes, berries, and beets [13]. The previous use of citric acid in wood was only as a crosslinker and was able to increase the physical and mechanical strength of wood [14]. The application of citric acid as a wood binder was first carried out by Umemura et al. in 2011. The results showed that citric acid can bind two hydroxyl groups of wood [15]. This statement also strengthens the research conducted by Menezzi [16]. Application of the use of citric acid binder will be used in the manufacture of com-ply. Com-ply is one of composite panels. Com-ply is a superior product because it can use low quality wood, small diameter, even wood residue in the core layer, as well as using a thin layer (veneer) on the front and back layers. Therefore, the utilization of high-quality wood will be minimized, while the com-ply appearance is the same as for plywood. This research aims to determine the effect of the manufacturing method and hot-pressing time on the physical and mechanical properties of composite plywood using citric acid adhesive. Research on compliance is still not done much. Whereas the potential for compliance as an alternative composite product is quite large.

2 Method

2.1 Material Preparation

Betung bamboo (*Dendrocalamus asper* (Schultes f.) Backer ex Heyne) is used in conditions of dry air with a water content of 12–15%. Bamboo stalks peeled clean up last in the number using the drum chipper. Furthermore, chopped bamboo is inserted to be the size of the particles. After that bamboo particles in the sieve to get the size of the past 14 mesh and smothered 20 mesh. The last particles of bamboo are put in a preheated oven at 80 °C, so the water content of bamboo particles is < 7% before used. Sengon veneers (*Paraserianthes falcataria*) used were obtained from PT. SGS that are in Balaraja, Banten. Size veneer used by 30 cm × 30 cm × 0.2 cm (length x width x thickness, respectively) with a moisture content of ≤ 10%.

Citric acid is derived from the PT. Budi Acid Jaya, Lampung, Indonesia. Adhesive made by dissolving citric acid in water with the concentration of the solution by 59%. The citric acid in powder form is mixed with water in a beaker glass to dissolve the citric acid. Once it's put in into the beaker glass that contains water with a temperature of 70 °C and stirred with a stirrer speed of 300 rpm for approximately 20 min. After that then removed and allowed to stand before use. The adhesive used for particle board of 20% of the weight of the particles used and 150 g/m^2 adhesive weight for coverage on the veneer.

2.2 *Com-Ply Manufactured*

The manufacture of com-ply consists of 3 layers, the front and back layers are sengon veneer and the core layer is betung bamboo particles. The dimensions of the com-ply made are 30 cm × 30 cm × 1.3 cm (length × width × thickness, respectively). First, the betung bamboo particles were sprayed with a 20% citric acid solution of the dry weight of the particles. After that, it was dried in an oven at 80 °C for 24 h. The manufacturing process is carried out in 2 methods, namely directly and gradually. The manufacturing process is carried out directly by smearing the sengon veneer with a solution of citric acid and placing it at the bottom of the mold. Furthermore, the top of the veneer is sprinkled with bamboo particles after being in the oven. After that, the top is covered with sengon veneer that has been smeared with citric acid solution. The manufacturing process is carried out in stages by first making the core of the betung bamboo particles, then it is coated with sengon veneer which has been coated with citric acid solution on the front and back. The glued spread used was 150 g/m^2. Finally, hot pressed at 200 °C for 8 min and 10 min.

2.3 *Physical and Mechanical Properties*

Testing of physical and mechanical properties refers to Japanese Industrial Standard Number A 5908-2003 [17] regarding veneered particleboard. Testing of physical properties includes measurement of water content, density, thickness expansion, and water absorption. Mechanical properties tests include flexural tests (MOR and MOE) and internal bonding tests.

(a) *Moisture content*

The size of the sample for the moisture content and density test is 100 mm × 100 mm in length and width.

$$MC(\%) = \frac{m_1 - m_0}{m_0} \times 100 \quad (1)$$

where:

M_0 mass (g) after drying
M_1 Initial weight of the test sample (g).

(b) *Thickness swelling and water absorption*

The size of the thick development test sample is 50 mm × 50 mm in length and width with a thickness of 14 mm. The test sample was first immersed in water at a temperature of 20 °C for 24 h. The calculation of thickness and water abosorbsion expansion is carried out using the formula:

$$ST(\%) = \frac{t_2 - t_1}{t_1} \times 100 \tag{2}$$

where:

T_2 Thickness after soaking (mm)
T_1 Thickness before soaking (mm).

$$WA(\%) = \frac{w_2 - w_1}{w_1} \times 100 \tag{3}$$

where:

W_2 Weight of test sample after immersion in water (g)
W_1 Weight of test sample before immersion in water (g).

(c) *Bending test*

Flexibility test using Universal Testing Machine (UTM) Shimadzu with three loading points. The test is carried out lengthwise at a load speed of 10 mm/min. The calculation of the flexibility test is carried out using the formula:

$$MOR\left(N/mm^2\right) = \frac{3P_{max}L}{2bt^2} \tag{4}$$

where:

P_{max} Maximum loading (N)
L Span length (mm)
b Width of test sample (mm)
t Thickness of test sample (mm).

$$MOE\left(N/mm^2\right) = \frac{\Delta PL^3}{4\Delta ybt^3} \tag{5}$$

where:

P The load difference is below the proportion limit (N)

L Span length (mm)
y Change in deflection below the limit of proportion (mm)
b Width of test sample (mm)
t Thickness of test sample (mm).

(d) *Internal bonding test*

Test samples measuring 50 mm × 50 mm × 14 mm. Internal bonding test using Shimadzu Universal Testing Machine (UTM). Loading speed of 2 mm/min. The calculation of the adhesive bond test is carried out using the formula:

$$IB\left(N/mm^2\right) = \frac{P\prime}{2bL} \quad (6)$$

where:

P′ Maximum loading (N)
L span length (mm)
b Width of test sample (mm).

3 Results and Discussion

A Physical Properties

The moisture content of composite plywood composed of particles of betung bamboo and sengon veneer ranged from 5.61 to 11.16% as shown in the Table 1. JIS A 5908-2003 requires a maximum water content of 13%, so that the water content value complied with in this study meets the standard. There is a difference in the value of moisture content in the appropriate manufacturing method between the direct manufacturing method and the stepwise manufacturing method with hot pressing times of 8 and 10 min. The largest moisture content was obtained by the direct manufacturing method with a hot-pressing time of 8 min. The density of each method of manufacture with variations in hot pressing time ranged from 0.438 to 0.448 g/cm^3. JIS A 5908-2003 requires a minimum density of 0.40 g/cm^3. The lowest density was obtained in the manufacture of com-ply in stages with a compression time of 8 min, while the highest density was obtained in the manufacture of com-ply in stages with a pressing time of 10 min. This density difference affects the water absorption and thickness swelling. The biggest thickness swelling occurred in the direct com-ply manufacturing method with a hot-pressing time of 8 min of 29.63% with a water absorption of 129.378%. This is due to the low density of the board having many void [18]. In this case it can be said that the density will affect the thickness swelling and water absorption [19]. The relationship between density and thickness of particleboard is directly proportional. The higher the density, the thicker the particle board tends to increase.

Table 1 Physical properties of com-ply

Method	Time (minute)	Moisture content (%)	Density (gram/ cm^3)	Thickness swelling (%)	Water absorption (%)
Directly	8	11.16	0.439	29.63	129.378
	10	6.96	0.443	12.35	101.610
Gradually	8	5.85	0.438	12.78	119.768
	10	5.61	0.448	11.31	100.341
JIS A 5908 2003		Maks: 13	Min: 0.40	Maks: 12	N/A

The thickness swelling of the stepwise manufacturing method with the hot-pressing time of 10 min (11.31%) is lower than the others. This thickness swelling is lower than that of lightweight particleboard expanded of polystyrene with a density of 0.45 [19].

B Mechanical properties

The MOR results show that the manufacture of com-ply in stages with a hot-pressing time of 10 min has a higher value than the others (Fig. 1). However, the MOE of making com-ply in stages with hot pressing time of 8 min is better than the others. The MOR and MOE values for the two types of compliance are still below the JIS A 5908-2003 standard which stipulates a minimum MOR value of 30 N/mm^2 and an MOE of 4000 N/mm^2. The MOR and MOE values are also lower than lightweight particleboard expanded of polystyrene at 22.4 N/mm^2 [19].

Internal bond values on adherence ranged from 0.1 to 0.23 N/mm^2 (Fig. 3). The greatest bonding value was obtained from the manufacture of panels in stages with a compression time of 10 min of 0.23 N/mm^2. However, all panel bond values are still below the minimum value required by JAS A 5908-2003, which is 0.3 N/mm^2. The low value of the mechanical properties is thought to be caused by the low density of the two types of composites made, which is around 0.44 g/cm^3 (Fig. 2).

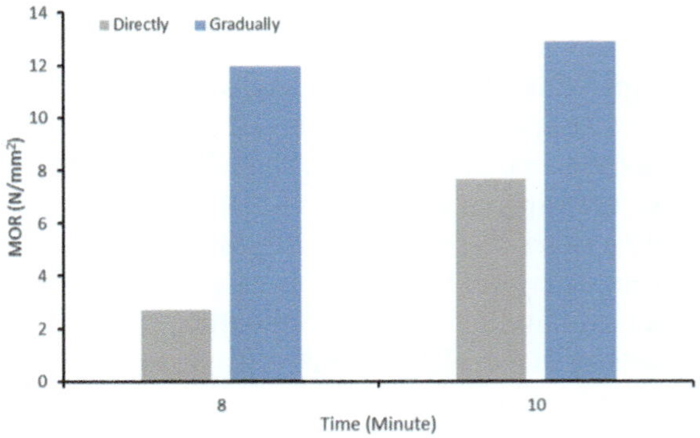

Fig. 1 MOR of hot press treatment com-ply

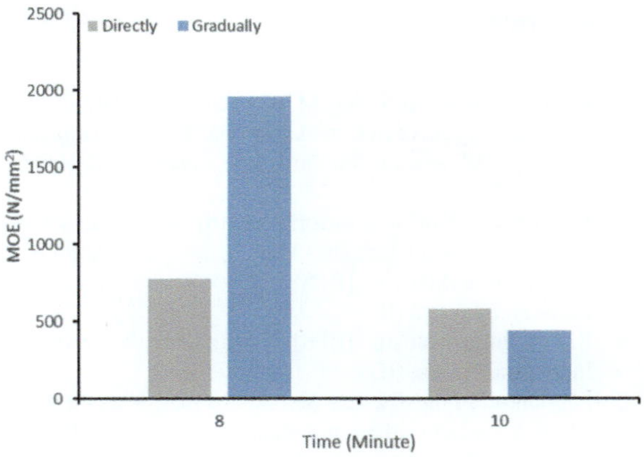

Fig. 2 MOE of hot press treatment com-ply

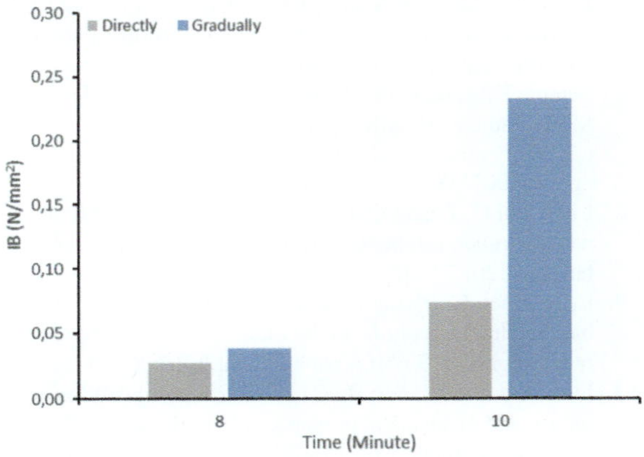

Fig. 3 Internal bonding of hot press treatment com-ply

4 Conclusion

The method of making com-ply in stages with a hot-pressing time of 10 min has good physical properties compared to making com-ply directly. This can be seen from the moisture content, density, thickness swelling value and better water absorption. In addition, the manufacture of com-ply gradually has better mechanical properties compared to the others, as evidenced by the better MOR and MOE values compared to the manufacture of another com-ply. However, the MOR and MOE values of com-ply are still below the value of JIS A 5908 2003.

Acknowledgements The author would like to thank the Center for Coaching, Education and Training LIPI for providing funding through the program by Research. In addition, the authors also thank JSPS for financial support.

References

1. Hadjar N, Marwah S, Sari M, Bana S, Uslinawaty Z, Pujirahayu N, Hamzah N, Kabe A, dan Agus (2020) Jenis dan keanekaragaman serta pemanfaatan bamboo oleh masyarakat kelurahan Tobimeita kecamatan Nambo kota Kendari. JKIC 1(1):53–63. http://ojs.uho.ac.id/index.php/CELEBICA
2. Hastuti RW, Yani AP, Ansori I (2018) Studi keanekaragaman jenis bambu di desa Tanjung Terdana Bengkulu Tengah. J Pend Pemb Biol 2(1):96–102
3. Badan Pusat Statistik [BPS] (2019) Statistik Produksi Kehutanan. Badan Pusat Statistik Indonesia, Jakarta (ID)
4. Badan Pusat Statistik [BPS] (2018) Statistik Produksi Kehutanan. Badan Pusat Statistik Indonesia, Jakarta (ID)
5. Febrianto F, Jang JH, Lee SH, Santosa IA, Hidayat W, Kwon JH, Kim NH (2015) Effect of bamboo species and resin content on properties of oriented strand board prepared of steam treated bamboo strands. BioResources 10(2):2642–2655. https://doi.org/10.15376/biores.10.2.2642-2655
6. Dixon PG, Malek S, Sample KE, Zhang PK, Smith GD, Gibson LJ (2017) Multiscale modeling of Moso bamboo-oriented strand board. BioResources 12(2):3166–3181. https://doi.org/10.15376/biores.12.2.3166-3181
7. Semple KE, Zhang PK, Smola M, Smith GD (2015) Hybrid oriented strand boards made from Moso bamboo (Phyllostachys pubescens Mazel) and Aspen (Populus tremuloidesMichx.): uniformly mixed single layer uni-directional boards. Eur J Wood Prod 73:515–525. https://doi.org/10.1007/s00107-015-0913-1
8. Li H, Wu G, Zhang Q, Deeks AJ, Su J (2018) Ultimate bending capacity evaluation of laminated bamboo lumber beams. Constr Build Mater 160:365–375. https://doi.org/10.1016/j.conbuildmat.2017.11.058
9. Li H, Chen G, Zhang Q, Ashraf M, Xu B, Li Y (2016) Mechanical properties of laminated bamboo lumber column under radial eccentric compression. Constr Build Mater 121:644–652. https://doi.org/10.1016/j.conbuildmat.2016.06.031
10. Lu Z, Zhou H, Liao Y, Hu C (2018) Effects of surface treatment and adhesives on bond performance and mechanical properties of cross-laminated timber (CLT) made from small diameter Eucalyptus timber. Const Build Mater 161:9–15. https://doi.org/10.1016/j.conbuildmat.2017.11.027
11. Nurhazwani O, Jawaid M, Paridah MT, Juliana AH, Hamid SA (2016) Hybrid Particleboard Made from Bamboo (Dendrocalamus asper) Veneer Waste and Rubberwood (Hevea brasilienses). BioResources 11(1):306–323. https://doi.org/10.15376/biores.11.1.306-323
12. Kanse NG, Deepali M, Kiran P, Priyanka B, Dhanke P (2017) A review on citric acid production and its applications. Int J Curr Adv Res 6(9):5880–5883. https://doi.org/10.24327/ijcar.2017.5883.0825
13. Kim KH, Jahan SA, Lee JT (2011) Exposure to formaldehyde and its potential human health hazards. J Environ Sci Health C 29:277–299
14. Hasan M, Jug M (2008) Biological durability of wood modified by citric acid. DRVNA Industrija 59(July):55–59
15. Umemura K, Ueda T, Munawar SS, Kawai S (2011) Application of citric acid as natural adhesive for wood. J Appl Polym Sci 123:1991–1996. https://doi.org/10.1002/app.34708
16. Menezzi CD, Amirou S, Pizzi A, Xi X, Delmonte L (2018) Reactions with wood carbohydrates and lignin of citric acid as a bond promoter of wood veneer panels. Polymers 833(10):1–13. https://doi.org/10.3390/polym10080833
17. Japanese Standards Association [JSA] (2003) Japanese Industrial Standard for Particleboards A 5908. Japanese Industrial Standards Committee Standards Board, Tokyo (JP)
18. Shalbafan A, Dietenberger MA (2013) Fire performances of foam core particleboards continuously produced in a one-step process. Eur J Wood Prod 71:49–59. https://doi.org/10.1007/s00107-012-0653-4

19. Subiyanto B, Raskita S, dan Efendy H (2003) Pemanfaatan Serbuk Sabut Kelapa sebagai Bahan Penyerap Air dan Oli Berupa Panel Papan Partikel. J Ilmu Teknologi Kayu Tropis 1

Description of Morphological Characterization and Dimensional Geometry for Several Prospective Bamboos

Adik Bahanawan, Teguh Darmawan, Wahyu Dwianto, Wida Banar Kusumaningrum, Ratih Damayanti, Malihatun Nufus, Rezky Lasekti Wicaksono, Yus Andhini Bhekti Pertiwi, and I. Putu Gede Parlida Damayanto

Abstract The aims of this study was to determine stem morphological and geometrical characteristics of several potential bamboo species. Bamboos were used in this study namely Ampel Kuning (*Bambusa vulgaris* Schrader ex Wendland (Var. Striata), Petung Hitam (*Dendrocalamus asper* Black), Andong (*Gigantochloa pseudoarundinaceae* (Steudel) Widjaja), Ater (*Gigantochloa atter* (Hassk.) Kurz), Betung (*Dendrocalamus asper* (Schultes f.) Backer ex Heyne), Sembilang (*Dendrocalamus giganteus* Wallich ex Munro) and Mayan (*Gigantochloa robusta* Kurz). Variable of internode length will produce a polynomial curve while variable of outer diameter, inner diameter and stem thickness will produce a negative linier curve. For internode length, bamboos shows an inverted U where base part (B) is shorter than middle part (M) but longer than top part (T). Outer diameter, inner diameter and stem thickness show trend where it will smaller to T part. Biggest internode length was Ater—B = 52.5 cm. Biggest outer diameter was Sembilang—B = 13.4 cm. Biggest inner diameter was Sembilang—B = 12.21 cm. Biggest stem thickness was Petung Hitam—B = 1.52 cm. Taper value for total length, most pointed bamboo culm was Sembilang (0.0060) while least pointed bamboo culm was Ampel Kuning (0.0028). Biggest slenderness ratio for B part was Andong (5.51), M part was Ater

A. Bahanawan (✉) · T. Darmawan · W. Dwianto · W. B. Kusumaningrum
Research Center for Biomaterials, Indonesian Institutes of Sciences, Cibinong, Bogor, Indonesia
e-mail: adikbahanawan@gmail.com

R. Damayanti
Forest Products Research and Development Center, Ministry of Environment and Forestry, Bogor, Indonesia

M. Nufus · R. L. Wicaksono · Y. A. B. Pertiwi · I. Putu Gede Parlida Damayanto
Forest Management Program, Faculty of Agriculture, Universitas Sebelas Maret, Surakarta, Indonesia

I. Putu Gede Parlida Damayanto
Research Center for Biology, Indonesian Institutes of Sciences, Bogor, Indonesia

M. A. Rahandi Lubis (ed.), *Proceedings of the 10th International Symposium for Sustainable Humanosphere*, Lecture Notes in Civil Engineering 255,
https://doi.org/10.1007/978-981-19-1422-5_11

(7.84) and T was Ater (10.97) while smallest slenderness ratio for B part was Ampel Kuning (2.95), M part was Sembilang (3.15) and T part was Ampel Kuning (4.80). Calculation of morphological uniformity based on smallest Standard Deviation (SD) value. Smallest SD value indicates morphological variable more uniform. Variable of internode length, Petung Hitam was bamboo with most uniform (SD = 3.38). Variable of outer diameter was Ampel Kuning (SD = 1). Variable of inner diameter was Ampel Kuning (SD = 0.87). Variable of stem thickness was Mayan (SD = 0.28).

Keywords Bamboo · Diameter · Length · Morphology · Standard deviation · Thickness · Uniformity

1 Introduction

Bamboo is a unique biological material. Some of its unique characteristics were various colors of bamboo stem, internode and node, and also the geometric of bamboo culm. Mass of bamboo stem was relatively light but has quite strong [1, 2]. Bamboo is fast growing plant. Sembilang (*Dendrocalamus giganteus* Wallich Ex Munro) was reported have growth rate of 6.06 cm/day [6]. Bamboo grows 3 times faster than other plant species [1]. Even, speed of growing bamboo was claimed about 25 times faster than fast-growing forest plant species like trees [4]. Bamboo can be developed as an alternative raw material to replace teh presence of wood [5]. Indonesia is one of the countries in Southeast Asia with high diversity of bamboo species with 161 species of which 126 species are native species which means about 15.5% of thr worlds total bamboo species [6]. These various types of bamboo can be used for various purposes such as raw materials for furniture, construction, and also pulp [1, 3, 7 and 8]. Various uses of bamboo raw materials for various purposes require us to know among others, their morphological characteristics because there are many variations of bamboo species with varying lenghts and diemeter. The morphological characteristics of bamboo stems are very unique with various variability in terms of length and diameter. Morphology in corridor of biology is the science of the external form and structure of living things [9]. In simple terms, morphology studies characteristics of shape from organs at a plant. Introduction of these morphological characteristics is important. This is done in order to maximize its potential benefits from point of view of unique and distinctive morphological characteristics of each bamboo. This study aims to describe morphological characteristics of bamboo stems consisting of length, number of internodes, diameter, thickness and to compile description stem geometri includes tapering and slenderness ratio parameters.

Fig. 1 Bamboos in this research. **a** Ampel Kuning = *B. vulgaris* Schrader ex Wendland (Var. Striata), **b** Petung Hitam = *D. asper* Black, **c** Andong = *G. pseudoarundinaceae* (Steudel) Widjaja, **d** Ater = *G. atter* (Hassk.) Kurz, **e** Betung = *D. asper* (Schultes f.) Backer ex Heyne, **f** Sembilang = *D. giganteus* Wallich ex Munro dan **g** Mayan = *G. robusta* Kurz. (photo by: Adik Bahanawan)

2 Materials and Methods

2.1 Materials

Seven types of bamboo were used as materials, namely Ampel Kuning (*Bambusa vulgaris* Schrader ex Wendland (Var. Striata)), Petung Hitam (*Dendrocalamus asper* Black), Andong (*Gigantochloa pseudoarundinaceae* (Steudel) Widjaja), Ater (*Gigantochloa atter* (Hassk.) Kurz), Betung (*Dendrocalamus asper* (Schultes f.) Backer ex Heyne), Sembilang (*Dendrocalamus giganteus* Wallich ex Munro) dan Mayan (*Gigantochloa robusta* Kurz) as shown in Fig. 1. Every type of bamboo used 6 replication. *Bambusa, Dendrocalamus,* and *Gigantochloa* are native Indonesian genera [6]. These bamboos grow in Indonesia's climate by growing in clumps. Bamboo used in this research have aged about 3–4 years and are collection of Research Center for Biomaterials—Indonesian Institutes of Sciences that located at 6°29′41.6″ South Latitude and 106°51′11.7″ East Latitude. This research was conducted from November 2020 until June 2021.

2.2 Methods

2.2.1 Measurements of Dimension, Geometry and Variable Equation Function

Several measurements were performed including total length, number of internodes, diameter (outer and inner), and thickness. Length, thickness and diameter are factors that affect dimensions of bamboo [10]. In addition, mean value of these measurement variable will also be obtained for each part of base (B), middle (T) and top (T). After obtaining measurement values, next step is visualization of variables on graphs and compiling an equation fuction for some measurements variables. Some of these equation include:

1) Equation of internode length, outer diameter, inner diameter and thickness
2) Equation of outer diameter and thickness
3) Equation of thickness to slenderness ratio
4) Equation of taper
5) Equation of slenderness ratio

In addition, geometric calculations, such as taper and slenderness ratio were performed. Taper was calculated for each type of bamboo with difference in length namely total length, length about 8 m and length about 6 m from base (B). Length about 8 m and 6 m are used because these 2 sizes are used in bamboo sales place especially in Java Island, Indonesia. Slenderness ratio was calculated for each internode and each type of bamboo. Formula of taper and slenderness ratio shown below:

$$\text{Taper} = \frac{\text{Db} - \text{Dt}}{\text{Lt}} \quad (1)$$

$$\text{Slenderness ratio} = \frac{\text{L internode x}}{\text{O diameter x}} \quad (2)$$

2.2.2 Data Analysis

Microsoft Excel 2013 was used for data analysis. Discussion presented (1) total length and total internodes, (2) biggest and smallest value for each measurement variable on base, middle and top part of bamboo, (3) morphological description for each measurement variable, (4) uniformity of morphology, (5) Linier equation for some variable, (6) taper and slenderness ratio (geometric calculation) and (7) morphological bamboo from other literature and potential utilization.

3 Results and Discussion

3.1 Total Length and Total Internodes

7 types of bamboos in this research have difference results of measurement. This phenomena because of difference in type and genetic for each bamboo. Table 1 shown total length and total internodes for each bamboos.

Average total length of seven bamboos was about 11.19–15.32 m with longest average bamboo was Ater (15.32 m). Average total internode was about 33.17–36.83 pieces with average biggest total internode was Sembilang (36.83 pieces).

Table 1 Total length and total internodes of each bamboos

No.	Bamboos	Total length (m)	Average total internode (pieces)
1	Ampel kuning	11.19**	35.83
2	Petung hitam	13.37	33.17**
3	Andong	14.40	34.17
4	Ater	15.32*	36.17
5	Betung	13.34	34.20
6	Sembilang	13.95	36.83*
7	Mayan	14.71	36.00

*biggest value, **smallest value

3.2 Length Internode, Outer Diameter, Inner Diameter and Thickness with Biggest and Smallest Value from Base (B), Middle (M) and Top (T) Part of Each Bamboos

Results showed that biggest internode length was Ater—B = 52.3 cm. Biggest outer diameter was Sembilang—B = 13.4 cm. Biggest inner diameter was Sembilang—B = 12.21 cm. Biggest thickness was Petung Hitam—B = 1.52 cm. Interesting thing among others, Sembilang with biggest diameter any others bamboo was not bamboo with biggest thickness. Therefore, diameter is not directly proportional to thickness [10]. Figure 2 below shows biggest and smallest values for all morphological measurement parameters.

3.3 Morphological Description of Length Internode, Outer Diameter, Inner Diameter and Thickness

Results show a relatively high morphological variation between species/type of bamboos. Species, genetic and environtment are factors that influence the character of growth. Figure 3 below explain how description of bamboo morphological parameters including internode length, outer diameter and inner diameter and also thickness for every species.

Figure 3 explain that there were variation of morphological variable of each bamboos. Trend formed for all variable of each bamboos are same. Trend of length internode was polynomial while for outer diameter, inner diameter and thickness were negative linier. Generally, internode length from base part (B) to top part (T) shown an inverted U shaped where B shorter than M but longer than T. This results are accordance with other research [3]. However, there is an anomaly where in certain section internode length will be shorter than normal length as seen in Ampel Kuning, Betung and Sembilang. Figure 3 explains that Petung Hitam was bamboo with relatively have same internode length from B to T when compared to another bamboos.

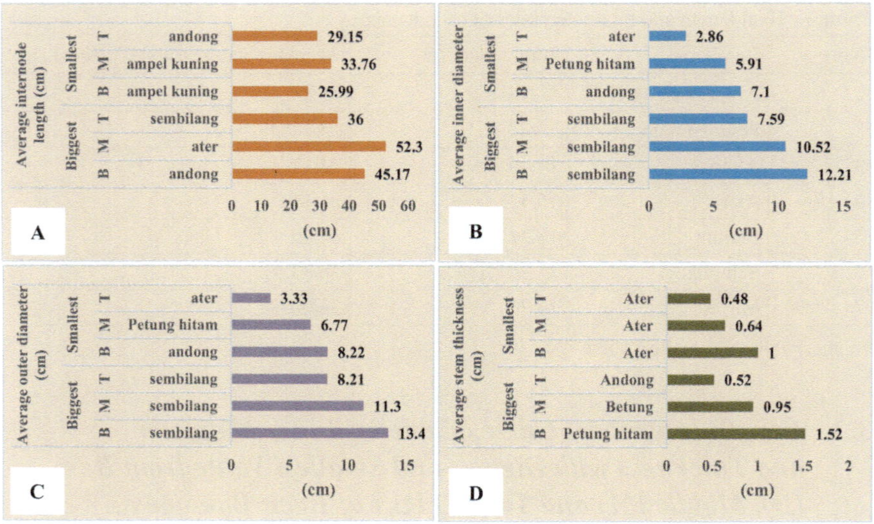

Fig. 2 Grapichs biggest and smallest value of some bamboos morphology. **a** Average internode length **b** average outer diameter **c** average inner diameter **d** average thickness. B = Bottom; M = Middle; T = Top

Outer diameter, inner diameter and thickness shown trend getting smaller values from B to T.

Figure 3 explains Mayan was bamboo with smallest difference values of outer diameter, inner diameter and thickness from B to T compared to other bamboos. Regarding imternode length, it can be concluded that internode length at M part has more similarity on size than B and T part, but there are also those that still involve end of B part. Internode length was relatively same size for Ampel Kuning was internode number 19–28, for Petung Hitam was number 7–22, for Andong was 7–16, for Ater was 13–25, for Betung was 19–25, for Sembilang was 4–16, and for Mayan was 10–22.

Outer diameter and inner diameter show trend where diameter smaller from B to T. This is occurred in all type of bamboos except Ampel Kuning where beginning at B part, outer diameter and inner diameter at internode number 3 until 7 relatively smaller than diameters internode number 1 and 2 and internode after number 7 up. Thickness for all type of bamboo shows same trend that smaller from B to T. Similarity or uniformity size of internode length, outer diameter, inner diameter and thickness will provide good raw materials for developing engineered bamboo product such as laminated bamboo or other.

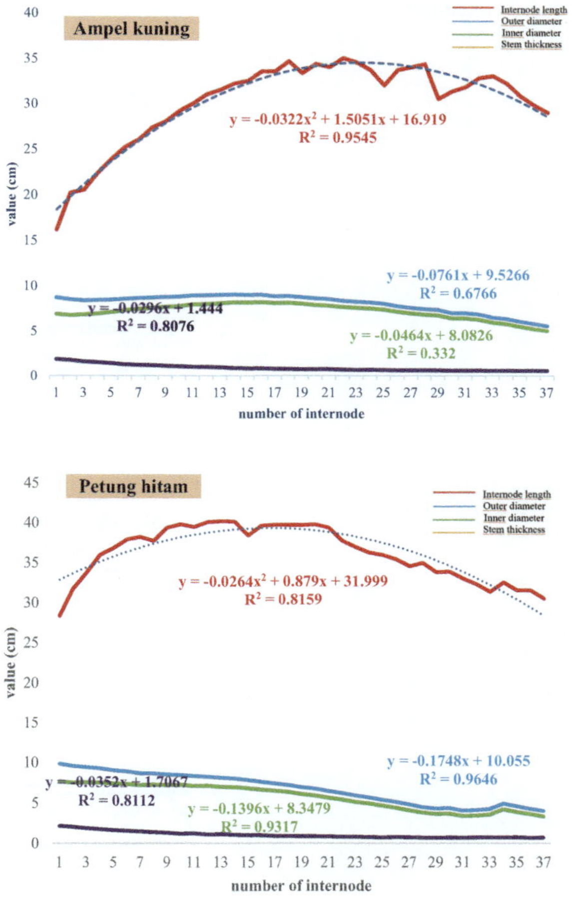

Fig. 3 Morphology description and equation of internode length, outer diameter, inner diameter and thickness

3.4 Uniformity of Morphological Bamboo

Table 2 explain uniformity values of morphological variable according to Standar Deviation (SD) value.

Smallest SD value shown more uniform morphological variable of bamboos. Internode length variable, Petung Hitam was bamboo that have uniformity than others with smallest SD value (3.38). Outer diameter variable, Ampel Kuning have uniformity than others with smallest SD value (1). Inner diameter variable, Ampel Kuning have uniformity than others with smallest SD value (0.87). Thickness variable, have uniformity than others with smallest SD value (0.28).

Fig. 3 (continued)

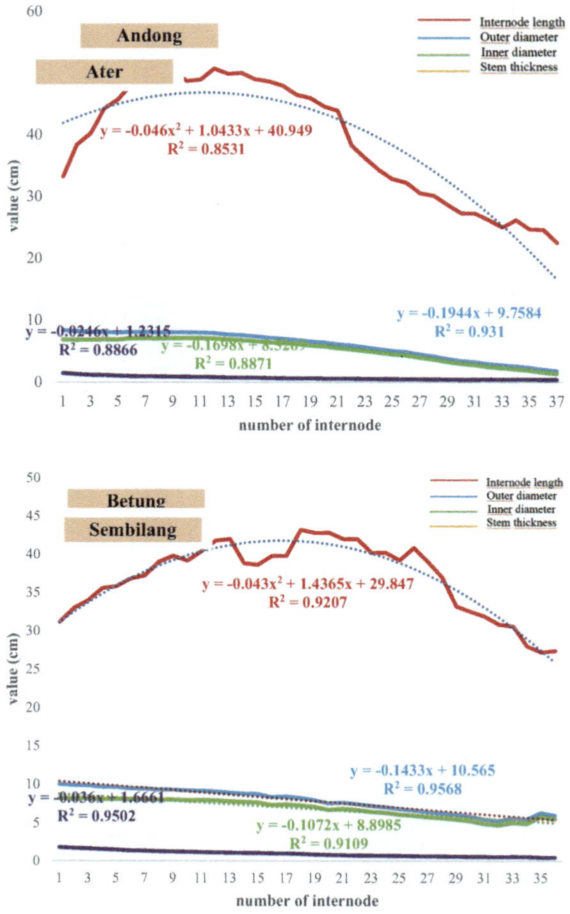

Fig. 3 (continued)

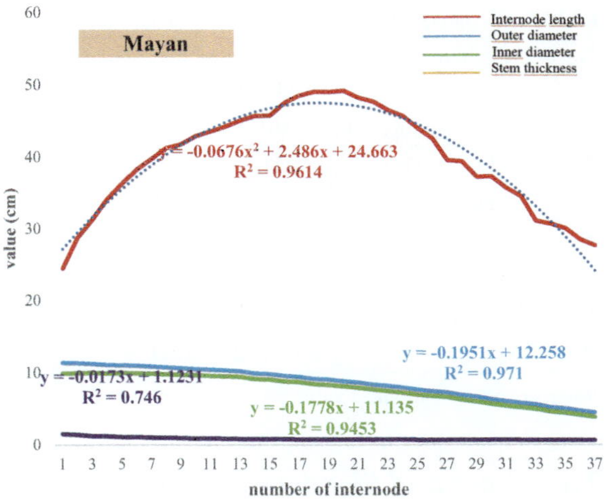

Table 2 Value of morphology unity based on SD

No.	Bamboos	Standard deviation			
		Internode length	Outer diameter	Inner diameter	Thickness
1	Ampel kuning	4.63	1*	0.87**	0.36
2	Petung hitam	3.38*	1.93	1.57	0.42*
3	Andong	9.73	2.18	1.95	0.28
4	Ater	10**	2.33	2.12*	0.28
5	Betung	4.69	1.54	1.18	0.39
6	Sembilang	3.99	2.37**	2.11	0.31
7	Mayan	7.18	2.14	1.98	0.22**

*biggest value, **smallest value

3.5 *Linier Equation Between Variable*

Equation function can be used to predict value of dependent variable if independent variable is already known. This will make it easier when to do a quick inventigate technique to find out value of measurement variable. Table 3 shows value of linier equation between outer diameter and inner diameter that can predict thickness. X value was independent variable namely outer diameter while y was dependent variable namely inner diameter. Generally, results shown value of R^2 was about 0.877–0.9952 that can conclude outer diameter and inner diameter have strong relationship.

Table 3 Persamaan dan nilai R^2 (koefisien determinasi)

No.	Bamboo	Linier equation between outer diameter with inner diameter	R^2 (*Coefficient of determination*)
1	Ampel kuning	$y = 0.8158x + 0.6078$	0.8770**
2	Petung hitam	$y = 0.8058x + 0.2704$	0.9839
3	Andong	$y = 0.8918x - 0.1077$	0.9936
4	Ater	$y = 0.9039x - 0.1181$	0.9937
5	Betung	$y = 0.7622x + 0.8824$	0.9877
6	Sembilang	$y = 0.8857x + 0.3659$	0.9943
7	Mayan	$y = 0.9212x - 0.1196$	0.9952*

*biggest value, **smallest value

3.6 *Taper and Slenderness Ratio*

3.6.1 Taper

Taper value for total length, length about 8 m and length about 6 m are presented in Tables 4, 5 and 6. More bigger taper value, bamboo culm mjore pointed. Results shows trend where tapper value getting smaller when bamboo culm also getting shorter. Taper value for total length, most pointed bamboo culm was Sembilang (0.0060) while least pointed bamboo culm was Ampel Kuning (0.0028). For about 8 m length, most pointed bamboo culm was Sembilang and Petung Hitam (0.0044) while least pointed bamboo culm was Ampel Kuning (0.0014). For about 6 m length, most pointed bamboo culm was Sembilang (0.0041) while least pointed bamboo culm was Ampel Kuning (0.0003).

Table 4 Taper value for total length

No.	Bamboos	Base diameter (cm)	Top diameter (cm)	Length (cm)	Taper
1	Ampel kuning	8.76	5.61	1.119	0.0028**
2	Petung hitam	9.82	4.01	1.337	0.0043
3	Andong	8.52	1.90	1.440	0.0046
4	Ater	8.61	1.80	1.532	0.0044
5	Betung	10.09	6.05	1.334	0.0030
6	Sembilang	14.48	6.14	1.395	0.0060*
7	Mayan	11.31	4.44	1.471	0.0047

*biggest value, **smallest value

Table 5 Taper value for about 8 length

No.	Bamboos	Base diameter (cm)	Top diameter (cm)	Length (cm)	Taper
1	Ampel kuning	8.76	7.61	803	0.0014**
2	Petung hitam	9.82	6.19	832.96	0.0044*
3	Andong	8.52	7.14	793.05	0.0017
4	Ater	8.61	7.01	826.07	0.0019
5	Betung	10.09	7.71	814.37	0.0029
6	Sembilang	14.48	10.88	827.38	0.0044*
7	Mayan	11.31	8.76	825.77	0.0031

*biggest value, **smallest value

Table 6 Taper value for about 6 length

No.	Bamboos	Base diameter (cm)	Top diameter (cm)	Length (cm)	Taper
1	Ampel kuning	8.76	8.57	599.94	0.0003**
2	Petung hitam	9.82	7.57	597.10	0.0038
3	Andong	8.52	7.81	597.44	0.0012
4	Ater	8.61	7.83	616.60	0.0013
5	Betung	10.09	8.43	603.58	0.0028
6	Sembilang	14.48	11.88	628.93	0.0041*
7	Mayan	11.31	9.49	629.97	0.0029

*biggest value, **smallest value

3.6.2 Slenderness Ratio

Bamboo culm with typical hollw column is material that has uniquess. Information of slenderness ratio value will be very helpfull in efforts to identify the potential mechanical strenth of this material, especially related to compression strenghness parallel to grain. By omitting value of thickness, Table 7 shown results of measurement of axial slenderness ratio for each bamboos.

Range value of slenderness ratio at B was 2.95–5.51; M was 3.15–7.84 and T was 4.80–10.97. Biggest slenderness ratio for B part was Andong (5.51), M part was Ater (7.84) and T was Ater (10.97) while smallest slenderness ratio for B part was Ampel Kuning (2.95), M part was Sembilang (3.15) and T part was Ampel Kuning (4.80). Slenderness ratio for each type of bamboo has a similarity in trend. It can be explained that this value shows an upward trend from B part to T part. Closer to T part, relative internode length will be longer and outer diameter will be getting smaller. Of course, this claim still has to pay attention to value of internode length and outer diameter each type bamboo in detail as presented in the results of measurement and visualization of these variables previously. Table 8 and Fig. 4

Table 7 Slenderness ratio

No.	Bamboos	Slenderness ratio of bamboo stem		
		Base	Middle	Top
1	Ampel kuning	2.95**	3.98	4.80**
2	Petung hitam	4.34	6.03	8.44
3	Andong	5.51*	6.64	8.22
4	Ater	4.9	7.84*	10.97*
5	Betung	3.88	5.62	6.45
6	Sembilang	2.98	3.15**	5.16
7	Mayan	4.23	6.6	6.66

*biggest value, **smallest value

Table 8 Equational function between thickness and slenderness ratio

No.	Bamboo	Linier equation between stem thickness with slenderness ratio	R^2 (Coefficient of determination)
1	Ampel kuning	$y = -2.2411x + 5.8473$	0.8715
2	Petung hitam	$y = -3.7552x + 10.113$	0.7593
3	Andong	$y = -4.648x + 10.538$	0.5566
4	Ater	$y = -9.4858x + 14.555$	0.8213
5	Betung	$y = -1.5851x + 6.3244$	0.5035**
6	Sembilang	$y = -1.9192x + 5.289$	0.6452
7	Mayan	$y = -5.0003x + 8.8587$	0.9122*

*biggest value, **smallest value

explain how relationship between thickness and slenderness ratio. Generally, it can be concluded that there was a negative linier relationship between those 2 variable where thickness upward slenderness ratio will be smaller. Value of R^2 (Coefficient of Determination) of all types of bamboo shows value > 0.6 which means X value have strong enaough relationship with Y value.

3.7 *Morphology of Bamboo from Another Reference and Potential Utilization*

Morphological of bamboo is srtongly influenced by species, genetics and environtment. Table 9 describes some reference related to bamboo species and their morphological information.

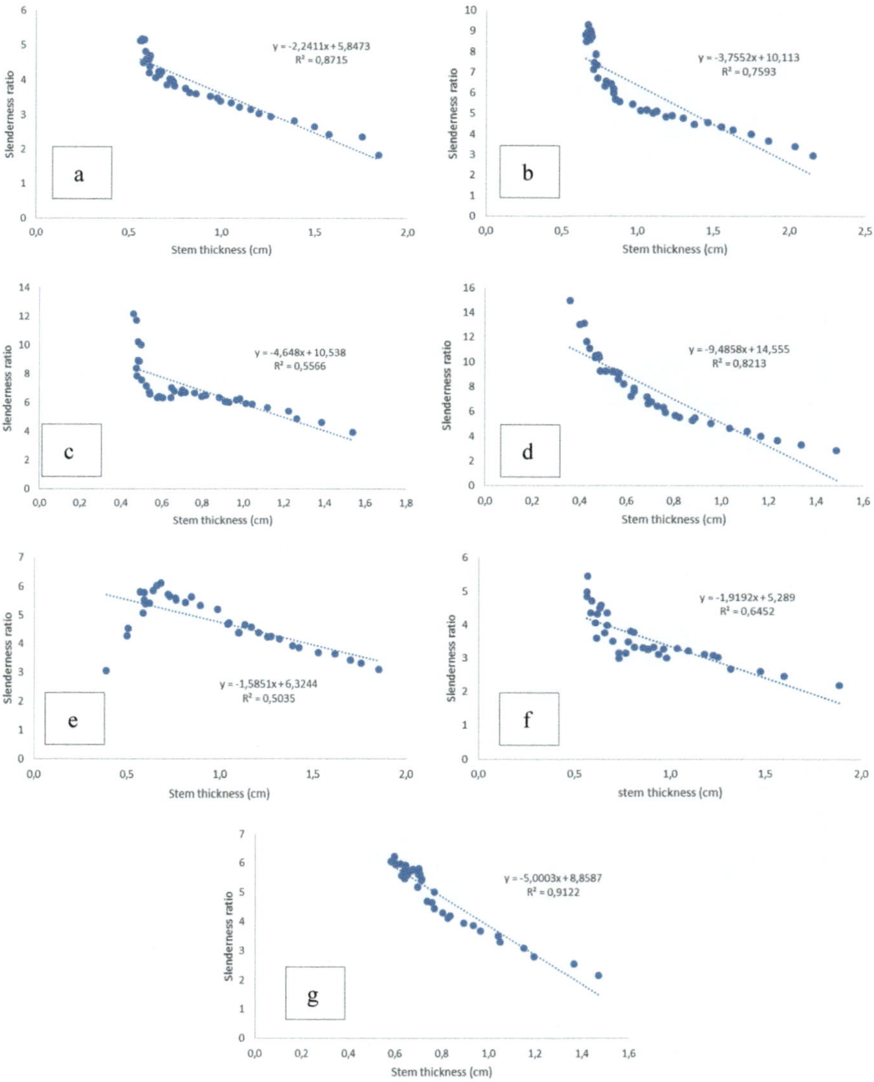

Fig. 4 Graphic between thickness and slenderness ratio. **a** Ampel Kuning, **b** Petung Hitam, **c** Andong, **d** Ater, **e** Betung, **f** Sembilang, **g** Mayan

3.8 Utilization of Bamboos

Several types of bamboos are classified for utilization including Andong for heavy construction, buildings, bridge, and laminated bamboo. Mayan for light construction, furniture and crafts. Betung for heavy construction, buildings, bridge, laminated bamboo and furniture. Ampel for light construction, furniture and crafts. Ater

Table 9 Morphology value of some bamboos

No.	Bamboos	Species	Sample origin	Total length (m)	Internodus length (cm)	Diameter (cm)	Thickness (cm)	References
1	Betung	*D. asper* (Schultes f.) Backer ex Heyne	Jasinga, Bogor, West Java, Indonesia	–	–	11.59	1.3	Rafif [10]
			–	–	–	8–20	2	Liese & Michael [11]
			Antirogo, Sumbersari, Jember, East Java, Indonesia	–	36	18	–	Murtodo & Setyati [12]
			Java Island	14.5–16.5	–	14.5–18.5 (B) 5–6 (T)	2.1–4 (B) 0.7 (T)	Rulliaty et al [13]
2	Ater	*G. atter* (Hassk.) Kurz	Antirogo, Sumbersari, Jember, East Java, Indonesia	–	27	11	–	Murtodo & Setyati [12]
			Java Island	9–15	–	5.4–8.7 (B) 4.2–6.1 (T)	1.1–1.6 (B) 0.3–0.5 (T)	Rulliaty et al [13]
3	Gombong/ andong	*Gigantochloa pseudoarundinaceae* (Steudel) Widjaja	Antirogo, Sumbersari, Jember, East Java, Indonesia	–	37	14	-	Murtodo & Setyati [12]
			Java Island	17–22	–	13.4 (B) 10.9 (M) 5.3 (T)	1.91 (B) 0.73 (M) 0.58 (T)	Rulliaty et al [13]

(continued)

Table 9 (continued)

No.	Bamboos	Species	Sample origin	Total length (m)	Internodus length (cm)	Diameter (cm)	Thickness (cm)	References
4	Mayan	*G. robusta* Kurz	Java Island	16.8	–	13 (B) 4–5 (T)	–	Rulliaty et al [13]
5	Sembilang	*D. giganteus* Wallich ex Munro	Cibinong, Bogor, Indonesia	–	–	22.12	–	Bahanawan et al [3]
6	*Bambusa* sp.	*Bambusa vulgaris* var. Striata (Lodd. Ex Lindl.)	Antirogo, Sumbersari, Jember, East Java, Indonesia	–	27	8	–	Murtodo & Setyati [12]
		Bambusa intermedia Hsueh et Yi	Bamboo forests in Ya'an and Yibin, Sichuan Province, China	14.80–16.60	–	6.93–7.69 (DBH)	–	Luo et al [14]
		Bambusa multiplex (Lour.) Raeusch. ex Schult	Bamboo forests in Huangshan, Anhui Province, China	6.30–8.70	–	2.20–2.79 (DBH)	–	Luo et al [14]
		Bambusa rigida Keng et Keng f	Bamboo forests in Ya'an and Yibin, Sichuan Province, China	12.90–14.30	–	5.09–5.82	–	Luo et al [14]

for construction, furniture and craft [13]. Petung Hitam, which is still same genus with Betung also has potential to be used as raw material for structural construction. Sembilang was species with large diameter but relatively thin thickness. Therefore, many people use this bamboo for gutters. Some people also known this species with Talang bamboo. However, it was explained that modulus of elastic (MOE) of Sembilang reaches 115723.5 kgf/cm^2 while modulus of rupture (MOR) reaches 1141.36 kgf/cm^2 so that it was claimed to be strong enough as raw material for structural components [15]. Natural resistance class of bamboo to subterranean termites (*Coptotermes curvighnatus* Holmgreen) reported including Mayan and Sembilnag classified as class II. Andong, Ater and Betung included class IV. Ampel included class V.

4 Conclussions

Morphological characters of bamboo species show variety. There are bamboos with unique and distinctive for total length and internode length, diameters and also thickness. Average total length of seven bamboos in this research were about 11.19–15.32 m with the longest average bamboo was Ater (15.32 m). Average number of internode were about 33.17–36.83 pcs with average number was Sembilang (36.83 pcs). Biggest internode length was Ater—B = 52.3 cm. Biggest outer diameter was Sembilang—B = 13.4 cm. Biggest inner diameter was Sembilang—B = 12.21 cm. Biggest thickness was Petung Hitam—B = 1.52 cm. Outer diameter, inner diameter and thickness getting smaller from B to T. Generally, internode length variable will produce polynomial curve while outer diameter, inner diameter and thickness variable will produce negatif linier curve.

Acknowledgements The authors would like thank to PN (National Priority) Research Program IPH (Departement of Biological Science) Indonesian Institutes of Sciences phase 2 year 2021 for funding support.

References

1. Sharma P, Dhanwantri K, Mehta S (2014) Bamboo as building material. Int J Civ Eng Res 5(3):249–254
2. Jiao J, Tang P Application of bamboo in a design-build course: Lianhuadang farm project. Front Architectural Res. https://doi.org/10.1016/j.foar.2019.09.003
3. Bahanawan A, Darmawan T, Sufiandi S, Dwianto W (2018) Getting to know the nine-leaf bamboo (*Dendrocalamus giganteus* Wallich ex Munro): a study of the stem characteristics of giant bamboo species. In: Proceedings of the 2018 lignocellulose seminar, Bogor, Indonesia, pp 88–91
4. Muchiri MN, Moga MO (2013) A preliminary yield model for natural Yushania Alpina bamboo in Kenya. J Nat Sci Res 3(10):77–84

5. Misdarti (2006) The quality of laminated bamboo from Toraja regency, South Sulawesi. Forest Prod Res J 24(3):183–189
6. Widjaja EA, Rahayuningsih Y, Rahajoe JS, Ubaidillah R, Maryanto I, Walujo EB, Semiadi G (eds) (2014) Current Indonesian biodiversity. LIPI Press
7. Barly, Ismanto A, Martono D, Abdurachman, Andianto (2012) Physical properties and dimensional stabilization of several types of commercial bamboo. Forest Prod Res J 30(3):163–170
8. Fatriasari W, Syafii W, Wistara N, Syamsu K, Prasetya B (2016) Lignin and cellulose changes of Betung bamboo (*Dendrocalamus asper*) pretreated microwave heating. Int J Adv Sci Eng Inf Technol 6(2):186–195
9. https://kbbi.kemdikbud.go.id/entri/morfologi. Accessed 19 Nov 2020 08.50 WIB
10. Rafif N (2018) Structural sorting of Betung bamboo (*Dendrocalamus asper*) based on its flexible properties. Thesis, IPB Bogor
11. Liese W, Michael K (2015) Bamboo the plant and its uses. Springer, Hamburg (DE)
12. Murtodo A, Setyati D (2015) Bamboo inventory in Antirogo village, Sumbersari District, Jember Regency. J Basic Sci 15(2):115–121
13. Rulliaty S, Nadjib N, Muslich M, Jasni, Sulastiningsih IM, Komaryati S, Suprapti S, Abdurrahman, Basri E (2015) Information science and technology package series on basic properties and possible uses of 10 types of bamboo. Center for Forest Products Research and Development, Research, Development and Innovation Agency, Ministry of Environment and Forestry, Bogor
14. Luo J, Lian C, Liu R, Zhang S, Yang F, Fei B (2019) Comparison of metaxylem vessels and pits in four sympodial bamboo species. Nat Sci Rep 9:10876
15. Bahanawan A, Darmawan T, Syamani FA, Adi DS, Amin Y (2017) Mechanical properties of sembilang bamboo (*Dendrocalamus giganteus* Wallich ex. Munro) culms and leaves. In: Proceedings of the 7th international symposium for sustainable humanosphere, Bogor, Indonesia, pp 37–43
16. Jasni, Damayanti R, Pari R (2017) Natural resistance of bamboo species growing in Indonesia to subterranean termites. J Forest Prod Res 35(4):289–301

Ovicidal Activity of *Brugmansia Candida* and *Cerbera Odollam* Extracts Against Dengue and Filarial Vector Mosquitoes

Dita Meisyara, Didi Tarmadi, Setiawan Khoirul Himmi, Anugerah Fajar, and Sulaeman Yusuf

Abstract *Brugmansia candida* and *Cerbera odollam* have been known as poisonous plants. The present study investigated the ovicidal activity of *B. candida* and *C. odollam* extracts against *Aedes aegypti* and *Culex quinquefasciatus*. *B. candida* leaves and *C. odollam* seed kernels were extracted by methanol, and then evaporated by rotary evaporator to obtain the dried extracts. The dried extracts with concentrations of 0.5 g/l, 1.0 g/l, and 2 g/l were diluted in the tap water and then eggs of *A. aedes* and *C. quinquefasciatus* were put on surface of those solutions. The hatch rate was observed until 5 days post treatment. The result showed that there was no significant difference of percent hatchability between *A. aegypti* and *C. quinquefaasciatus* at the end of observation for both extarcts, *B. candida* and *C. odollam* extracts. The percent hatchability of both eggs decreased with increasing the concentration of extracts. Both extracts exhibited high ovicidal activity against dengue and filarial vectors. However, the percent hatchability of *A. aegypti* and *C. quinquefasciatus* eggs for *B. candida* was significantly higher than that for *C. odollam*, suggesting that *C. odollam* is more toxic for mosquitoes' eggs than *B. candida*. Overall, the data clearly suggest that *B. candida* and *C. odollam* are promising as an ovicidal agent against mosquitoes, *A. aegypti* and *C. quinquefasciatus*.

D. Meisyara · D. Tarmadi (✉) · S. K. Himmi · A. Fajar · S. Yusuf
Research Center for Applied Zoology, National Research and Innovation Agency (BRIN), Cibinong, Bogor, Indonesia
e-mail: didi019@brin.go.id

D. Meisyara
e-mail: dita002@brin.go.id

S. K. Himmi
e-mail: khoi003@brin.go.id

A. Fajar
e-mail: anug005@brin.go.id

S. Yusuf
e-mail: sula002@brin.go.id

M. A. Rahandi Lubis (ed.), *Proceedings of the 10th International Symposium for Sustainable Humanosphere*, Lecture Notes in Civil Engineering 255,
https://doi.org/10.1007/978-981-19-1422-5_12

Keywords *Aedes aegypti* · *Brugmansia candida* · *Cerbera odollam* · *Culex* quinquefasciatus · Ovicidal activity

1 Introduction

Plants have been widely known to have anti-insect activities including anti-mosquitoes, such as *Vitex agus castus* [1], *Ageratina adenophlora* [2], *Gmelina asiatica* [3], *Ricinus communis* [4], *Aristolochia bracteata* [5]. Utilization of plant as insecticide due to two reasons, i.e.: (1) plants have chemical substances, and (2) the fact that, commercial insecticides cause hazard for humans and environment [6, 7]. Therefore, some researchers are trying to develop bioinsecticides based on plant extracts to substitute the commercial insecticides.

Mosquitoes are one of the insects that disturb humans due to their habit of baiting the skin and they are vector for several dangerous diseases. In general, there are two mosquitoes species that are commonly found in urban habitats, i.e.: *A. aegypti* and *C. quinquefasciatus*. *A. aegypti* is mosquito that can be found in early morning or in the evening before sunset. *A. aegypti* serves as vector for dengue, zika viruses, and yellow fever. In Indonesia, dengue is one of serious diseases that can causes death. While, *C. quinquefasciatus* serves as vector for filariasis disease. This mosquito can be easily recognized from its colour, *C. quinquefasciatus* is brown in colour and is commonly found in the night.

Based on such adverse reasons, mosquitoes must be controlled. One way to control mosquitoes is to prevent they hatch eggs (ovicidal). Earlier studies reported that some plant extracts exhibited high ovicidal activity against mosquitoes, i.e.: *Cassia fistula* [8]; *Cymbopogan citratus* [9]; *Eclipta alba* [10]; *Ervatamia coronaria* and *Caesalpinia pulcherrima* [11]; *Sphaeranthus indicus* and *Limonia acidissima* [12]. Therefore, the present study investigated the ovicidal activity of *B. candida* and *C. odollam* extracts against *A. aegypti* and *C. quinquefasciatus*.

B. candida is belonging to a member of Solanaceae and is well known a shrub plant and commonly used as ornamental plants in tropical and subtropical areas including Indonesia. This plant growths in in humid and warm areas in forest edges and lowland rainforest. While *C. odollam* is belonging to a member of Apocynacea and is found mainly in tropical area, especially in Asia countries [13]. In Indonesia, *C. odollam* is including to beneficial plant, this plant is commonly planted as shade plant due to it has shady and dense leaves. We chose these plants (*B. candida* and *C. odollam*) because based on our previous studies suggested that *B. candida* and *C. odollam* exhibited anti-mosquitoes against *C. quinquefasciatus* and *A. aegypti* [6, 7, 14], and showed anti-termites activity against lower termites, *Coptotermes gestroi* and *Cryptotermes cynocephalus* [15, 16], suggesting that *B. candida* and *C. odollam* can be developed as anti-ovicidal agent against mosquitoes.

2 Materials and Methods

2.1 Extract Materials

Plant materials, i.e.: *B. candida* leaves and *C. odollam* seed kernels were collected from Bogor, West Java, Indonesia. Plant materials were dried in the oven at 60 °C for 2 days. The dried plant materials were powdered using a mechanical blender and were then sieved using sieve sizes of 40 mesh. Approximately 1 kg of powdered plant materials were prepared for extraction process.

2.2 Extraction

The extraction process referred to our previous studies [7, 13]. 1 kg of powdered plant samples were extracted using methanol solvent and maceration procedure with three cycles (24 h for one cycle). The liquid from filtration was collected and polled in the glass vials, and evaporated by rotary evaporator (RV 10 Digital, IKA Works GmbH & Co., Germany) to obtain dry extract. The dry extracts were then dried in the oven at 40 °C for 2 days to release remaining methanol solvent in the dry extracts.

2.3 Ovicidal Bioassay

The extract solutions with concentrations 0.5 g/l, 1.0 g/l, and 2.0 g/l were prepared by dissolving those extracts in the tap water with tween 80 as surfactant. Untreated solutions (control) were prepared from combination of tap water and tween 80. Ovicidal bioassay referred to Jabenasen et al. [8] and Reegan et al. [12]. Approximately 145–150 freshly laid *A. aegypti* and *C. quinquefasciatus* eggs obtained from National Research and Innovation Agency (BRIN) were separately subjected to extract solutions with different concentrations (0.5 g/l, 1.0 g/l, and 2.0 g/l). The percent ovicidal activity was recorded until 5 days post treatment by the following formula:

$$(\text{Number of hatched eggs} / \text{Total number of eggs}) \times 100\%$$

2.4 Statistical Analysis

The difference in the number of hatched eggs was tested by ANOVA-SPSS ver. 23 (IBM, Armonk, NY, USA). Tukey's post hoc test ($P < 0.05$) was used to calculate the significant difference.

3 Results and Discussion

Figure 1 shows the ovicidal activity of *B. candida* extract against *A. aegypti* and *C. quinquefasciatus*. The percentage of hatched eggs of *A. aegypti* decreased by increasing the concentration (Fig. 1). We detected the significant difference among concentrations, suggesting that the increasing of concentration gives the positive effect on the ovicidal activity of *B. candida* extract. The numbers of hatched eggs of *A. aegypti* were 51%, 21%, and 6% for 0.5 g/l, 1.0 g/l, and 2.0 g/l, respectively. Similar results were recorded from *C. quinquefasciatus*, in which percent hatchability decreased with increasing the concentration. The values of hatched eggs were 47%, 17%, and 4% for 0.5 g/l, 1.0 g/l, and 2.0 g/l, respectively. Based on statistical analysis, the percent hatchability between *A. aegypti* and *C. quinquefasciatus* was not significantly different ($P > 0.05$). Earlier study reported *Cassia fistula* extract at concentration 200 ml/l showed no eggs hatch from *C. quinquefasciatus* and *Anopheles stephensi* eggs [8]. Govindarajan and Karuppannan [10] suggested that *Eclipta alba* extract resulted in zero hatchability at 300 ppm against *A. aegypti* eggs. Despite *B. candida* extract used in this present study generated lower ovicidal activity compared with *Cassia fistula*, and *Aclipta alba*, however *B. candida* is still promosing as an ovicidal agent at 2 g/l.

Figure 2 shows the ovicidal activity of *C. odollam* extract against *A. aegypti* and *C. quinquefasciatus*. Similar with results from *B. candida*, we found that the percent hatchability of *A. aegypti* and *C. quinquefasciatus* decreased by increasing the concentration (Fig. 2). The increasing of concentration gave positive effect on the percent hatchability of both mosquitoes eggs, *A. aegypti* and *C. quinquefasciatus* due to this study recorded the significant difference among concentrations. Based on the results, the percent of hatched eggs of *A. aegypti* was 26%, 9%, and 0.7% for 0.5 g/l, 1.0 g/l, and 2.0 g/l, respectively. While the percent of hatched eggs of *C. quinquefasciatus* was 18%, 5%, and 0.2% for 0.5 g/l, 1.0 g/l, and 2.0 g/l, respectively. In addition, statistical analysis showed that no significant difference between the percentage of hatched eggs *A. aegypti* and *C. quinquefasciatus* for all concentration ($P > 0.05$). This finding is in line with previous study by Govindarajan et al. [11], they showed that the numbers of hatched *C. quinquefasciatus* and *A. aegypti* eggs were similar after exposure to by *Ervatamia coronaria* plant extract. In addition, Govindarajan et al. [11] revealed that *Ervatamia coronaria* plant extract resulted in

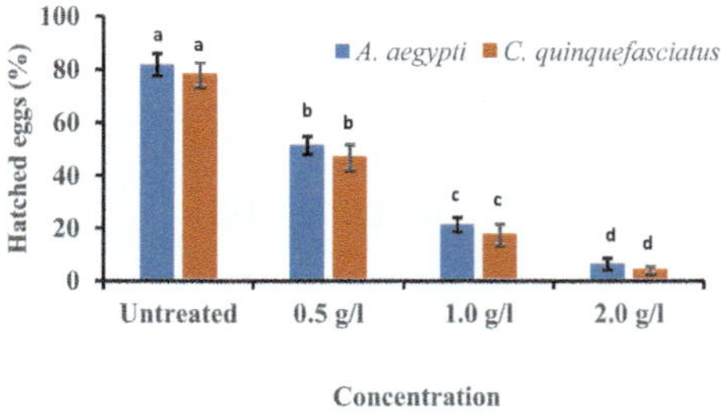

Fig. 1 The percentage of hatched eggs of *A. aegypti* and *C. quinquefasciatus* after exposure to *B. candida* extract. Error bars represent standard deviations. Same letters do not differ statistically

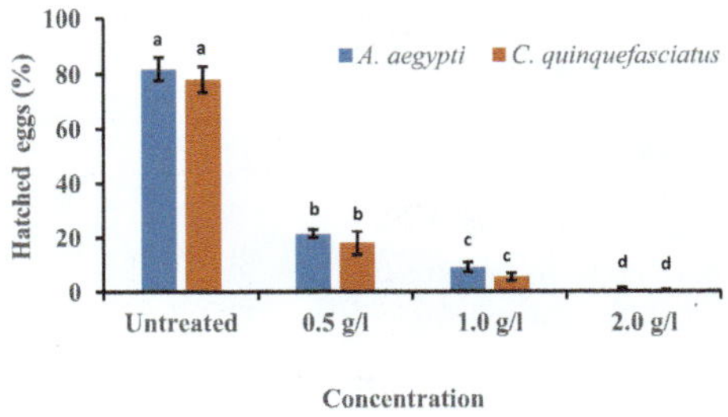

Fig. 2 The percentage of hatched eggs of *A. aegypti* and *C. quinquefasciatus* after exposure to *C. odollam* extract. Error bars represent standard deviations. Same letters do not differ statistically

zero hatchability of *C. quinquefasciatus* and *A. aegypti* eggs at 0.3 g/l. *Terminalia chebula* plant extract also showed zero hatchability at concentrations of 0.3 g/l and 0.25 g/l for *C. quinquefasciatus* and *A. aegypti* eggs, respectively [17].

To know which one of two extracts has higher ovicidal activity against mosquitoes, *A. aegypti* and *C. quinquefasciatus*, we employed the statistical analysis from data both extracts, *B. candida* and *C. odollam*. Figures 3 and 4 show the comparation of hatched eggs of *A. aegypti* dan *C. quinquefasciatus* after exposure to *B. candida* and *C. odollam* extracts. The results showed that the percent of hatched eggs of both mosquitoes, *A. aegypti* and *C. quinquefasciatus* after exposure to *B. candida* extract were significantly higher than that after exposure to *C. odollam* extract ($P < 0.05$), suggestimg that *C. odollam* showed higher ovicidal activity against both mosquitoes than *B. candida*. This finding is consistent with our previous studies suggested that *C. odollam* extract generated higher larvicidal activity against *C. quinquefasciatus* and *A. aegypti* than *B. candida* [6, 7, 14]. It has been generally accepted that the ovicidal activity of plant extracts is determined by chemical components in the plant extracts. Based on the data of ovicidal activities from *C. odollam* and B. candida, it indicates that chemical components in *C. odollom* have higher ability to inhibit hatching eggs of *A. aegypti* and *C. quimquefasciatus* than those in *B. candida*. Laphookhieo et al. [18] reported that *C. odollam* contains neriifolin, cerberin, and other compounds. Cerberin and neriifolin are toxic against insects. While, *B. candida* is also well known as a poisonous plant because contain tropane alkaloid [19, 20]. The toxic chemical components in the *C. odollam* and *B. candida* prevent *A. aegypti* and *C. quiquefasciatus* eggs to hatch. In next study, it is needed to check the condition of eggs after treatment by extract using a microscope to investigate the damage of eggs.

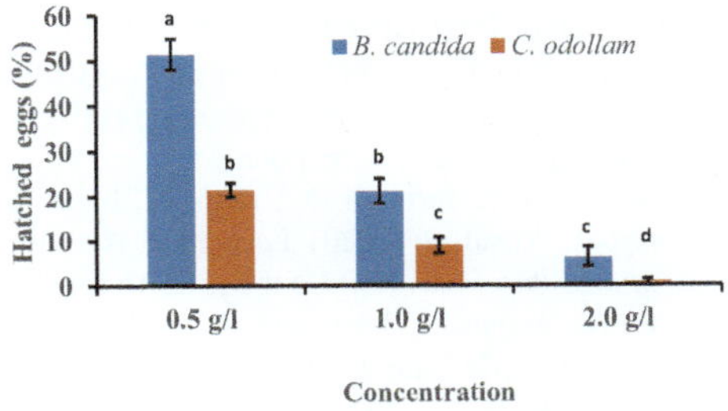

Fig. 3 Comparation of hatched eggs of *A. aegypti* after exposure to *B. candida* extract. Error bars represent standard deviations. Same letters do not differ statistically.

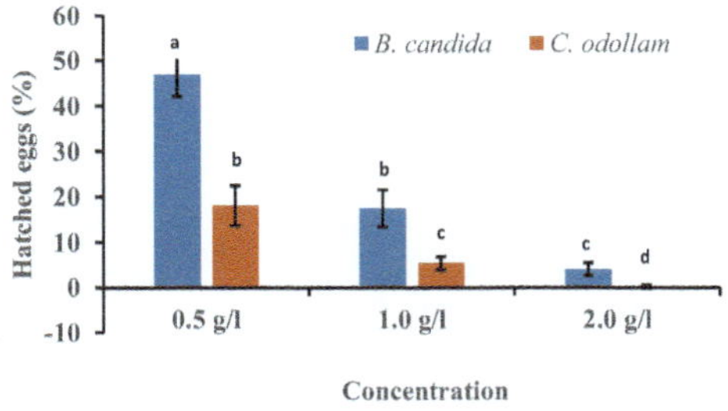

Fig. 4 Comparation of hatched eggs of *C. quinquefasciatus* after exposure to *C. odollam* extract. Error bars represent standard deviations. Same letters do not differ statistically.

4 Conclusion

C. odollam and *B. candida* showed high ovicidal activity at concentrations of 1.0 g/l and 2.0 g/l, respectively, suggesting that *C. odollam* extract exhibited higher ovicidal activity than *B. candida*. Overall, our data suggested that both extracts, *C. odollam* and *B. candida,* are promosing as ovicidal agenst against mosquitoes, *A. aegypti* and *C. quinquefasciatus*.

Acknowledgements The authors thank JSPS Alumni Association of Indonesia (JAAI) for supporting the funding.

References

1. Mehlhorn H, Schmahl G, Schmidt J (2005) Extract of the seeds of the plant *Vitex agnus castus* proven to be highly efficacious as a repellent against ticks, fleas, mosquitoes and biting flies. Parasitol Res 95:363–365
2. Raj Mohan D, Ramaswamy M (2007) Evaluation of larvicidal activity of the leaf extract of a weed plant, *Ageratina adenophora*, against two important species of mosquitoes, *Aedes aegypti* and *Culex quinquefasciatus*. Afr J Biotech 6(5):631–638
3. Muthukumaran U, Govindarajan M, Rajeswary M, Hoti L (2015) Synthesis and characterization of silver nanoparticles using *Gmelina asiatica* leaf extract against filariasis, dengue, and malaria vector mosquitoes. Parasitol Res 114:1817–1827
4. Sogan N, Kapoor N, Singh H, Kala S, Nayak A, Nagpal BN (2018) Larvicidal activity of *Ricinus communis* extract against mosquitoes. J Vector Borne Dis 55(4):282–290
5. Krishnappa K, Elumalai K (2012) Toxicity of *Aristolochia bracteata* methanol leaf extract against selected medically important vector mosquitoes (Diptera: Culicidae). Asian Pacific Journal of Tropical Disease 2:S553–S557
6. Tarmadi D, Gunandini DJ, Yusuf S (2018) Larvicidal activity of *Cerbera odollam* Gaertn against a dengue vector, *Aedes aegypti* (Diptera: Culicidae). In: Sustainable future for human security. Springer, Singapore
7. Meisyara D, Tarmadi D, Zulfitri A, Fajar A, Ismayati M, Himmi SK, Kartika T, Guswenrivo I, Yusuf S (2020) Larvicidal Activity of Bintaro (*Cerbera odollam*) against *Culex quinquefasciatus*. IOP Conf Ser Earth Environ Sci 591
8. Govindarajan M, Jebanesan A, Pushpanathan T (2008) Larvicidal and ovicidal activity of *Cassia fistula* Linn. Leaf extract against filarial and malarial vector mosquitoes. Parasitol Res 102:289–292

9. Pushpanathan T, Jebanesan A, Govindarajan M (2006) Larvicidal, ovicidal and repellent activities of *Cymbopogan citratus* Stapf (Graminae) essential oil against the filarial mosquito *Culex quinquefasciatus* (Say) (Diptera:Culicidae). Trop Biomed 23(2):208–212
10. M Govindarajan M, Karuppannan P (2011) Mosquito larvicidal and ovicidal properties of *Eclipta alba* (L.) Hassk (Asteraceae) against chikungunya vector, *Aedes aegypti* (Linn.) (Diptera: Culicidae). Asian Pac J Trop Med, 24–28
11. Govindarajan M, Mathivanan T, Elumalai K, Krishnappa K, Anandan A (2011) Mosquito larvicidal, ovicidal, and repellent properties of botanical extracts against *Anopheles stephensi, Aedes aegypti,* and *Culex quinquefasciatus* (Diptera: Culicidae). Parasitol Res 109:353–367
12. Reegan AD, Gandhi MR, Paulraj MG, Ignacimuthu S (2015) Ovicidal and oviposition deterrent activities of medicinal plant extracts against *Aedes aegypti* L. and *Culex quinquefasciatus* Say Mosquitoes (Diptera: Culicidae). Osong Pub Health Res Perspect 6(1):64–69
13. Gillard Y, Ananthasankaran K, Fabien B (2004) *Cerbera odollam*: a 'suicide tree' and cause of death in the state of Kerala, India. J Ethnopharmacol 95:123–126
14. Himmi SK, Tarmadi D, Meisyara D, Fajar A, Kartika T, Guswenrivo I, Yusuf S (2020) Larvicidal activity of *Brugmansia candida* against *Aedes aegypti* and *Culex quinquefasciatus* (Diptera: Culicidae). *Biosaintifika:* J Biol Biol Educ 12(3):363–369
15. Tarmadi D, Himmi SK, Yusuf S (2014) Termiticidal activity of an extract of *Brugmansia candida* leaf against a subterranean termite *Copotermes gestroi* Wasmann and a drywood termite *Cryptotermes cynocephalus* Light. In: Proceedings of the 10th Pacific-termite research group conference, Kuala Lumpur, Malaysia, 26–28 Feb 2014, pp S3:4
16. Tarmadi D, Prianto AH, Guswenrivo I, Kartika T, Yusuf S (2007) The efficacy of *Carbera odollam* Gaertn and *Brugmansia candida* Pers against subterranean termite *Coptotermes* sp. (in Indonesian). J Trop Wood Sci Technol 5:38–42
17. Veni T, Pushpanathan T, Mohanraj J (2017) Larvicidal and ovicidal activity of *Terminalia chebula* Retz. (Family: Combretaceae) medicinal plant extracts against *Anopheles stephensi, Aedes aegypti* and *Culex quinquefasciatus*. J Parasitic Dis 41(3):693–70
18. Laphookhie S, Cheenpracha S, Karalai C, Chantrapromma S, Ponglimanont C, Chantrapromma K (2004) Cytotoxic cardenolide glycoside from the seeds of *Cerbera odollam*. Phytochemistry 65:507–510
19. Cardillo AB, Otálvaro AM, Busto VD, Talou JR, Velásquez ME, Giulietti AM (2010) Scopolamine, anisodamine and hyoscyamine production by *Brugmansia candida* hairy root cultures in bioreactors. Process Biochem 45(9):1577–1581
20. Kerchner A, Farkas A (2020) Worldwide poisoning potential of Brugmansia and Datura. Forensic Toxicol 38:30–41

Synthesis of Bio-Silica from Bamboo Waste and Its Applications as an Adsorbent

Nur Adi Saputra and Lisna Efiyanti

Abstract The potential silica utilization as an adsorbent continues to develop. In this study, silica was extracted from bamboo leaf ash using the sol-gel method with NaOH and HCl as the main chemicals. The formation of Na-silicate was obtained by refluxing bamboo leaf ash with NaOH at a temperature of 80 °C for 2 h and then adding HCl until the pH was neutral to obtain silica gel. The silica gel is then filtered and heated to obtain bio-silica. The bio-silica is characterized using Scanning Electron Microscope (SEM), Fourier-Transform Infrared Spectroscopy (FTIR), surface area analyzer (SAA), and X-Ray Fluorescence (XRF). The bio-silica application process was carried out on the adsorption of iodine, methylene blue, chloroform, benzene, and ammonia compounds to produce adsorption capacities of 107.6 mg/g, 70.28 mg/g, 2.51%, 14.22%, 15.7%, and 19.46. %. In the future, silica from natural material waste such as bamboo can be an alternative adsorbent for various adsorption processes.

Keywords Adsorption · Silica · Bamboo leaf ash

1 Introduction

One of the main issues in the world is environmental pollution caused by various compounds such as dyes, volatile organic compounds, and even metal compounds. Iodine, methylene blue, chloroform, benzene, and ammonia compounds are chemical compounds produced from various human and industrial activities. Iodine is generally used in the health sector, especially as antiseptics. The hospital waste usually produces a high iodine content, which will be harmful to the environment and the waters around the hospital [1]. Methylene blue compounds are water-soluble and produced from various industries such as the cosmetic and textile industries.

N. A. Saputra · L. Efiyanti (✉)
Research Center for Biomass and Bioproduct, BRIN, Jl. Raya Jakarta-Bogor No. Km. 46, Bogor, West Java 16911, Indonesia
e-mail: lisn002@brin.go.id

M. A. Rahandi Lubis (ed.), *Proceedings of the 10th International Symposium for Sustainable Humanosphere*, Lecture Notes in Civil Engineering 255,
https://doi.org/10.1007/978-981-19-1422-5_13

The high quantities of methylene blue (more than 5–10 mg/L) will be dangerous for the environment, humans, and aquatic organisms [2]. Chloroform, toluene, and benzene are commonly used as solvents in the industry. The excessive release of these compounds brings environmental damage, such as water contamination. Meanwhile, ammonia is a kind of waste produces from industry and domestic waste which generally smells and toxic for aquatic biota and human health [3, 4]. The threshold level of ammonia in industrial and domestic waste according to the Minister of Environment and Forestry Regulation No. 68 of 2016 is 5–10 mg/L, so this needs to be considered that industrial liquid waste or domestic waste does not release excess ammonia levels into the environment.

The reduction effort to remove the environmental waste is by adsorption mechanism. The researchers develop technology to overcome pollution problems and improve environmental quality. The adsorption process can be an alternative method because it is generally inexpensive and can bind pollutant compounds well because it involves the accumulation of binding of adsorbate to the adsorbent. There are two types of the adsorption process, namely liquid and solid adsorption. Some of the shortcomings in the liquid adsorption process are difficult and require high energy in the separation process, while in the solid adsorption mechanism, the separation process is easier because of the phase difference between the adsorbent and the adsorbate. Other advantages include lower energy use, more cost-effective and easy to handle [5]. Moreover, solid adsorbents derived mainly from natural materials are expected to be a non-toxic, non-corrosive approach and can adsorb adsorbate well. The solid adsorbent needs to have good characteristics such as has a high surface area and pore structure that can adsorb chemical compounds (adsorbate) optimally.

Some porous solid materials commonly used as adsorbents are zeolite, activated carbon, bentonite, and silica. Silica is an inorganic material that is widely used in chemical processes because it provides several advantages including good chemical and thermal stability during the reaction process, has a high surface area, and low cost [6]. Silica can be obtained from the extraction of tetraethyl orthosilicate (TEOS) [7], sand [8, 9], olivine [10] as well as agricultural and forestry waste. Silica with higher yield and purity is generally extracted from rice husks [11, 12]. Several research previously used a silica from rice husk as adsorbent of Pb and Au [11, 13], silica gel commercial for Congo red adsorption [14] and sound adsorption [15], silica from TEOS for Hg, Au, nitrobenzene, phenol, and methylene blue adsorption [7, 16, 17], and silica from tetramethyl orthosilicate (TMOS) for CO_2 adsorption [18].

Biomass waste is considered more likely as a silica raw material because it contains high silica, is inexpensive, abundantly available, and can provide added value from the waste. One kind of silica source is bamboo. Bamboo is a fast growing and abundant species in Indonesia with limited utilization [19]. Bamboo produces leaves waste and unutilized whereas it is contained of Si element. Silica from bamboo used as materials for many purposes such as for water glass [20], cement [21, 22], ceramics [23], catalyst [24], Al metal matrix composite [25], nano-silica material [26], and PDMS membrane [27]. Other research found that removal uranium by silica from bamboo leaves is about 45% [28]. An excellent performance of mesoporous silica bamboo for the removal methyl orang were found about 92.48% with 0.4 g/L adsorbent dosage at pH 3 [29]. However, natural-based silica development from bamboo leaves waste is still attractive. Therefore, in this study, silica was extracted from bamboo waste, then characterized and used as an adsorbent for various adsorbate types. This research is expected to produce useful and effective bio-silica adsorbents for future needs.

2 Materials and Method

2.1 Materials

The raw material used is bamboo leaf. For chemical compounds used are NaOH 10% (Merck, Germany), HCl 1 M (Merck, Germany), KI (Merck, Germany), Na_2SO_3 0.1 N (Merck, Germany), starch, methylene blue (Merck, Germany), ammonia (Merck, Germany), chloroform (Merck, Germany), benzene (Smart Lab, Indonesia), distilled water. Several equipment used are glass equipment (Pyrex, Iwaki), filter paper, pH indicator, mortar, oven, analytical balance, crucibles, hotplate magnetic stirrer. The instrument for analysis used are UV-Visible spectrophotometer (Shimadzu-1700, Japan), SEM (Evo 50, ZEISS, Germany), SAA (JWGB Meso 112-China), FTIR (Thermo Scientific Nicolet iS5-United States), and XRF (MA 01,876, Thermo Fisher Scientific-United States).

2.2 Methods

Preparation and silica extraction from bamboo leaf [30]. Bamboo leaves are dried and cleaned to remove impurity particles. After that, the samples were burned in a conventional furnace for 5–6 h to obtain bamboo leaf ash as a silica precursor. The ash was then extracted using 10% NaOH with a reflux system at a temperature of 80 °C for 2 h. The extraction results were then filtered and centrifuged until the filtrate was collected in the form of a Na-silicate solution. Na silicate is then reacted with 1 M HCl until the pH is neutral and forms a gel. The gel formed was then aged for 24 h and filtered, then the residue was collected and dried in the oven for about 24 h. The solid obtained was then ground and calcined at a temperature of 500 °C for 1 h to form silica solids (Si/BW).

Characterization and silica application as an adsorbent. The silica was analyzed using SEM for the surface morphology analysis, FTIR to determine functional groups contents, SAA to determine surface area, pore distribution, and pore size of silica, and XRF to analyze silica and other elements content. The adsorption process with silica as an adsorbent is carried out by several methods.

Iodine adsorption methods in accordance with SNI 06-3730-1995. The silica sample was weighed as much as 0.25 g and put into an Erlenmeyer. Then 25 mL of

0.1 N iodine solution was added and shaken for 15 min. The mixture was filtered and 10 ml of the filtrate was added with three drops of starch solution and then titrated with 0.1 N Na2SO3 until the blue color disappeared. The calculations followed the equation:

$$\text{Iodine adsorption (mg/g)} = \frac{A\frac{BxNNa_2SO_3}{NIod}x12.693xdf}{a}$$

A filtrate volume (mL).
B Na_2SO_3 volume (mL).
df dilution factor.
a silica weight (g).
12.693 iodine number according with 1 mL Na_2SO_3 0.1 N.

Methylene blue adsorption. The sample was weighed as much as 0.25 g and then put into an Erlenmeyer. Then add 25 mL of 1200 mg/L methylene blue solution and shake for 30 min. This method in accordance to previous method with slight modification [2]. Measure the absorption at the maximum absorption wavelength of 664 nm (after the standard curve is calculated).

$$X/m = \frac{V(Co - Ca)}{m}$$

X/m adsorption capacity (mg/g).
V methylene blue solution (L).
Co methylene blue initial concentration (mg/l).
Ca methylene blue final concentration (mg/l).

Chloroform, toluene, benzene and ammonia adsorption process. The ± 0.5 g sample was put into a cup (a) and then placed in a desiccator (previously saturated with ammonia vapor for 24 h: adsorption equilibrium was reached). Then the cup containing the sample is weighed again (b). This analysis was carried out in 2 replications. Determination of adsorption using the following equation:

$$\text{Adsorption (\%)} = \frac{b - a}{a}x100\%$$

a sample weight before adsorption (g)
b sample weight after adsorption (g)

The adsorption process for benzene, toluene, and chloroform similarly conducted with ammonia adsorption [31].

3 Result and Discussion

3.1 Physicochemical Properties of Bio-Silica from Bamboo Waste

The process of extracting silica from bamboo waste begins with the preparation of bamboo leaf waste through cleaning and drying processes. Then, the bamboo leaf is burned in a furnace for 5–6 h to obtain ash which is an inorganic material leftover from burning organic matter [32] with a higher silica content [33]. Bamboo leaf ash then becomes the precursor of the silica extraction process through the reflux method using NaOH as a reagent to produce Na-silicate, then HCl solution is added to separate the silica. The reaction mechanism in the process of binding silica in bamboo ash follows the reaction.

$$SiO_2 + 2\,NaOH \rightarrow Na_2SiO_3 + H_2O$$

$$Na_2SiO_3 + 2\,HCl \rightarrow SiO_2 + 2\,NaCl + H_2O$$

The silica resulted were then characterized for their surface morphology using SEM instrument as shown in Fig. 1. Based on this figure, the silica surface morphology was observed as irregular, and some form aggregates. Previously research found the similar morphology surface silica with rough and irregular shape [34].

Chemically, silica contains siloxane (Si–O–Si) and silanol Si–OH groups on its surface and it can be modified according to specific targets. As with l-arginine, organic compounds with COOH and NH_2 groups, and Al metal. The functional group spectra of silica were analyzed using FTIR, which is shown in Fig. 2. From this figure, it can be seen that several spectra at certain wavenumbers. At a wavelength of 3459.66 cm^{-1}

Fig. 1 Surface morphology of bio-silica (Si/BW) from bamboo leaf ash

and 1635. 34 cm^{-1}, it shows the presence of a silanol group (Si–OH) on the surface and an OH group that comes from the absorption of H_2O [12, 35, 36]. Wave numbers 1099.22, 802.24, and 478.25 cm^1 respectively indicate the presence of Si–O, Si–OH, and Si–O–Si groups [16, 37, 38].

One of the factors that affect the adsorption strength of an adsorbent is the surface area. The higher surface area of adsorbent will give more active sites to adsorb the adsorbate. The surface area analyzer with BET and BJH method was used to determine the surface area, pore-volume, and pore size. Figure 3 shows the results of silica analysis on the surface area analyzer so that the isotherm pattern of nitrogen adsorption on silica and the distribution of silica pores is obtained.

Figure 3a, it can be seen that the type of silica adsorption isotherm follows the type IV isotherm with an H2 hysteresis pattern due to the capillary condensation phenomenon, where this type of isotherm indicates multilayer adsorption with the formation of molecular layer channels. The results of the BET calculation show that

Fig. 2 FTIR spectra of bio-silica (Si/BW) from bamboo leaf ash

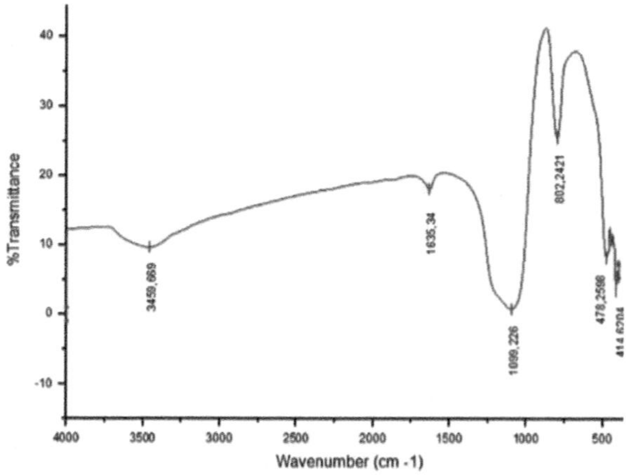

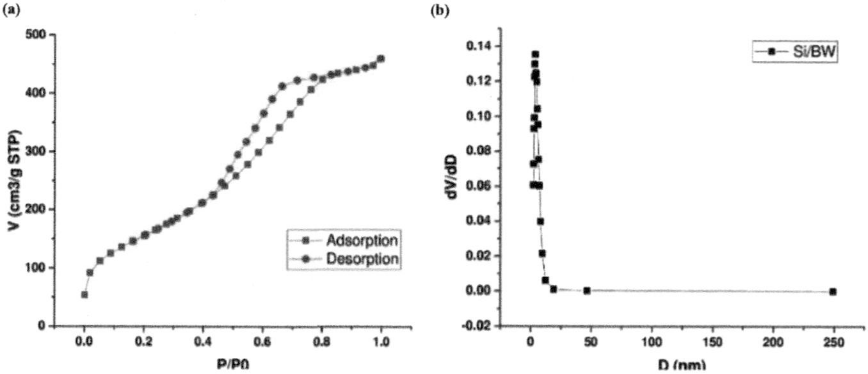

Fig. 3 N_2 adsorption-desorption graph **a** and pore distribution **b** of bio-silica (Si/BW) from bamboo leaf ash

the silica surface area is 572.833 m^2/g, and this high surface area is expected to provide good adsorption power on silica. Figure 3b also shows the pore distribution of silica with an average pore size of 4.945 nm, the silica obtained in this study can be classified into the mesoporous class. This result is higher than in previous studies with a surface area of only 330–380 m^2/g on silica extracted from sand [8] but smaller than the surface area of silica aerogel from TEOS with a surface area of 888.73 m^2/g [39].

In this research, the silica and other element content analyzed using XRF instrument, as seen in Table 1.

From Table 1, it is seen that the elemental content in silica is very varied, with the main content being Si at 493,403,938 ppm or 49.34%. The result in this work higher than silica content from bamboo leaf using hydrothermal reaction [40]. Similar result reported that 52.26% Si from bamboo leaf ash produced by [21]. The silica content in this work tends to be impure because it is still mixed with other elements, which may interfere with the adsorption process. After all, these elements can close or fill the silica pores. For example, if the element is positively charged, it will interfere with the adsorption process on the positively charged adsorbate (methylene blue). The

Table 1 Element composition of bio-silica (Si/BW) from bamboo leaf ash

Elements	ppm
Cd	16.765
Pd	8.509
Ag	7.260
Bal	440,834.844
Mo	15.023
Nb	10.364
Zr	25.119
Rb	47.677
Bi	28.065
Zn	108.897
Cu	227.625
Fe	1122.964
Ti	1753.748
Ca	2348.088
K	425.046
Al	1726.807
P	20,740.609
Si	493,403.938
Cl	24,684.418
S	4527.275
Mg	7883.158

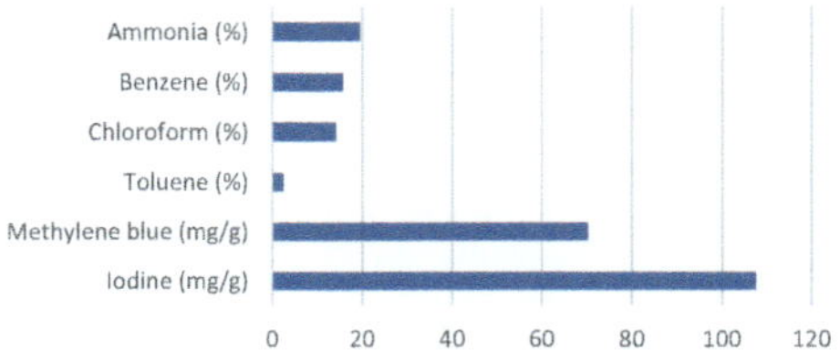

	Iodine (mg/g)	Methylene blue (mg/g)	Toluene (%)	Chloroform (%)	Benzene (%)	Ammonia (%)
■	107.6	70.28	2.51	14.22	15.7	19.46

Fig. 4 Adsorption capacity of bio-silica from bamboo leaf as toward iodine, methylene blue, toluene, chloroform, benzene, and ammonia

diversity of silica material content in this study is probably due to the absence of a purification process at the beginning of the stage so that the impurity content cannot be removed. It is well known that the extraction yield can be affected by various factors, including purification, pretreatment, sample preparation, and operating conditions. Silica with a purity level of 81.75% can be obtained from natural sand raw materials with a material purification process using magnetic materials and ultrasonic cleaner [41]. The magnetic separation process and acid leaching also affect increasing the silica content by up to 12% compared to no treatment [42]. However, the Si content in this bio-silica is higher than the Si content by 25.66% from rice husk silica [12].

3.2 *Bio-Silica from Bamboo Leaf Ash Application as Adsorbent*

The adsorption process was conducted with various adsorbates, namely iodine compounds, methylene blue, toluene, chloroform, benzene, and ammonia. Figure 4 presents the bio-silica adsorption data on these various compounds.

From Figure 4, it is seen that the adsorption capacity of bamboo silica in this study was different for each compound that was adsorbed. The adsorption process is affected by many factors, including the properties of the adsorbent and adsorbate as well as adsorption conditions such as pH, adsorbent doses, adsorbate concentration, and contact time reaction [43]. The adsorption capacity of silica against iodine compounds is 107.6 mg/g, which means that each gram of silica can only adsorb 107.6 mg of iodine compounds. Meanwhile, in general, the activated carbon provides a higher adsorption capacity, reached until 750 mg/g according to the requirements of the Indonesian National Standard (SNI). The low iodine adsorption capacity of silica is probably due to the silica characteristic in this study which tends to be hydrophilic, so that it is difficult to adsorb iodine compounds even though the surface area of the silica is high. This proved in the FTIR spectra of silica which did not show any hydrophobic contributing groups, such as CH from CH_3 stretching vibration, symmetric vibration of CH_3, and in-plane vibration CH_3 at wavenumbers 2968 cm^{-1}, 1387 cm^{-1}, and 928 cm^{-1}, respectively [17].

In the methylene blue adsorption, the adsorption capacity of 70.28 mg/g with a removal percentage of 58.56% at a high initial concentration (1200 mg/L) with a low adsorbent dose (0.25 gr). Methylene blue is a water-soluble organic compound that is more easily adsorbed into silica. Then, this process is affected by the electronegativity of methylene blue and silica, where methylene blue is a cationic dye with a positive charge while silica has a negative charge due to the presence of Si–O groups, especially at high reaction pH or not acidic. In higher pH conditions, the Si–OH group will be deprotonated into Si–O and induce the cations adsorbate attached to the adsorbent [44]. However, the negative charge on the material can provide a repulsive force when the adsorption process is carried out on an anionic adsorbate or with a negative charge. Some negatively charged metal oxides in silica, namely Al_2O_3, Fe_2O_3, CaO, and MgO can reduce the adsorption capacity, therefore the purification process through acid leaching is needed.

In a previous study, the adsorption yield of methylene blue using commercial silica is 93% at 25 mg/L solutions and 94.6% at 12.5 mg/L [45]. Meanwhile, silica from rice husk ash and silica aerogel from TEOS can remove methylene blue by 95% and 45.35 mg/g, respectively [38, 39]. Therefore, the higher the initial concentration used, the removal percentage tends to be smaller and stagnant. This work is in line with [46] that there is a decrease in phenol removal when the initial concentration of adsorbate was increased (for 20 mg/L adsorbate resulted in 99% removal, while 500 mg/L adsorbates only resulted in 31.5% removal). On the other hand, the removal percentage will increase when the adsorbent dose is increased. The adsorption capacity will be decreased as an increase of removal percentage (0.4 g/L doses resulted in about 23.3% removal with an adsorption capacity of 58.3 mg/g while 0.8 g/L doses could produce a 98.8% removal with an adsorption capacity of 12.5 mg/g). Silica gel adsorption capacity of indigo carmine is 25,707 mg/g with a 21.82% removal probably indicates that the adsorbate covers the active site of the adsorbent and then becomes saturated [37].

Figure 4 also shows that the adsorption capacity of silica on toluene, chloroform, benzene, and ammonia is 2.51, 14.22, 15.7, and 19.46%, respectively. The adsorption capacity of silica in these compounds is relatively small compared to other studies, but many factors influence the low absorption capacity. In this study, an adsorbent weight of 0.5 g was used, while in this study [47] an adsorbent weight of 10 g of activated carbon was used at a volume of 0.5 mL of benzene to produce an adsorption capacity of 55.85 mg/g. Silica aerogel also has adsorption capacity on chlorobenzene, DMSO, xylene, methanol, and toluene of 10.12; 9.6; 8.92;7.24, and 7.0 mg/g [48]. Priyanto [49] reported that ammonium removal is about 35% using silica from bamboo. Ammonia adsorption using activated carbon can also result in a reduction in ammonia concentration of 49–75% [50] and 71–84% at doses of 3 and 4 g [51].

Silica is a non-flammable material with good safety for volatile organic compounds (VOCs) adsorption, but the silica produced in this study tends to be hydrophilic, therefore it is easy to absorb water vapor. This condition has implications for causing adsorption competition between VOCs and water vapor. As well as the effect of VOCs recovery causing difficulties in adsorption process efficiency. The solubility properties of the adsorbate also affect the adsorption strength of silica. As is known, ammonia is soluble in water, chloroform is slightly soluble in water, while benzene and toluene are insoluble in water. For the adsorption process improvement, a modification character of silica was conducted, by adding TMCS, CTAB, L-arginine, certain metal oxides, and pH conditioning so that the surface properties of the silica change and suitable for the adsorbate to be adsorbed. Silica from sugarcane ash modified using CTAB resulting in 88–98% purity with a surface area of 131 m^2/g and pore size of 22 nm has an adsorption capacity of 230 mg/g for acid orange adsorbate [52]. The hexane adsorption resulted in 93% removal by silica gel with TMCS modification [53]. The silanol group in silica acts as a weak acid and interacts with polar compounds through hydrogen bonds and dipole-dipole interactions known as Bronsted acid. Therefore, silica can absorb many types of compounds such as cyclic amines, amides, proteins, and platinum metals [6]. So that to produce a maximum adsorption process, special attention is needed on the interaction between the adsorbent surface and the adsorbate which is influenced by chemical bonds, hydrogen bonds, hydrophobic/hydrophilic bonds, and van der Waals forces [54].

4 Conclusion

The results showed that silica synthesized from bamboo leaf waste with a mesoporous structure and high surface area. The functional groups in silica are OH, Si–O–Si and Si–O, with a silica content of 49%. The adsorption capacities of iodine, methylene blue, toluene, chloroform, benzene, and ammonia compounds were 107.6 mg/g, 70.28 mg/g, 2.51%, 14.22%, 15.7%, and 19.46%, respectively. Silica from bamboo waste can potentially be an alternative green adsorbent in the future.

Acknowledgements Author(s) is expressing grateful to the technicians of Chemistry Laboratory of Forestry Research and Development Center (Nela R. Sari, S.Si and Dadang Setiawan) who have provided assistance in this work. We are also grateful to Japan Society for the Promotion of Science (JSPS) for providing partial funding of this publication fee.

References

1. Rahmi R, Fachruddin S, Nurmalasari N (2018) Pemanfaatan Limbah Serat Sagu (Metroxylon sago) Sebagai Adsorben Iodin. J Rekayasa Kim Lingkung 13:70–77. https://doi.org/10.23955/rkl.v13i1.10072
2. Efiyanti L, Indrawan DA, Hastuti N, Darmawan S (2020) The activated carbon produced from mayan bamboo (Gigantochloa robusta Kurz) and its application as dye removal. IOP Conf Ser Mater Sci Eng
3. Kosim H, Arita S, Hermansyah H (2015) Pengurangan Kadar Amonia dari Limbah Cair Pupuk Urea dengan Proses Adsorpsi Menggunakan Adsorben Bentonit. J Penelit Sains 17:168291

4. Nurhidayanti N, Ardiatma D, Anggriawan B (2020) Pemanfaatan Karbon Aktif Dari Tempurung Kelapa Dalam Menurunkan Kadar Amonia total Dalam Air Limbah Industri. J Pelita Teknol 15:68–76
5. Tahari MNA, Yarmo MA (2014) Adsorption of CO_2 on silica dioxide catalyst impregnated with various alkylamine. AIP Conf Proc 1614:334–341. https://doi.org/10.1063/1.4895218
6. Sulastri S, Nuryono N, Kartini I, Kunarti ES (2011) Adsorption of Ca(II), Pb(II) and Ag(I) on sulfonato-silica hybrid prepared from rice hull ash. Indones J Chem 11:273–278. https://doi.org/10.22146/ijc.21392
7. Rabie AM, Abd El-Salam HM, Betiha MA, El-Maghrabi HH, Aman D (2019) Mercury removal from aqueous solution via functionalized mesoporous silica nanoparticles with the amine compound. Egypt J Pet 28:289–296. https://doi.org/10.1016/j.ejpe.2019.07.003
8. Efiyanti L, Trisunaryanti W, Bahri S, Lailun Y, Meisa N, Satriyo W (2021) Synthesis of mesoporous silica from beach sand using variation of cetyl trimethyl ammonium bromide (CTAB). In:14, pp 374–381
9. Trisunaryanti W, Larasati S, Bahri S, Ni'mah YL, Efiyanti L, Amri K, Nuryanto R, Sumbogo SD (2020) Performance comparison of Ni-Fe loaded on NH_2-functionalized mesoporous silica and beach sand in the hydrotreatment of waste palm cooking oil. J Environ Chem Eng 8:104477. https://doi.org/10.1016/j.jece.2020.104477
10. Chen YX, Sepahvand S, Gauvin F, Schollbach K, Brouwers HJH, Yu Q (2021) One-pot synthesis of monolithic silica-cellulose aerogel applying a sustainable sodium silicate precursor. Constr Build Mater 293:123289. https://doi.org/10.1016/j.conbuildmat.2021.123289
11. Yusmaniar, Purwanto A, Putri EA, Rosyidah D (2017) Adsorption of Pb(II) using silica gel composite from rice husk ash modified 3-aminopropyltriethoxysilane (APTES)-activated carbon from coconut shell. AIP Conf Proc 1823. https://doi.org/10.1063/1.4978107
12. Ragadhita R, Nandiyanto ABD, Nugraha WC, Mudzakir A (2019) Adsorption isotherm of mesopore-free submicron silica particles from rice husk. J Eng Sci Technol 14:2052–2062
13. Istiningrum RB, Tiwow CD (2015) Nuryono, Narsito: Au(III) selective adsorption of quaternary ammonium-silica hybrid in Au/Cu system. Procedia Chem 17:132–138. https://doi.org/10.1016/j.proche.2015.12.126
14. de Farias RS, Buarque HL de B, da Cruz MR, Cardoso LMF, Gondim T de A, de Paulo VR (2018) Adsorption of congo red dye from aqueous solution onto amino-functionalized silica gel. Eng Sanit e Ambient 23:1053–1060. https://doi.org/10.1590/s1413-41522018172982
15. Mazrouei-Sebdani Z, Begum H, Schoenwald S, Horoshenkov KV, Malfait WJ (2021) A review on silica aerogel-based materials for acoustic applications. J Non Cryst Solids 562:120770. https://doi.org/10.1016/j.jnoncrysol.2021.120770
16. Hastuti S, Nuryono, Kuncaka A (2015) L-arginine-modified silica for adsorption of gold(III). Indones J Chem 15:108–115. https://doi.org/10.22146/ijc.21203
17. Yi Z, Tang Q, Jiang T, Cheng Y (2019) Adsorption performance of hydrophobic/hydrophilic silica aerogel for low concentration organic pollutant in aqueous solution. Nanotechnol Rev 8:266–274. https://doi.org/10.1515/ntrev-2019-0025
18. Wörmeyer K, Smirnova I (2013) Adsorption of CO_2, moisture and ethanol at low partial pressure using aminofunctionalised silica aerogels. Chem Eng J 225:350–357. https://doi.org/10.1016/j.cej.2013.02.022
19. Noverliana N, Asmi D (2014) Sintesis Keramik Silika dari Daun Bambu dengan Teknik sol-gel dan Karakteristik Pada Suhu Kalsinasi 500 °C–700 °C. J Chem Inf Model 03:17–23
20. Kow KW, Yusoff R, Aziz ARA, Abdullah EC (2014) From bamboo leaf to aerogel: preparation of water glass as a precursor. J Non Cryst Solids 386:76–84. https://doi.org/10.1016/j.jnoncrysol.2013.11.041
21. Lwin PP, Nu T, May KA (2020) Application of extracted silica samples from bamboo leaves and rice husk in the formulation of cement. Dagon Univ Res J 11:174–181
22. Nura Diana AI, Fansuri S, Deshariyanto D (2020) Penambahan Abu Daun Bambu Sebagai Substitusi Material Semen Terhadap Kinerja Beton. Padur J Tek Sipil Univ Warmadewa 9:172–182. https://doi.org/10.22225/pd.9.2.1788.172-182

23. Olawale O (2020) Bamboo leaves as an alternative source for silica in ceramics using Box Benhken design. Sci African 8:e00418. https://doi.org/10.1016/j.sciaf.2020.e00418
24. Chen H, Wang C, Zhou H, Tao R (2016) Isolation of Sulfuretin and Butin from Rhus verniciflua stokes using medium-pressure liquid chromatography and their Tyrosinase inhibitory effects. BioResources 11:759–771
25. Olaniran O, Uwaifo O, Bamidele E, Olaniran B (2019) An investigation of the mechanical properties of organic silica, bamboo leaf ash and rice husk reinforced aluminium hybrid composite. Mater Sci Eng Int J 3:129–134. https://doi.org/10.15406/mseij.2019.03.00103
26. Dirna FC, Rahayu I, Maddu A, Darmawan W, Nandika D, Prihatini E (2020) Nanosilica synthesis from betung bamboo sticks and leaves by ultrasonication. Nanotechnol Sci Appl 13:131–136. https://doi.org/10.2147/NSA.S282357
27. Sethy NK, Arif Z, Mishra PK, Kumar P (2019) Synthesis of SiO_2 nanoparticle from bamboo leaf and its incorporation in PDMS membrane to enhance its separation properties. J Polym Eng 39:679–687. https://doi.org/10.1515/polyeng-2019-0120
28. Megasari K, Herdiyanti H, Nurliati G, Kadarwati A, Swantomo D (2019) Sintesis Silika Xerogel Dari Abu Daun Bambu Sebagai Adsorben Uranium. J Forum Nukl 13:27–36
29. Mohd Nazri M, Sapawe N (2020) Removal of methyl orange over low-cost silica nanoparticles extrated from bamboo leaves ash. Mater Today Proc 31. https://doi.org/10.1016/j.matpr.2020.10.969
30. Wulandari NM, Efiyanti L, Trisunaryanti W, Oktaviano HS, Bahri S, Ni'mah YL, Larasati S (2021) Effect of CTAB ratio to the characters of mesoporous silica prepared from rice husk ash in the pyrolysis of a–cellulose. Bull Chem React Eng Catal 16:632–640. https://doi.org/10.9767/bcrec.16.3.10828.632-640
31. Wibowo S, Efiyanti L, Pari G (2020) Catalytic and thermal cracking of bio-oil from oil-palm empty fruit bunches, in batch reactor. Indones J Chem 20:1000–1009. https://doi.org/10.22146/ijc.44076
32. Riandani U (2014) Preparasi dan Karakterisasi keramik silika dari daun bambu hasil leaching asam sitrat dan suhu pembakaran 800 °–1000 °C. Thesis, Departement of Phisics, Lampung University
33. Maulidiyah A (2017) Sintesis dan karakterisasi nanozeolit y dari abu sekam padi dengan variasi suhu hidrotermal. Thesis, Departement of Chemistry, Maulana Malik Ibrahim University
34. Budnyak TM, Pylypchuk IV, Tertykh VA, Yanovska ES, Kolodynska D (2015) Synthesis and adsorption properties of chitosan-silica nanocomposite prepared by sol-gel method. Nanoscale Res Lett 10:1–10. https://doi.org/10.1186/s11671-014-0722-1
35. Nandiyanto ABD, Oktiani R, Ragadhita R (2019) How to read and interpret ftir spectroscope of organic material. Indones J Sci Technol 4:97–118. https://doi.org/10.17509/ijost.v4i1.15806
36. Nuruddin A, Yuliarto B, Saputro AG, Badruzaman B, Ramelan A (2020) Preparation of polycrystalline silicon from rice husk by thermal decomposition and aluminothermic reduction. Molekul 15:26–33. https://doi.org/10.20884/1.jm.2020.15.1.568
37. Sulistiyo YA, Rofi'ah F, Suwardiyanto, Nugraha AS, Zulfikar, Sunnardianto GK (2020) Isothermal and kinetic adsorption of anionic dye onto impregnated silica gels with aluminum. Environ Eng Manage J 19:1299–1308. https://doi.org/10.30638/eemj.2020.123
38. Rengga WDP, Sediawan WB, Imani NAC, Harianingsih H, Salsabiil KA (2020) Adsorption studies of rice husk-based silica/carbon composite. EKSAKTA J Sci Data Anal 1:98–104. https://doi.org/10.20885/eksakta.vol1.iss2.art1
39. Wei W, Hu H, Ji X, Yan Z, Sun W, Xie J (2018) Selective adsorption of organic dyes by porous hydrophilic silica aerogels from aqueous system. Water Sci Technol 78:402–414. https://doi.org/10.2166/wst.2018.313
40. Hasri H, Muharram M, Nadwi F (2020) Sintesis Nanosilika Daun Bambu (*Bambusa* sp.) Menggunakan Metode Hidrotermal. J Kartika Kim 3:96–100
41. Silvia L, Zainuri M (2020) Analisis silika (SiO_2) Hasil Kopresipitasi Berbasis Bahan Alam menggunakan Uji XRF dan XRD. J Fis dan Apl 16:12. https://doi.org/10.12962/j24604682.v16i1.5322

42. Sutal Y, Pingak RK, Ahab AS, Baunsele SD (2019) Seminar nasional sains dan teknik fst undana (sainstek). In: Kajian awal ekstraksi silika dari pasir noeltoko menggunakan x-ray fluorescence. Seminar Nasional sains dan Teknik fst Undana (sainstek) Hotel Swiss-Belinn Kristal Kupang, Kupang-25 Oktober 2019, pp 278–289
43. Efiyanti L, Wati SA, Maslahat M (2020) Pembuatan dan Analisis Karbon Aktif dari Cangkang Buah Karet dengan Proses Kimia dan Fisika. J Ilmu Kehutan 14:94. https://doi.org/10.22146/jik.57479
44. Ikhsan J, Sulastri S, Priyambodo E (2015) Pengaruh pH pada adsorpsi kation unsur hara Ca^{2+} dan K^+ oleh silika dari lumpur Lapindo. J Penelit Saintek 20:10–18. https://doi.org/10.21831/jps.v20i1.5603
45. Gaikwad RW, Misal SA (2010) Sorption studies of methylene blue on silica gel. Int J Chem Eng Appl 1:342–345. https://doi.org/10.7763/ijcea.2010.v1.59
46. Rahmani A, Nazemi F, Barjasteh Askari F, Davoudi M (2018) Preparation, characterization, and application of silica aerogel for adsorption of phenol: an in-depth isotherm study. Health Scope (In Press). https://doi.org/10.5812/jhealthscope.15115
47. Holle RB, Wuntu AD, Sangi MS (2013) Kinetika Adsorpsi Gas Benzena Pada Karbon Aktif Tempurung Kelapa. J MIPA 2:100. https://doi.org/10.35799/jm.2.2.2013.2997
48. Shi HX, Cui JT, Shen HM, Wu HK (2014) Preparation of silica aerogel and its adsorption performance to organic molecule. Adv Mater Sci Eng https://doi.org/10.1155/2014/850420
49. Priyanto A (2015) Sintesis Dan Aplikasi Silika Dari Abu Daun Bambu Petung (*Dendrocalamus asper* (Schult f.) Backer ex Heyne) Untuk Mengurangi Kadar Ammonium Dan Nitrat Pada Limbah Cair Tahu
50. Pradana AA, Pujiono P, Yulianto B, Ruhmawati T (2019) Perbedaan waktu kontak karbon aktif terhadap penurunan kadar amonia pada limbah cair domestik. J Ris Kesehat Poltekkes Depkes Bandung 11:215
51. Solichah A, Rachmaniyah, Rokhmalia F (2018) Pemanfaatan ampas tebu sebagai karbon aktif terhadap penurunan kadar cod dan amonia (NH_3) (Studi Pada Limbah Cair Industri Tahu Dinoyo Kota Surabaya). Gema Lingkung Kesehat 16. https://doi.org/10.36568/kesling.v16i3.894
52. Rovani S, Santos JJ, Corio P, Fungaro DA (2018) Highly pure silica nanoparticles with high adsorption capacity obtained from sugarcane waste ash. ACS Omega 3:2618–2627. https://doi.org/10.1021/acsomega.8b00092
53. Huang W, Xu J, Tang B, Wang H, Tan X, Lv A (2018) Adsorption performance of hydrophobically modified silica gel for the vapors of n-hexane and water. Adsorpt Sci Technol 36:888–903. https://doi.org/10.1177/0263617417728835
54. Parida SK, Dash S, Patel S, Mishra BK (2006) Adsorption of organic molecules on silica surface. Adv Colloid Interface Sci 121:77–110. https://doi.org/10.1016/j.cis.2006.05.028

Effect of Densification and Steam Treatment on the Quality of Laminated Bamboo

Teguh Darmawan, Dimas Triwibowo, Muhammad Adly Rahandi Lubis, Wahyu Dwianto, and Nareswor o Nugroho

Abstract This research is related to the optimization of the amount of epoxy adhesive used in the manufacture of laminated bamboo from curved cross-sectional blades and efforts to homogenize the characteristics of laminated bamboo through the use of bamboo slats that have been compressed by steam treatment. In this study, there were two stages of making laminated bamboo. The first was laminated bamboo without a compression process. Laminated bamboo was made with a target thickness of 5 cm using epoxy adhesive with a glue spread weight of 100 g/m^2, 120 g/m^2, and 150 g/m^2. The second was laminated bamboo from compressed bamboo slats. The bamboo slats were compressed at the ratio of 50% for the bottom, 30% for the middle, and without compression for the upper part. Then the slats were subjected to steam treated at 160 °C for 60 min. The lamination process was carried out with the best amount of glue spread weight in the previous stage. The results showed that based on the value of delamination and shear strength, the use of adhesive with a glue spread weight of 120 g/m^2 could be chosen as the optimal mass of adhesive. Laminated bamboo with compression has good adhesive qualities indicated by the absence of delamination and the value of shear strength that meets the standards. The position factor has no significant effect on the value of the density and shear strength of laminated bamboo. The values of density and shear strength were not different or the same between samples made from the bottom, middle, and upper part. This showed that the compression of the bamboo slats at the bottom and middle by 50% and 30% respectively succeeded in making the characteristics of the laminated bamboo uniform.

Keywords Laminated bamboo · Curved cross section · Bamboo densification · Steam treatment · Epoxy

T. Darmawan · N. Nugroho (✉)
Faculty of Forestry and Environment, IPB University, Bogor, Indonesia
e-mail: nares@apps.ipb.ac.id

T. Darmawan · D. Triwibowo · M. A. R. Lubis · W. Dwianto
Research Center for Biomaterials, Indonesian Institute of Sciences, Jakarta, Indonesia

M. A. Rahandi Lubis (ed.), *Proceedings of the 10th International Symposium for Sustainable Humanosphere*, Lecture Notes in Civil Engineering 255,
https://doi.org/10.1007/978-981-19-1422-5_14

1 Introduction

Bamboo is known as a lignocellulosic material with excellent mechanical properties. Hence, bamboo can be used as a material for various products, including structural and non-structural building materials. Sun et al. [27] explained that in general the stages of making bamboo products consist of the process of cutting/splitting, gluing, and pressing. The cutting and splitting process is the stage of preparing materials in the form of slats, zephyr, strand, fiber or other shapes according to the product made. Various types of research and development of bamboo-based products include bamboo oriented strand board [7, 14, 26], bamboo oriented strand lumber [13], bamboo zephyr board [17, 18], laminated bamboo (Sulastiningsih et al. 2009, [10, 16, 29], ply-bamboo [2] and cross laminated bamboo [1, 19].

From the above engineered bamboo products, laminated products are relatively more efficient in the use of adhesives. Types of adhesives commonly used for laminated bamboo with large sizes and thicknesses are poly vinyl acetate (PVA), isocyanate (EPI), epoxy (EP), polyurethane (PUR) and other types of adhesives that can be set at room temperature or cold pressing processes. Each type of adhesive used has an influence on the characteristics of the laminated product. Lu et al. [11] reported that EP, EPI, and PUR adhesives can be used to produce cross laminated timber (CLT) because they have good values for shear strength in dry and wet conditions, delamination ratio values, and percentage of damage.

The manufacture of square-section slats has several drawbacks, including a lot of discarding the cylindrical part of the bamboo. The size of the slats depends on the thickness of the bamboo used, square cutting will be effective only on bamboo species that have thick walls [25, 28]. In addition, bamboo has an uneven density distribution, where the density distribution increases in the position from the bottom of the stem to the upper, as well as from the inner part out of the bamboo stem [20, 31]. Therefore, making a square slat will cut the part near the skin which actually has the highest density.

The shortcomings in the manufacture of bamboo can be reduced by the use of bamboo slats that still maintain the curvature of the bamboo. Bamboo slats with a significant curved cross section increase the yield of bamboo slats by 7–13% higher than those of square slats [3]. Curved sections have better mechanical advantages than square sections because of the wider modulus of the section. This was proven in the experiment by Mujiman et al. [15] where laminated bamboo beams from curved slats have better mechanical properties than laminated beams from square slats. Laminated beams of curved slats are more resistant to loading, more rigid and tough.

The strength characteristics at each position of the bamboo stem have the same pattern as the density distribution, because the strength of a material is strongly influenced by its density [5]. On the inner part it has a low density and is soft so that it will become an area of weakness when laminated bamboo is made. One of the efforts to increase the density can be done by densification, where the deformation starts from the weakest part of the cell walls [9] so that the softer part of the bamboo will be compressed first. Thus densification can regulate the uniformity of the density of bamboo slats from the bottom and middle of the stem as well as on the inner part.

Previous research has carried out the development of laminated bamboo from curved bamboo slats and also carried out densification of bamboo slats at a compression ratio of 20% and steam treatment of 120 °C for 60 min. The results showed that the compression process and steam treatment of curved bamboo slats increased the value of shear strength, and improved the delamination ratio. However, the delamination that occurred has still quite high, namely 16.8%, it indicated that the deformation due to compression has not yet occurred fixation. Compressed bamboo tended to return to its original shape or experience spring back due to the influence of immersion or boiling, this was one of the causes of the release of adhesive on the laminate line [4]. The next development related to homogenizing the density of bamboo slats through densification at a compression level of 50% for the bottom and 30% for the middle and steam treatment until permanent fixation achieved. The results showed that the density of the bamboo slats became more uniform near the upper, and the fixation could be achieved by steam treatment at 160 °C for 60 min.

Bamboo will experience changes in physical and mechanical properties when subjected to heat or steam treatment. Steam treatment of bamboo can increase its strength, but at a high temperature and for a long time it will degrade its strength. After steam treatment at 140 °C the strength of bamboo will increase, but at 180 °C for 30 min it can significantly reduce the strength of bamboo by 20–47% [30]. Therefore, in addition to determining the weight of the epoxy adhesive used in the manufacture of curved laminated bamboo, this study also aims to optimize the amount of epoxy adhesive used in the manufacture of laminated bamboo from curved cross-sectional blades and efforts to homogenize the characteristics of laminated bamboo through the use of bamboo slats that have been compressed by steam treatment.

2 Method

The bamboo used in this study was the Sembilang bamboo (*Dendrocalamus giganteus* Munro). The bamboo stems were cut at the bottom and upper part with a length based on an inner diameter (d) which was set at least 60 mm, then the number of segments was calculated and grouped into 3 parts at the bottom, middle and upper part based on the thickness of the bamboo. The segment used was cut into a cylinder with a length of 50 mm, then split with a width of 50 mm and cleaned the outer and inner part of the bamboo using a sandpaper machine.

The first group was laminated bamboo without a compression process. The bamboo used in this study had an air-dry moisture content (< 12%). Laminated bamboo is made with a target thickness of 5 cm, so the number of layers of laminated bamboo at the bottom, middle and upper part were 4, 6 and 10 layers, respectively. The test samples were glued with Qbond HQ EP 501 epoxy adhesive produced by PT. Konishi Lemindo Indonesia. Adhesive consists of two components in the form of resin and hardener (1: 0.8). The adhesive is applied in a double spread with a glue spread weight of 100 g/m^2, 120 g/m^2, and 150 g/m^2 and prepared on a mold. The mold was made with a curved shape on the upper and bottom surfaces, so that bamboo slats with a curved cross section can be compressed without damage. The compression was carried out for 24 h at room temperature.

The second group was laminated bamboo from compressed bamboo slats. The test sample used in the wet condition was saturated with fiber. The bamboo slats were arranged in a mold and compressed with a compression ratio of 50% for the bottom, 30% for the middle, and no compression for the upper parts. Furthermore, the test sample in the mold was inserted into the chamber and steam treated at a temperature of 160 °C for 60 min. After that it was dried in an oven at a temperature of 100 °C for 24 h and then glued with the total weight of the glue spread according to the best results in the previous process.

Physical properties testing included measurement of moisture content, density, delamination and for testing of mechanical properties in the form of shear strength test. The method of measuring moisture content, delamination test and shear strength test referred to the Japanese Agricultural Standard (JAS) for Glued laminated timber 1152 [12]. The test sample for shear strength testing of laminated bamboo was made with the shape and size as shown in Fig. 1. The shear plane was selected in the middle of the bamboo laminate arrangement that has been made. At the time of the test, the load was applied in the direction of the longitudinal bamboo and in this study, no shear test was carried out in the direction of the width of the laminated bamboo. In the axial plane, the test sample was cut following the glue line thickness so that a curved shear plane was obtained.

The shear strength testing of laminated bamboo using a universal testing machine (UTM) of the Shimadzu brand, Autograph AG-IS 50 kN Models, made in Japan. The loading was carried out to the maximum limit and damage occurred, the loading speed was 9800 N per minute. To determine the characteristics of the adhesive and the quality of the adhesive, the adhesive viscosity test was carried out, observing the adhesive line using a digital microscope. The microscope used was manufactured by Keyence Corporation USA with model number VHX-6000, followed by measuring the width of the adhesive line using the image-J software. The research design in this study was a completely randomized factorial design and each experiment was

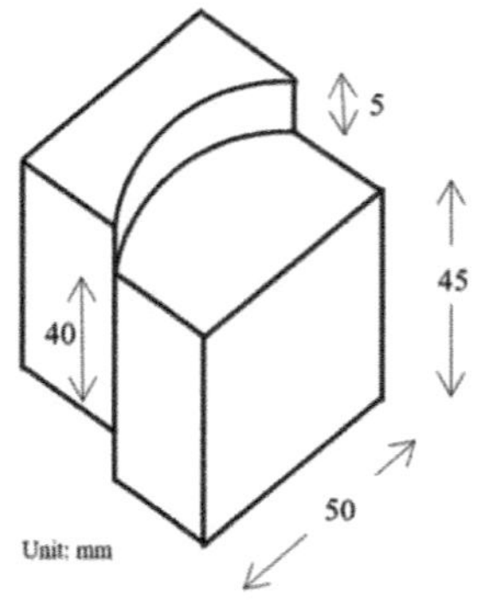

Fig. 1 The shape and size of the shear strength test of laminated bamboo from curved cross-section slats

repeated 3 times. The data obtained were analyzed by means of variance (ANOVA) at a 95% confidence interval, if there was a significant difference followed by Duncan's test [24].

3 Results and Discussion

3.1 Characteristics of Laminated Bamboo Without Compression

The results of the measurement of the physical properties of laminated bamboo from curved cross-sectional slats are shown in Table 1. The moisture content of the samples ranged from 7.95 to 9.38% so that all of them were included in the values required in the JAS 1152-2007 standard, which was no more than 12%. The position of sampling from bamboo determined the density value of laminated bamboo, the lowest to the highest values were shown in the sample from the position of the bamboo at the bottom, middle and upper part with around 0.5, 0.7, 0.9 g/cm^3 respectively. The weight of glue spread has little effect on the density value, giving more glue spread weight did not make all samples increase in density. Based on the analysis of variance in Table 1, it was known that the position and weight factors of the coating have an effect on the density value of laminated bamboo. Further analysis in Table 2 showed that the density value of the sample was influenced by the position of the bamboo used, samples from the bottom, middle, and upper part have different density values. This was related to the nature of bamboo which has a non-uniform density distribution, where at the bottom it has a low density and the highest density was at the upper part of the bamboo [20, 31]. The density of wood or bamboo depended on the anatomical structure, which was mainly influenced by the composition and density of the vascular bundle of the bamboo [5, 32]. In the glue spread weight factor, it can be seen that the glue spread weight of 150 g/m^2 has a density value that was different from the sample with 100 g/m^2 and 120 g/m^2 glue spread weight, but these two glue spread weights have no different density values.

The value of delamination after the immersion test showed that all samples at the bottom with the treatment of the bulk weight did not experience delamination,

Table 1 ANOVA physical and mechanical properties of laminated bamboo without compression

Parameter		Position (A)	Glue spread weight (B)	Interaction (AB)
Density (g/cm^3)		*(0.000)	*(0.011)	NS (0.505)
Delamination (%)	Soaking	*(0.000)	*(0.000)	*(0.000)
	Boiling	*(0.000)	*(0.014)	NS (0.185)
Shear strength (MPa)		*(0.000)	NS (0.095)	NS (0.918)

*Significant ($P \leq 0.05$); NS = not significant

Table 2 Duncan's test of the physical and mechanical properties of laminated bamboo without compression

Parameter		Duncan test of physical and mechanical properties base on factors					
		Position (A)			Glue Spread g/m^3 (B)		
		Bottom	Middle	Upper	100	120	150
Density (g/cm^3)		0.559^a	0.727^b	0.922^c	0.727^a	0.728^a	0.753^b
Delamination (%)	Soaking	0.000^a	2.310^b	12.26^c	6.787^a	3.561^a	4.22^{a,b}
	Boiling	4.553^a	5.372^a	24.54^b	13.041^a	11.31^{a,b}	10.11^b
Shear strength (MPa)		5.600^a	7.333^b	6.672^b	6.204^a	6.431^{a,b}	6.970^b

Note The values followed by the same letter in one line in each factor are not significantly different

in the middle delamination occurred in the sample with a volume of 100 g/m^2, and at the end all samples were delaminated. The allowable delamination value of the JAS 1152-2007 standard is not more than 5%, so that in this immersion test the samples were from the bottom and middle, except for the 100 g/m^2 glue spread weight which met the standard. ANOVA shows that the position factor, the weight of the glue spread and their interaction have an influence on the value of the soak test delamination (Table 1). Further analysis (Table 2) showed that the position of the bamboo sample has an influence on the delamination value, where the delamination value was different at each position. The difference in the weight of the glue spread did not all have an effect on the delamination value, the 100 g/m^2 glue spread weight had a different delamination value, while the samples with 120 g/m^2 and 150 g/m^2 glue spread weights did not make any difference (Table 3).

The results of the boiling test showed that all samples were delaminated with various values. It can be seen that there was an influence of the sampling position and the difference in the weight of the glue spread used, where the sample from

Table 3 Physical properties of laminated bamboo without compression

Position	Glue spread	Moisture content	Density	Delamination	
				soaking	Boiling
	(g/m^2)	(%)	(g/cm^3)	(%)	(%)
Bottom (P)	100	9.17(0.27)	0.55(0.02)	0.00(0.00)	6.52(1.89)
	120	9.22(0.17)	0.55(0.02)	0.00(0.00)	4.33(3.81)
	150	9.38(0.40)	0.57(0.02)	0.00(0.00)	2.80(1.14)
Middle (T)	100	8.92(0.50)	0.71(0.03)	6.93(1.61)	8.29(1.44)
	120	9.04(0.45)	0.73(0.00)	0.00(0.00)	4.77(0.24)
	150	8.66(0.39)	0.74(0.02)	0.00(0.00)	3.06(1.14)
Upper (U)	100	8.38(0.13)	0.91(0.01)	13.43(2.22)	24.31(1.16)
	120	8.43(0.66)	0.91(0.01)	10.68(0.65)	25.69(2.97)
	150	7.95(0.37)	0.94(0.02)	12.67(0.24)	24.47(1.90)

the upper part experienced higher delamination than the bottom and middle parts. The use of more lamination weight tended to reduce the delamination that occurred. The lowest delamination occurred in the sample at the bottom with a volume of 150 g/m^2 of 2.80%, and the highest delamination occurred at the upper part of the sample with a mass of 120 g/m^2 of 25.69%. The allowable delamination value of the JAS 1152-2007 standard in the boiling test is not more than 10%, so that in this boiled test the samples from the bottom and middle met the standard. Based on the ANOVA in Table 1, the position factor and the weight of the glue spread have an effect on the delamination value while the interaction of the two has no significant effect, further Duncan's test (Table 2) it was known that the weight treatment of the glue spread produced a delamination value that was not different, namely at the glue spread weight of 100 g/m^2 and 120 g/m^2 and also between the glue spread weights of 120 g/m^2 and 150 g/m^2. The value of delamination was different in the treatment of 100 g/m^2 and 150 g/m^2 glue spread weight.

The shear strength of laminated bamboo from curved cross-sectional slats without compression is shown in Fig. 2. The value of shear strength varied due to the treatment position of the bamboo stem sampling and the given weight of the coating. In general, laminated bamboo from the bottom of the slats has a lower bond strength and in the middle has a higher shear strength value than the samples from other parts, but was still included in the JAS 1152-2007 standard, which was not less than 5.4 MPa. The value of the shear strength of the bottom has a value that was close to the value of the shear strength of laminated bamboo made using a tannin resorcinol formaldehyde adhesive and the material using *G. robusta* bamboo, which is 5.58 MPa (Sulastiningsih et al. 2009). Based on ANOVA (Table 1) only the position factor has an effect, for the glue spread weight factor and the interaction of the two factors, it did not have a significant effect. Based on Duncan's test (Table 2), it was known that the samples from the middle and upper part of the bamboo have shear strength values that were not significantly different, only at different bottoms. The effect of the weight of the glue spread on the value of the adhesive strength was known to have a difference in the weight of 100 g/m^2 and 150 g/m^2, but the weight of 120 g/m^2 has a shear strength value that was not different from the weight of other glue spreads.

Based on the analysis of the results of the immersion, boiling and shear strength tests that have been carried out, it can be seen that the effect of 120 g/m^2 and 150 g/m^2

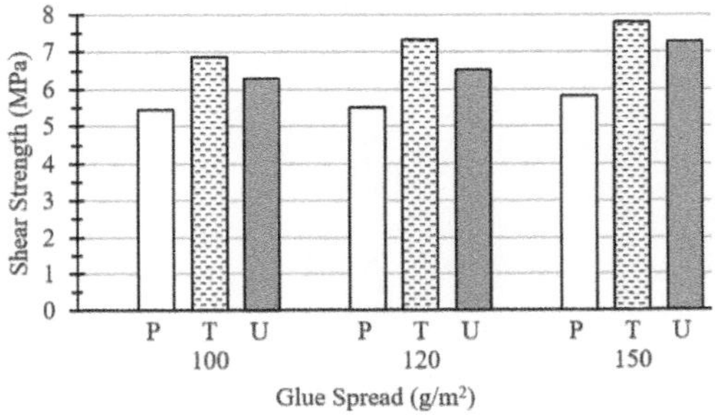

Fig. 2 Shear strength of laminated bamboo without compression

glue spread weight on the delamination value and shear strength was not different, with consideration of the efficiency of the use of adhesives, the weight of 120 g/m^2 can be chosen as the optimal mass of glue spread.

3.2 Characteristics of Laminated Bamboo with Compression

Laminated bamboo with compression using adhesive with a glue spread of 120 g/m^2 in accordance with the optimization stage of the total weight of laminated bamboo without prior compression. The physical properties of laminated bamboo from compressed curved slats are shown in Table 5. The sample has a moisture content of 5.6–6.39% so that it was in accordance with the JAS 1152-2007 standard, namely the moisture content was not more than 12%. The density of laminated bamboo that has been compressed from the position of the bamboo at the bottom, middle and upper part has an average value that was almost uniform, which was 0.96–0.99 g/cm^3. The results of the immersion and boiling test showed that the adhesiveness of laminated bamboo was very good where there was no delamination in all samples.

The results of the shear strength test showed that at the bottom it increased so that it was not much different from the value of the shear strength of the middle and upper part. The shear strength values from the lowest values were the upper, bottom, and middle, respectively, which were 6.2, 7.07, 7.35 MPa (Fig. 3). The minimum acceptable shear strength value for JAS 1152-2007 is 5.4 MPa, so that all samples have shear strength that meets the standard. The absence of delamination and the value of shear strength that met the standard indicated that laminated bamboo from curved cross-sectional slats that had been compressed and glued with 120 g/m^2 epoxy had good adhesive qualities.

ANOVA (Table 4) shows that the position factor has no significant effect on the value of the density and shear strength of laminated bamboo. Furthermore, Duncan's test showed that the density and shear strength values were not different or the same between samples made from the bottom, middle, and ends. This showed that the compression of the bamboo slats at the bottom and middle by 50% and 30% succeeded in making the characteristics of the laminated bamboo uniform.

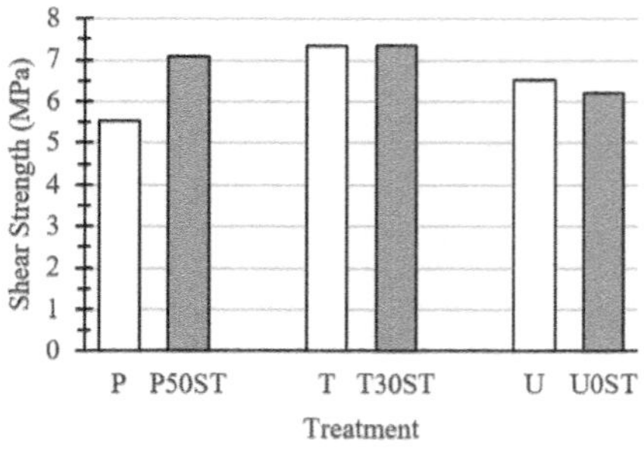

Fig. 3 Shear strength laminated bamboo with compression

Table 4 ANOVA and Duncan's test of the value of density and shear strength of laminated bamboo by compression

Parameter	ANOVA	Duncan's test by position		
	Position factor	P 50 ST	T 30 ST	U 0 ST
Density	NS (0.086)	0.980[a]	0.967[a]	0.977[a]
Shear strength	NS (0.273)	6.428[a]	7.989[a]	6.203[a]

Note *Significant ($P \leq 0.05$); NS = not significant. Values followed by the same letter in a row are not significantly different. P;T;U, Bottom; Middle; Upper. 50;30;0, compression level. ST, Steam Treatment

Table 5 Physical properties of laminated bamboo by compression

Position	Glue spread	Moisture content	Density	Delamination	
				Soaking	Boiling
	(g/m^2)	(%)	(g/cm^3)	(%)	(%)
P 50% ST	120	6.31 (0.92)	0.98 (0.01)	0.00 (0.00)	0.00 (0.00)
T 30% ST	120	6.39 (0.87)	0.96 (0.01)	0.00 (0.00)	0.00 (0.00)
U 0% ST	120	5.60 (0.38)	0.98 (0.01)	0.00 (0.00)	0.00 (0.00)

P; T; U, Bottom; Middle; Upper. 50%; 30%; 0%, compression level. ST, Steam Treatment

3.3 Microscopic Observation of Laminated Bamboo Sections

Microscopic observation of the cross-section of laminated bamboo in the area of the glue line is shown in Fig. 4. We can see the glue line between the bamboo slats, which glue the outside and the inner part of the bamboo. A greater number of vascular bundles indicates the outer part of the slats near the skin, whereas on the inner part it showed a sparse number of vascular bundles and large cavities.

The difference in the weight of the glue spread can be seen in its effect on the thickness of the glue line that occurred, where the more weight of the glue spread produced a thicker glue line. The size of the thickness of the glue line was shown in Table 6. The average size of the glue line based on the lowest weight of the glue spread was 31.00–38.73 μm; 59.22–67.95 μm; and 75.57–91.97 μm.

Observation of the glue line plane at a higher magnification showed that the adhesive and the bamboo slats were fused together on the surface of the bamboo slats, where there was a contrast between the adhesive and the bamboo slats (Fig. 5). In laminated bamboo without and with compression, there was no adhesive that was able to penetrate into the cavity or other bamboo components, so that the bamboo gluing occurred was more likely to be a surface bond strength without penetration of the adhesive into the pores on the bamboo slats. The ability of the adhesive to penetrate the wood pores was strongly influenced by its viscosity value [22]. The higher the viscosity, the more difficult the ability to wet and penetrate into the voids of the bonded surface [6], but if the adhesive has a low viscosity, it caused over-penetration of the voids on the surface and lack of glue lines formed [21]. The viscosity of the

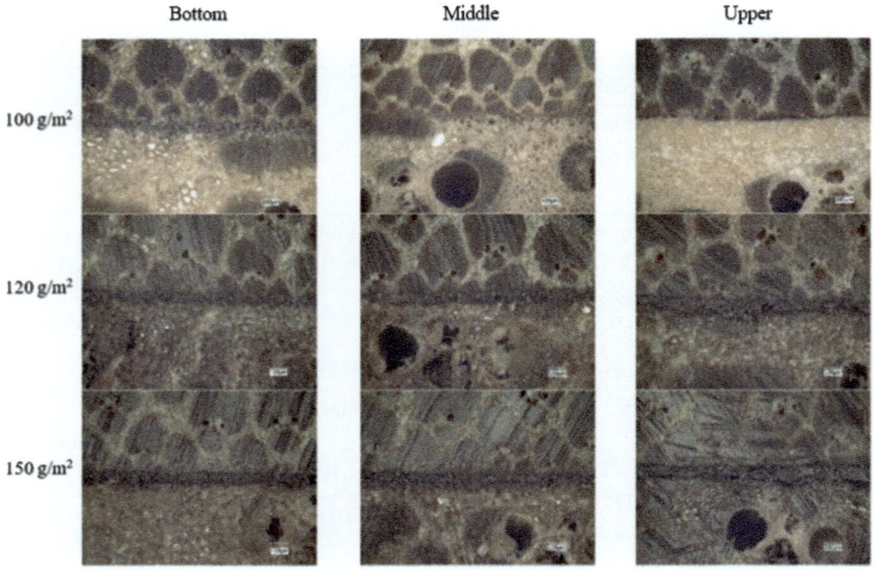

Fig. 4 Cross section of laminated bamboo without compression at 250 × magnification

Table 6 Laminated bamboo glue line thickness without compression

Glue spread (g/m^2)	Glue line thickness (μm)		
	Bottom	Middle	Upper
100	38.73 (4.46)	31.00 (6.23)	33.76 (17.55)
120	65.65 (10.92)	59.22 (5.57)	67.95 (14.54)
150	75.68 (11.75)	75.57 (11.36)	91.97 (15.52)

epoxy adhesive used was similar to that of PVA adhesive and polyurethane which has a viscosity value of 134–150 ps [8, 23] this value was higher when compared to the Isocyanate type of wood adhesive which has a viscosity of around 2.94 ps [6].

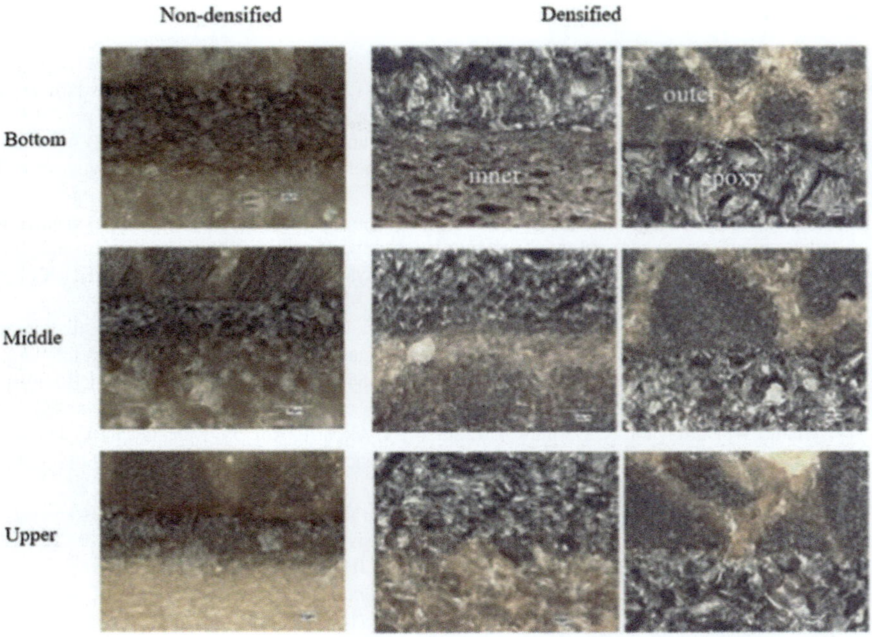

Fig. 5 Cross section of laminated bamboo without and with compression at 1000 × magnification

4 Conclusion

Based on the analysis of the results of the immersion, boiling and shear strength tests, it was known that the effect of 120 g/m^2 and 150 g/m^2 of glue spread weight on the delamination value and shear strength was not different. Therefore, the use of adhesive with a glue spread weight of 120 g/m^2 can be chosen as the optimal weight of the glue spread.

Laminated bamboo with compression has a moisture content of 5.6–6.39%, an average density of 0.96–0.99 g/cm^3, no delamination after immersion or boiling testing and has an average shear strength of 6.2–7.35 MPa. The absence of delamination and the value of shear strength that met the standard indicated that laminated bamboo from curved cross-sectional slats that had been compressed and glued with 120 g/m^2 epoxy had good adhesive qualities. The position factor has no significant effect on the value of the density and shear strength of laminated bamboo. The values of density and shear strength were not different between samples made from the bottom, middle, and upper part. This showed that the compression of the bamboo slats at the bottom and middle by 50% and 30% succeeded in making the characteristics of the laminated bamboo uniform.

Acknowledgements We thank the Japan Society for the Promotion of Science (JSPS) for partial funding of the publication fee.

References

1. Agustina A, Nugroho N, Bahtiar ET, Hermawan D (2015) Characteristic of cross laminated bamboo as structural composite material. J Teknol Indust Pert 25(2):174–181
2. Anwar UMK, Paridah MT, Hamdan H, Zaidon A, Hanim AR, Nordahlia AS (2012) Adhesion and bonding properties of low molecular weight phenol formaldehyde-treated plybamboo. J Trop Forest Sci 24(3):379–386
3. Budiana IBGP, Kusmawan D, Rusli (2014) Bambu laminasi sistem bilah lengkung dan penambahan air suling sebagai optimasi polymer isocyanate. Di dalam: Suwinarti W, Kusuma IW, Erwin, Ismail (ed) Prosiding Seminar Nasional Masyarakat Peneliti Kayu (MAPEKI) XVI, 6 Nov 2013. Balikpapan, Indonesia. Balikpapan (ID): MAPEKI. hlm 103–112
4. Darmawan T, Sudarmanto, Jayadi, Amin Y, Lestari E, Dwianto W (2016) Karakteristik bambu laminasi dai bilah berpenampang lengkung dan pengaruh perlakuan uap pada bilah lengkung terpadatkan. Laporan Teknik (Technical Report) 2016 Pusat Penelitian Biomaterial LIPI. Pusat Penelitian Biomaterial LIPI, Bogor
5. Dwianto W, Damayanti R, Darmawan T, Sejati PS, Akbar F, Adi DS, Bahadawan A, Amin Y, Triwibowo (2020) Bending strength of lignocellulosic materials in softening condition. IJFR 7(1):59–70
6. Fatriasari W, Ruhendi S (2010) Fortification of natural rubber lateks-styrene with isocyanate: characteristics and its application for plywood adhesive. JITKT 8(2):121–132
7. Febrianto F, Sahroni HW, Bakar ES, Kwon GJ, Kwon JH, Hong SI, Kim NH (2012) Properties of oriented strand board made from Betung bamboo (*Dendrocalamus asper* (Schultes.f) Backerex Heyne). Wood Sci Technol 46:53–62. https://doi.org/10.1007/s00226-010-0385-8
8. Kusumah SS, Subiyanto B, Massijaya MY (2011) Optimizing of composite board production made from wood waste and bamboo. Widyariset 14(2):415–422
9. Laine K, Segerholm K, Wålinder M, Rautkari L, Hughes M (2016) Wood densification and thermal modification: hardness, set-recovery and micromorphology. Wood Sci Technol 50(5):883–894. https://doi.org/10.1007/s00226-016-0835-z
10. Li H, Zhang Q, Huang D, Deeks AJ (2013) Compressive performance of laminated bamboo. Compos Part B 54:319–328
11. Lu Z, Zhou H, Liao Y, Hu C (2018) Effects of surface treatment and adhesives on bond performance and mechanical properties of cross-laminated timber (CLT) made from small diameter *Eucalyptus* timber. Const Build Mater 161:9–15. https://doi.org/10.1016/j.conbuildmat.2017.11.027
12. (MAFF) Minister of Agriculture Forestry and Fisheries (2007) Japanese agricultural standard for glued laminated timber 1152 Tokyo (JP)
13. Malanit P, Barbu MC, Frühwald A (2011) Physical and mechanical properties of oriented strand lumber made from an Asian bamboo (*Dendrocalamus asper* Backer). Eur J Wood Wood Prod 69:27–36
14. Maulana S, Busyra I, Fatrawana A, Hidayat W, Sari RK, Sumardi I, Wistara INJ, Lee SH, Kim NH, Febrianto F (2017) Effects of steam treatment on physical and mechanical properties of bamboo oriented strand board. J Korean Wood Sci Technol 45(6):872–882
15. Mujiman PH, Sulistyo D, Prayitno TA (2014) Influence of shape and dimensions of lamina on shear and bending strength of vertically glue laminated bamboo beam. Procedia Eng 95(2014):22–30
16. Ni L, Zhang X, Liu H, Sun Z, Song G, Yang L, Jiang Z (2016) Manufacture and mechanical properties of glued bamboo laminates. BioResources 11(2):4459–4471
17. Nugroho N, Ando N (2000) Development of structural composite products made from bamboo I: fundamental properties of bamboo zephyr board. J Wood Sci 46:68–74
18. Nugroho N, Ando N (2001) Development of structural composite products made from bamboo II: fundamental properties of laminated bamboo lumber. J of Wood Sci 47(3):237–242
19. Qingfang L, Wang W, Liu Y (2019) Flexural performance of cross-laminated bamboo (CLB) slabs and CFRP grid composite CLB slabs. Adv Civil Eng 2019:1–17

20. Rosalita Y, Nugroho N, Subiyanto B, Kusumah SS (2009) Pola distribusi kekuatan mekanik bambu betung *(Dendrocalamus Asper)* dan bambu Sembilang *(Dendrocalamus Gigantochloa)*. Di dalam: (editor tidak diketahui). Prosiding Seminar Nasional Masyarakat Peneliti Kayu (MAPEKI) XII; 23–25 Juli 2009; Bandung, Indonesia. Bandung (ID): MAPEKI. hlm 78–85
21. Ruhendi S, Koroh DN, Syamani FA, Yanti H, Nurhaida, Saad S, Sucipto T (2007) Analisis Perekatan Kayu. Fakultas Kehutanan. Institut Pertanian Bogor, Bogor
22. Santoso A, Latifah HH (2019) Formulation optimization on tannin phenol formaldehyde adhesive by XRD and DTA methods. J ITEKIMA 5(1):48–59
23. Sihombing RP, Ngatin A (2019) Modifikasi homopolimer poli (vinil asetat) dengan variabel hidrofobisitas emulsifier untuk aplikasi perkayuan. J Fluida 12(2):72–77
24. Steel RGD, Torrie JH (1995) Prinsip dan Prosedur Statistika. Ed ke–2. Terjemahan: B. Sumantri. Gramedia Pustaka Utama, Jakarta (ID)
25. Sulastiningsih IM, Nurwati H (2009) Physical and mechanical properties of laminated bamboo board. J Trop For Sci 21(3):51–246
26. Sumardi I, Suzuki S, Rahmawati N (2015) Effect of board type on some properties of bamboo strandboard. J Math Fund Sci 47(1):51–59
27. Sun X, He M, Li Z (2020) Novel engineered wood and bamboo composites for structural applications: state-of-art of manufacturing technology and mechanical performance evaluation. Const Build Mater 249:1–22. https://doi.org/10.1016/j.conbuildmat.2020.118751
28. Sutiyono (2010) Penggunaan efektif batang enam jenis bambu sebagai bahan baku bambu lamina. Di dalam: [editor tidak diketahui]. Prosiding Seminar Nasional Masyarakat Peneliti Kayu (MAPEKI) XIII; 10–11 Nopember 2010; Denpasar, Indonesia. Denpasar (ID): MAPEKI. hlm 555–560
29. Verma CS, Chariar VM (2012) Stiffness and strength analysis of four layered laminate bamboo composite at macroscopic scale. Compos Part B 45(1):369–376. https://doi.org/10.1016/j.compositesb.2012.07.048.
30. Wang X, Cheng D, Huang X, Song L, Gu W, Liang X, Li Yand XuB (2020) Effect of high-temperature saturated steam treatment on the physical, chemical, and mechanical properties of moso bamboo. J Wood Sci 66(52):1–9. https://doi.org/10.1186/s10086-020-01899-8
31. Yu HQ, Jiang ZH, Hse CY, Shupe TF (2008) Selected physical and mechanical properties of moso bamboo *(Phyllostachys pubescens)*. J Trop Forest Sci 20(4):258–263
32. Zhang Q, Shenxue J, Yongyu T (2002) Industrial utilization on bamboo. Technical Report No. 26. The International Network for Bamboo and Rattan (INBAR), Beijing

Prediction of Physical and Mechanical Properties of 5-years Old Platinum Teakwood by Near Infrared Spectroscopy

Danang Sudarwoko Adi, Setiawan Khoirul Himmi, and Wahyu Dwianto

Abstract This study investigated the possible use of Near Infrared (NIR) spectroscopy in combination with Partial Least Square Regression (PLS-R) to build a prediction model of density, Modulus of Rupture (MOR), and Modulus of Elasticity (MOE) of 5-years old Platinum teakwood. The results showed that the model for density and MOR is adequate to predict the value in the future, but the MOE prediction model is relatively fair. Root Means Square Error validation (RMSEv) and coefficient determination of validation (R^2 validation) were 0.028, and 0.77 for density, 149.41 and 0.71 for MOR, and 2.01×10^4 and 0.53 for MOE, respectively. The bands at (1) 7315 cm^{-1} cellulose; (2) 6281 cm^{-1} crystalline bonds in cellulose; (3) 5980 cm^{-1} lignin; (4) 5795 cm^{-1} lignin or C–H stretching of CH_2-groups; (5) 4686 cm^{-1} lignin and extractive; (6) 4288 cm^{-1} hemicellulose and xylans were the main chemical affects the density. The peaks at (1) 7000 cm^{-1} cellulose/amorphous region in cellulose; (2) 5980 cm^{-1} lignin; (3) 5795 cm^{-1} lignin; (4) 5464 cm^{-1} semi-crystalline/crystalline region in cellulose; (5) 4195 cm^{-1} Lignin were the main factor for MOR. While cellulose is the main factor for the MOE prediction models. There were: (1) 6281 cm^{-1} crystalline bonds in cellulose; (2) 5800 cm^{-1} furanose/pyranose due to hemicellulose; (3) 5618 cm^{-1} cellulose; (4) 4808 cm^{-1} semi-crystalline or crystalline region in cellulose; (5) 4795 cm^{-1} cellulose.

Keywords Platinum teak · NIR spectroscopy · PLS-R · MOR · Density

---

D. S. Adi (✉)
Research Center for Applied Botany, National Research and Innovation Agency (BRIN), Cibinong, Bogor, Indonesia
e-mail: danang.sudarwoko.adi@brin.go.id

S. K. Himmi
Research Organization for Life Sciences and Environment, National Research and Innovation Agency (BRIN), Cibinong, Bogor, Indonesia

W. Dwianto
Research Center for Biomass and Bioproducts, National Research and Innovation Agency (BRIN), Cibinong, Bogor, Indonesia

M. A. Rahandi Lubis (ed.), *Proceedings of the 10th International Symposium for Sustainable Humanosphere*, Lecture Notes in Civil Engineering 255,
https://doi.org/10.1007/978-981-19-1422-5_15

1 Introduction

Timber has higher economic benefits to function as a structural material than other utilizations, such as for pulp and paper making, wood charcoal, composite, etc. Among thousands of wood species, *Tectona grandis*, namely teakwood, is one of the valuable raw materials for structural timber in Indonesia. Not only it has a beautiful pattern on its surface, but it also provides good mechanical quality as well as good durability against termites and fungi [1, 2]. However, the growth of teak species is categorized as a slow-growing wood since the rotation of teak plantation reaches more than 60-years old [3]. On the other hand, the increasing demand for this wood for wood industries over the years cannot be fulfilled due to the limited availability in the field.

The studies to overcome the issues regarding growth rotation, harvesting time, and quality of teakwood as a structural timber have been conducted by many scientists, such as using the genetic improvement [4, 5] and utilizing the short-harvested teak from both conventional and community forest [3, 6]. A previous study on the Platinum teakwood, a fast-growing teakwood variety developed by the Indonesian Institute of Sciences (LIPI) through tissue cultured wood, had revealed that the 5-years old Platinum teak was promising as structural timber since its quality was almost similar to that of the conventional one [7].

As a new categorized fast-growing wood, the development and evaluation of Platinum teakwood is needed. The monitoring of wood quality properties is required even since there are in the early growing stage as guidance to manage the stands [8] and to preserve the quality for future forest plantations [9]. Generally, quality testing is determined by destructive analysis, which means that the wood should be felled to get the data. This approach is difficult to employ since there are limiting numbers of newly developed wood. Therefore, studying a non-destructive method to analyze the wood quality is urgently required.

NIR spectroscopy is well known as a tool to analyze a material without causing damage to the sample [10]. It has a unique characteristic since it has analytical capabilities that can be used for qualification and quantification [11]. In recent years, NIR spectroscopy is widely used for monitoring and predicting the properties of wood and wood-derived products to provide quality assurance in the wood industries, such as the chemical, physical, mechanical, and anatomical properties [12–21]. One of the methods to build a prediction model, which is widely applied on the wood properties, is Partial Least Square Regression (PLS-R) [14, 22]. All prediction model of the wood properties by using the combination of NIR spectroscopy and PLS-R as an analytical method is satisfied with the good prediction to build a model. Therefore, this study investigated the possibility of NIR spectroscopy to predict the physical and mechanical traits of 5-years old Platinum teakwood and to determine the regression model of their quality.

2 Materials and Method

2.1 Materials

This study used 5-years old Platinum teakwood collected from teak plantations, established at the experimental area of the Cibinong Science Center—Botanical Garden, Bogor, Indonesia. The data of physical and mechanical properties, such as density, Modulus of Rupture (MOR), and Modulus of Elasticity (MOE), were adopted from the previous study [7] since this analysis was a series to monitor the growth and the quality of Platinum teakwood. The parameter of the sample was radial and axial direction. The destructive test was conducted to get the standard or as initial data for density, MOR, and MOE.

2.2 NIR Spectroscopy Analysis

This research used some part of the teakwood samples from physical and mechanical analysis. NIR spectra recording was conducted on both radial wood surfaces by using the Fourier Transform Near-Infrared machine, PerkinElmer Spectrum 100 N (PerkinElmer Inc., Massachusetts, USA). The spectra were collected at the wavenumber of 10,000–4000 cm^{-1}, 32 scans, and a resolution of 16 cm^{-1} per record process [23].

2.3 Spectra Pretreatment and Prediction Model

The further analysis used only the spectra at the wavenumber of 8000–4000 cm^{-1} since there was no significant information on the range more than 8000 cm^{-1} [24]. In addition, overlapping band spectra on the original spectra contains featureless due to the light scattering effect during scanning processes [13]. Thus, the original spectra were derivatized by the Savitzky-Golay method [25] to get the second derivative spectra, and then, it was smoothened at nine points with fifth polynomial order. The PLS-R was used to build a prediction model of physical and mechanical properties of 5-years old Platinum teakwood. Multivariate analysis was performed on the Unscrambler X (CAMO Software, Inc., USA). Only the best model and factor will be discussed further in this study that is maximizing the R^2v (coefficient determination of validation) and minimizing the value of Root Means Square Error of validation (RMSEv).

3 Results and Discussion

Figure 1 showed the original spectra of all teakwood samples at the wavenumber of 10,000–4000 cm^{-1}. At first glance, the original spectra were identical and there were also difficult to find the valuable peaks due to the bands overlapping. However, the absorbance level was different due to the varied density of the samples [26]. The relationship of density and absorbance was in the linear correlation. The denser wood has a small light scattering pattern; thus the absorbance is high [27]. Therefore, preprocessing on the original spectra is required to reveal the important peak of the sample through reducing the effect of light scattering, such as using spectral derivatives [13]. The derivatives have also helped the visualization of the important bands [20]. The results showed that the second derivative spectra demonstrated the discrepancies of the band spectra, which may contribute to the prediction model (Fig. 2). Previous research on the 5-years old Platinum teakwood proposed that the trend of density on the radial and axial direction were raised from the pith to outer part as well as from the bottom to the upper part of the tree.

The best combination of prediction models for the density of 5-years old Platinum teakwood was reached at factor 4. The value of RMSE calibration (RMSEc), RMSE validation (RMSEv), R^2 calibration, R^2 validation were 0.021, 0.028, 0.86, and 0.77, respectively. The RMSEc represents the goodness of fit at the model, while the RMSEv is a tool to show the predictive ability of the model [28]. Figure 3 showed the result of the prediction model for the density. The coefficient of determination of the building model was 0.77, indicated that the model was adequate to predict the value of density.

The best building model of MOR and MOE were achieved at factors 5 and 6, respectively. The value of RMSEc, RMSEv, R^2 calibration, R^2 validation for were 73.77, 149.41, 0.88, and 0.71, respectively, while for the MOE, those value were

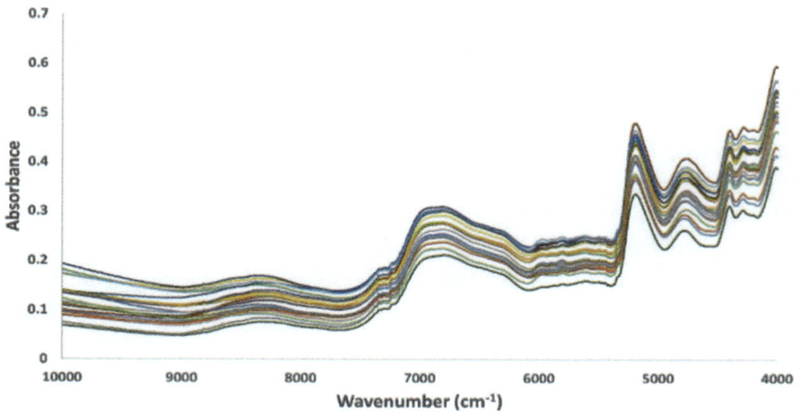

Fig. 1 The original NIR spectra of all samples on the radial and axial direction of 5-years old Platinum teakwood at the wavenumber of 10,000–4000 cm^{-1}

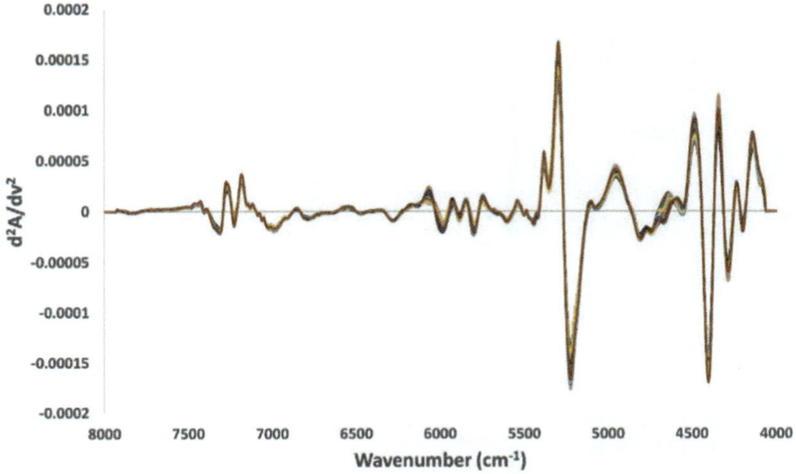

Fig. 2 The second derivative NIR spectra of radial and axial 5-years old Platinum teakwood at the wavenumber of 8000–4000 cm^{-1}

Fig. 3 Predicted versus measured density based on the data validation of PLS-R at the wavenumber of 8000–4000 cm^{-1}

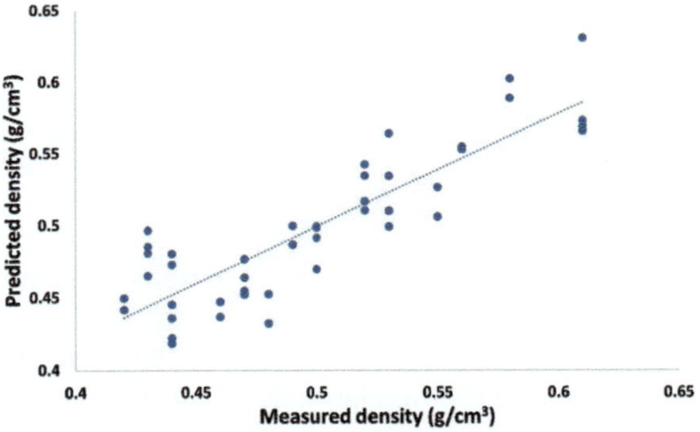

1.11 × 10^4, 2.01 × 10^4, 0.85, and 0.53, respectively. The prediction models of the MOR were almost similar to those densities, and it was a good model to predict the MOR value. However, MOE predictive models were relatively low due to the determination coefficient was only 0.53 (Figs. 4 and 5).

The results suggested that the best combination model for the density was reached at factor 4, and thus, the loading of factor 4 could be used to estimate the chemical component of wood that affects the prediction (Fig. 6). The relationship between the second derivative spectra and the loading revealed that there were 6 band spectra which had effect to the model based on the band spectra proposed by Schwanninger et al [24]: (1) 7315 cm^{-1} assigned to cellulose; (2) 6281 cm^{-1} assigned crystalline bonds in cellulose; (3) 5980 cm^{-1} assigned lignin; (4) 5795 cm^{-1} assigned lignin or C–H stretching of CH_2-groups; (5) 4686 cm^{-1} assigned lignin and extractive; and (6) 4288 cm^{-1} assigned hemicellulose and xylans. The main chemical component that

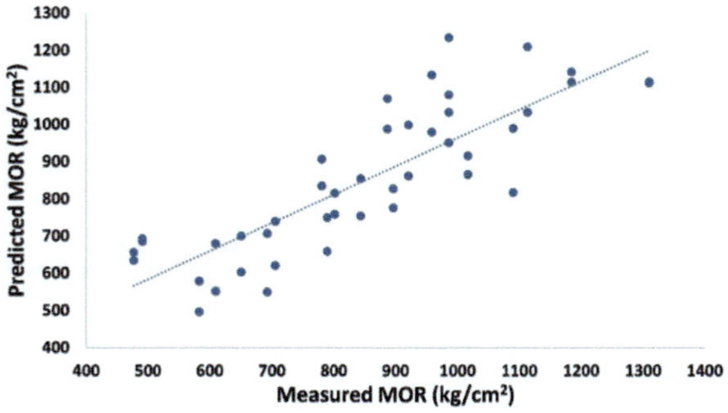

Fig. 4 Predicted versus measured MOR based on the data validation of PLS-R at the wavenumber of 8000–4000 cm^{-1}

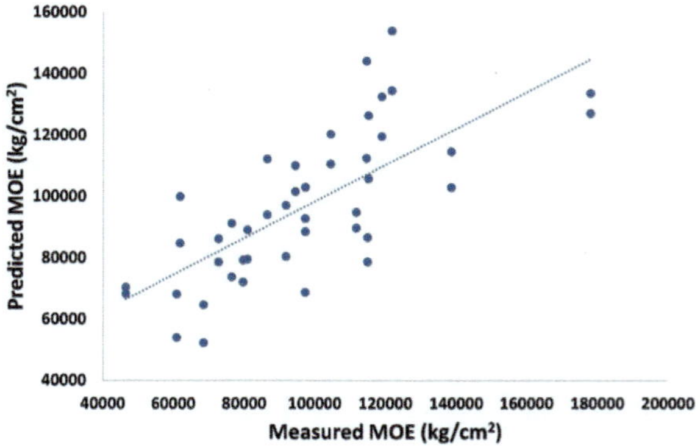

Fig. 5 Predicted versus measured MOE based on the data validation of PLS-R at the wavenumber of 8000–4000 cm^{-1}

affects wood density was lignin. The trend in the radial was increased from pith to bark, but at the axial direction, the density was decreased to the top of trees [29, 30].

The loading of the best model for MOR and MOE was achieved at factors 5 and 6, respectively. Similar to those densities, these loadings were used to find the wood chemical component that affects the prediction model (Fig. 7). At least 5 peaks of the spectra were assigned to the wood chemicals that were found on the loading factor 5 for MOR prediction model: (1) 7000 cm^{-1} cellulose/amorphous region in cellulose; (2) 5980 cm^{-1} lignin; (3) 5795 cm^{-1} lignin; (4) 5464 cm^{-1} semi-crystalline/crystalline region in cellulose; and (5) 4195 cm^{-1} lignin. The results indicated that lignin and semi-crystalline, crystalline, and amorphous region of the cellulose were the main factors to the MOR. Fiber frequency and the distribution of the soft part of the cell have an effect on the static bending [31]. Cellulose was the main factor for the MOE prediction models. There were: (1) 6281 cm^{-1} crystalline bonds in cellulose; (2) 5800 cm^{-1} furanos/pyranose due to hemicellulose; (3) 5618 cm^{-1} cellulose; (4) 4808 cm^{-1} semi-crystalline or crystalline region in cellulose; and (5) 4795 cm^{-1} cellulose.

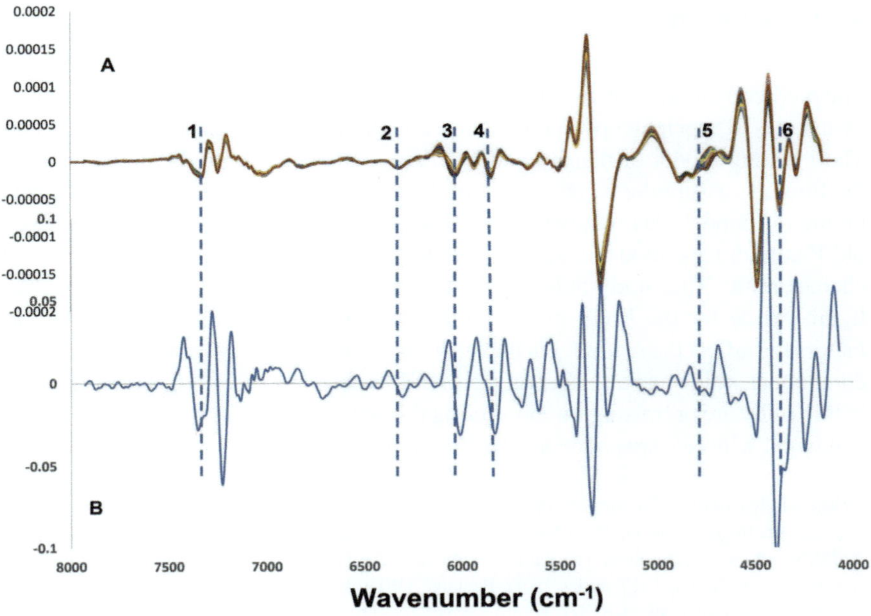

Fig. 6 The relationship between the second derivative spectra of all samples of 5-years old Platinum teakwood **a** and the loading at the best prediction model of density at factor 4 **b**

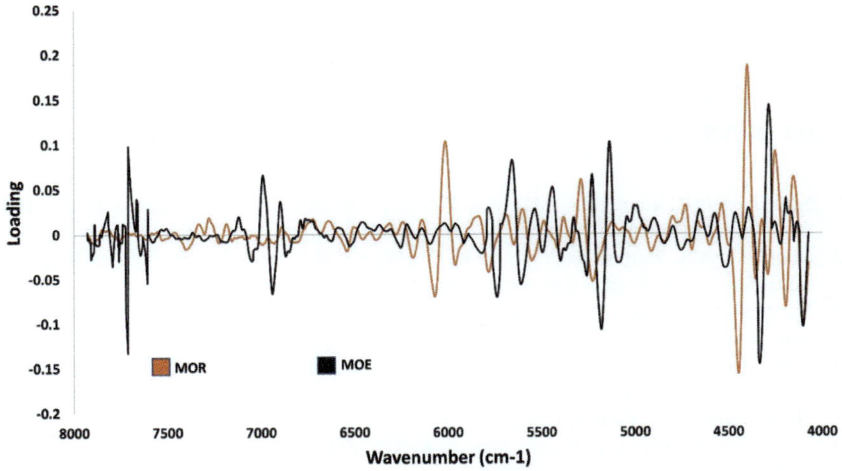

Fig. 7 The loading at the best combination of MOR and MOE prediction models

4 Conclusion

The prediction model for density, MOR, and MOE of 5-years old Platinum teakwood by using NIR spectroscopy in combination with PLS-R was introduced in this study. The building model of density and MOR was adequate to predict those values in the future to assure the wood quality of the young plantation. In addition, the MOE prediction model was relatively low. However, our previous study on the 5-years old Platinum teakwood showed that MOR has a linear correlation with their MOE characteristic. This study indicated that the main factor for the density model was lignin, while for the MOR prediction model, lignin and cellulose were the main factors that affect the models. On the other hand, holocellulose was the main chemical component which had an effect on the MOE models. In general, the building models in this study are promising for monitoring the newly developed species like Platinum teakwood with low specimen availability in the field.

Acknowledgements The authors thank the Indonesian Institute of Sciences (LIPI) for the financial support through the DIPA of the Deputy of Life Sciences. The authors also thank the Japan-ASEAN Science Technology Innovation Platform (JASTIP) Program, a collaborative project of Kyoto University and LIPI under Work-Package (WP) 3: Bioresources and Biodiversity for the support on data analysis at RISH, Kyoto University. The authors are indebted to Professor Junji Sugiyama, Professor Tomoya Imai, and DR. Sung-Wook Hwang for their assistance during the laboratory work at RISH, Kyoto University. The authors extend their gratitude to the Japan Society for the Promotion of Sciences (JSPS) and JSPS Alumnae Association Indonesia (JAAI) for their support on the conference grant.

References

1. Soerianegara I, Lemmens RHMJ (1994) Prosea, plant resources of South-East Asia 5(1). Timber trees: major commercial timbers. Prosea, Bogor
2. Martawijaya A, Kartasujana I, Kadir K, Prawira S (2005) Atlas kayu Indonesia Jilid I. Balai Penelitian dan Pengembangan Kehutanan. Departemen Kehutanan
3. Darmawan W, Nandika D, Sari RK, Sitompul A, Rahayu I, Gardner D (2015) Juvenile and mature wood characteristics of short and long rotation teak in Java. IAWA J 36:428–442
4. Solorzano S, Moya R, Murillo O (2012) Early prediction of basic density, shrinking, presence of growth stress, and dynamic elastic modulus based on the morphological tree parameters of *Tectona grandis*. J Wood Sci 58:290–299
5. Hidayati F, Ishiguri F, Iizuka K, Makino K, Takashima Y, Danarto S, Winarni WW, Irawati D, Na'iem M, Yokota S (2013) Variation in tree growth characteristics, stress-wave velocity, and Pilodyn penetration of 24-year-old teak (*Tectona grandis*) trees originating in 21 seed provenances planted in Indonesia. J Wood Sci 59:512–516
6. Rizanti DE, Darmawan W, George B, Merlin A, Dumarcay S, Chapuis H, Gérardin C, Gelhaye E, Raharivelomanana P, Kartika Sari R, Syafii W, Mohamed R, Gerardin P (2018) Comparison of teak wood properties according to forest management: short versus long rotation. Ann For Sci 75
7. Adi DS, Sudarmanto, Ismadi, Gopar M, Darmawan T, Amin Y, Dwianto W, Witjaksono (2016) Evaluation of the wood quality of platinum teak wood. Teknol Indones 39:36–44

8. Kojima M, Yamamoto H, Marsoem SN, Okuyama T, Yoshida M, Nakai T, Yamashita S, Saegusa K, Matsune K, Nakamura K, Inoue Y, Arizono T (2009) Effects of the lateral growth rate on wood quality of *Gmelina arborea* from 3.5-, 7- and 12-year-old plantations. Ann For Sci 66:507–507
9. Nocetti M, Rozenberg P, Chaix G, MacChioni N (2011) Provenance effect on the ring structure of teak (*Tectona grandis* L.f.) wood by X-ray microdensitometry. Ann For Sci 68:1375–1383
10. Ito N, Okubo N, Kurata Y (2019) Nondestructive near-infrared spectroscopic analysis of oils on wood surfaces. Forests 10:1–7
11. Tsuchikawa S, Kobori H (2015) A review of recent application of near infrared spectroscopy to wood science and technology. J Wood Sci 61:213–220
12. Ban M, Inagaki T, Ma T, Tsuchikawa S (2018) Effect of cellular structure on the optical properties of wood. J Near Infrared Spectrosc 26:53–60
13. Fahey LM, Nieuwoudt MK, Harris PJ (2018) Using near infrared spectroscopy to predict the lignin content and monosaccharide compositions of *Pinus radiata* wood cell walls. Int J Biol Macromol 113:507–514
14. Elle O, Richter R, Vohland M, Weigelt A (2019) Fine root lignin content is well predictable with near-infrared spectroscopy. Sci Rep 9:1–11
15. Hodge GR, Acosta JJ, Unda F, Woodbridge WC, Mansfield SD (2018) Global near infrared spectroscopy models to predict wood chemical properties of *Eucalyptus*. J Near Infrared Spectrosc 26:117–132
16. Kokutse A-D, Brancheriau L, Chaix G (2010) Rapid prediction of shrinkage and fibre saturation point on teak (*Tectona grandis*) wood based on near-infrared spectroscopy. Ann For Sci 67:403–403
17. Saha UK, Sonon L, Kane M (2017) Prediction of calorific values, moisture, ash, carbon, nitrogen, and sulfur content of pine tree biomass using near infrared spectroscopy. J Near Infrared Spectrosc 25:242–255
18. Segla KN, Rabiou H, Adjonou K, Guibal D, Bationo AB, Mahamane A, Kokou K, Langbour P, Kokutse AD, Chaix G (2017) Useful near infrared spectroscopy model calibrations on solid wood samples of *Pterocarpus erinaceus* (Poir.) for physical, mechanical and colour properties. J Near Infrared Spectrosc 25:256–266
19. Inagaki T, Schwanninger M, Kato R, Kurata Y, Thanapase W, Puthson P, Tsuchikawa S (2012) *Eucalyptus camaldulensis* density and fiber length estimated by near-infrared spectroscopy. Wood Sci Technol 46:143–155
20. Sandak J, Sandak A, Meder R (2016) Assessing trees, wood and derived products with near infrared spectroscopy: hints and tips. J Near Infrared Spectrosc 24:485–505
21. Costa LR, Trugilho PF, Hein PRG (2018) Evaluation and classification of eucalypt charcoal quality by near infrared spectroscopy. Biomass Bioenerg 112:85–92
22. Inagaki T, Yonenobu H, Asanuma Y, Tsuchikawa S (2018) Determination of physical and chemical properties and degradation of archeological Japanese cypress wood from the Tohyamago area using near-infrared spectroscopy. J Wood Sci 64:347–355
23. Hwang SW, Horikawa Y, Lee WH, Sugiyama J (2016) Identification of *Pinus* species related to historic architecture in Korea using NIR chemometric approaches. J Wood Sci 62:156–167
24. Schwanninger M, Carlos J, Fackler K (2011) A review of band assignments in near infrared spectra of wood and wood. J Near Infrared Spectrosc 19:287–308
25. Savitzky A, Golay MJE (1964) Smoothing and differentiation of data by simplified least squares procedures. Anal Chem 36:1627–1639
26. Via B, McDonald T, Fulton J (2012) Nonlinear multivariate modeling of strand density from near-infrared spectra. Wood Sci Technol 46:1073–1084
27. Ma T, Inagaki T, Ban M, Tsuchikawa S (2019) Rapid identification of wood species by near-infrared spatially resolved spectroscopy (NIR-SRS) based on hyperspectral imaging (HSI). Holzforschung 73:323–330
28. Schwanninger M, Rodrigues JC, Gierlinger N, Hinterstoisser B (2011) Determination of lignin content in Norway spruce wood by Fourier transformed near infrared spectroscopy and partial least squares regression. Part 1: wavenumber selection and evaluation of the selected range. J Near Infrared Spectrosc 19:319–329

29. Panshin AJ, de Zeeuw C (1980) Textbook of wood technology. Structure, identification, properties, and uses of the commercial woods of the United States and Canada. McGraw-Hill Company, New York (1980)
30. Bowyer JL, Haygreen JG, Schmulsky R (2003) Forest products and wood science: an introduction. Iowa State Press, USA
31. Tsoumis G (1991) Science and technology of wood. Structure, properties, utilization. Van Nostrand Reinhold, New York

Durability Characteristics of *Alstonia scholaris* Wood Treated with *Cerbera manghas* Seed Ethanol Extract Against *Schizophyllum commune* Attack

Riana Anggraini, Jauhar Khabibi, Albayudi, and Via Elva Riani

Abstract *Alstonia scholaris* wood has a high potential as a raw material of the forestry industry, such as veneer, crates, matches, heels, masks, concrete molds, and pulp. The low durability of *A. scholaris* wood is a weak point that must be improved. One technique for improving the *A. scholaris* wood durability is giving it a treatment using ethanol extract of *Cerbera manghas* seed, which has antifungal and antibacterial abilities. The purpose of this study was to analyze the durability characteristics of *A. scholaris* wood treated with *C. manghas* seed ethanol extract. The experiment was carried out with a combination of *C. manghas* seed ethanol extract concentration with 5 levels (0%, 20%, 40%, 60%, and 80%). The durability analysis using the *Schizophyllum commune* for 12 weeks (SNI 01-7207-2006). The result shows that the differences in concentration of *C. manghas* seed ethanol extract had a significant effect on retention, absorption, and weight loss of *A. scholaris* wood. The giving of *C. manghas* seed ethanol extract concentrations both of 40% and 80% was able to increase the durability of *A. scholaris* wood. These concentrations increase the durable class into III and I, respectively. GC-MS analysis showed that the *C. manghas* seed ethanol extract contained 35 types of compounds. The five dominant types of compounds are: Octadecane (CAS) n-Octadecane, Nonadecane (CAS) n-Nonadecane, Heneicosane (CAS) n-Heneicosane, Tetratetracontane (CAS) n-Tetratetracontane, and Tetracontane, 3, 5, 24-trimethyl—(CAS).

Keywords *Schizophyllum commune* · *Cerbera manghas* seed · Durability · *Alstonia scholaris* · Ethanol extract

---

R. Anggraini (✉) · J. Khabibi · Albayudi · V. E. Riani
Forestry Department, Jambi University, Jambi, Indonesia
e-mail: nanuk_onra@yahoo.co.id

J. Khabibi
Center of Excellence for Land-Use Transformation Systems, Universitas Jambi, Jambi, Indonesia

M. A. Rahandi Lubis (ed.), *Proceedings of the 10th International Symposium for Sustainable Humanosphere*, Lecture Notes in Civil Engineering 255,
https://doi.org/10.1007/978-981-19-1422-5_16

1 Introduction

Alstonia scholaris wood is one type of wood that has the potential as an industrial raw material. Its utilization is quite large and it's a type of plant that has economic value in Indonesia. *A. scholaris* wood is used for the manufacture of veneers, crates, matches, heels, handicrafts such as wayang golek, masks, concrete molds, and pulp. *A. scholaris* wood belongs to low quality wood class with durable class IV-V [6]. Therefore, *A. scholaris* wood is easily attacked wood-destroying organisms such as fungi, insects and sea borers so that theis use is not optimal. One way to prevent the damage of the wood caused by wood-destroying organisms is by means of preservation. Preservatives from the *C. manghas* seed extract can be used as antifungal and antibacterial because they contain extractive substances that are toxic. Extractive substances are represented in almost all parts of plants such as leaves, stems, seeds and flowers. But the most extractive content contained in seeds and flowers, extractive substances contained in it, in the form of cerberin compounds, secondary metabolite compounds such as saponins, polyphenols, alkaloids, tannins and flavonoids [25].

This study used five levels of treatment concentration of *C. manghas* seed extract. This concentration variation also refers to [7] research which states that the best concentration of coarse extract of *C. manghas* leaves is 50.953% in killing mealybugs with coarse extract of *C. manghas* leaves. Determination of this concentration is done to see the difference in the durability of the wood after it is preserved and prove what concentration has the highest effectiveness in withstanding the attack of the fungus *Schizophyllum commune*.

2 Research Methods

2.1 Wood Sample Preparation

The study design used a complete randomized design with one factor, the concentration of *Cerbera manghas* seed extract using five levels 0%, 20%, 40%, 60%, and 80% [7]. The wood sample size used in this method is 5 cm × 2.5 cm × 1.5 cm [5]. The number of samples needed is 25 for analysis of retention, absorption, and weight loss, respectively. The wood samples were air-dried for 1 week, and then the moisture content was analyzed following [14, 23].

2.2 Extraction of C. manghas *Seed*

C. manghas seed extract is taken from the *C. manghas* fruit which has slightly reddish green color on the outside. After the fruit has been splited, the core seed part was taken and washed. After that, the core seeds were oven-drying at 50 °C temperature for 48 h and grinded into 40–60 mesh size [14]. Then, this fine size was extracted using the ethanol solvent (1:3) w/w in the black bottle for 3 days, and then it was filtering used filter paper. Afterward, the ethanol extract of *C. manghas* seed was stirring used a rotary evaporator at 50 °C temperature, and 50 rpm speed to increase its concentration which is called *C. manghas* seed extract [8]. Moreover, 12 ml of *C. manghas* seed extract was prepared for GC-MS (gas chromatography mass spectrometry) analysis [27].

2.3 Wood Preservation and Analysis

C. manghas seed extract was diluted into five concentrations 0%, 20%, 40%, 60%, and 80% for preserving *Alstonia scholaris* wood. Cold soaking method was used to preserve *A. scholaris* wood for 48 h [4]. Afterward, the treated wood was fed to *Schizophyllum commune* for 12 weeks [22]. The *S. commune* was bred using PDA (potato dextrose agar) [9]. After that, the weight loss analysis of *A. scholaris* wood was done following [1, 18].

Retention is the amount of *C. manghas* seed extract retained in the wood. This analysis calculated from initial wood weight (W1), treated wood weight (W2), wood volume (V), and concentration (C) of *C. manghas* seed extract using this Eq. 1 [19].

$$\text{Retention} = \left(\frac{W2 - W1}{V}\right)xC \tag{1}$$

Absorption is the amount of *C. manghas* seed extract fulfilling the wood void. This analysis calculated using modified Eq. 2 from initial wood weight (W1), treated wood weight (W2), and wood volume (V) [2].

$$\text{Absorption} = \left(\frac{W2 - W1}{V}\right) \tag{2}$$

3 Results and Discussion

3.1 Moisture Content

The moisture content of the *A. scholaris* wood shows a similarity value before it is preserved (Fig. 1). The water content ranges from 12.22% to 12.75%. The moisture content value of the material to be preserved is a very important factor to know. This is related to the successful determination of the wood preservation process. Beside that

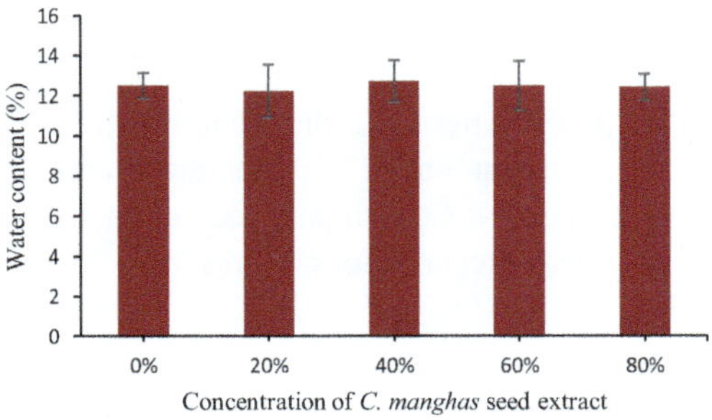

Fig. 1 Moisture content of *A. scholaris*

natural durability is affecting by moisture content, which also corelate with fungus attack [17].

Water in wood is in the cell cavity called free water and bound water in the cell walls [6]. The high moisture content condition resulted in restricted other materials or compounds to enter the wood cell wall. It can be affected the retention and absorption of *C. manghas* seed extract [3]. Therefore, the low moisture content in the wood will further support the retention and absorption of *C. manghas* seed extract. In addition, the moisture content can be an important factor for its susceotibility to fungus growth [16].

3.2 Retention of C. manghas *Seed Extract*

The results showed differences in concentrations of *C. manghas* seed extract had a significant effect on retention values (Fig. 2). Optimal treatment is at a concentration of 80%. This is because an 80% concentration provides a different retention value to the control (0%) while the other concentrations do not give a different effect. This condition, in line with [11], increasing the concentration of preservatives will increase the retention value. Previous research [13] has also reported similar results that increase in concentration of preservatives resulted in increased retention in all wood species.

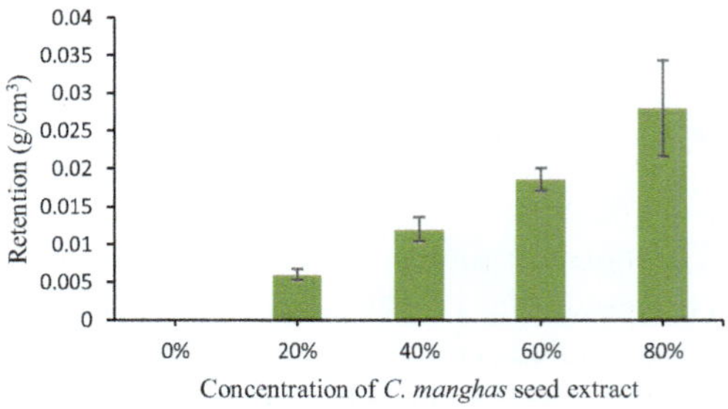

Fig. 2 Correlation between concentration of *C. manghas* seed extract toward retention

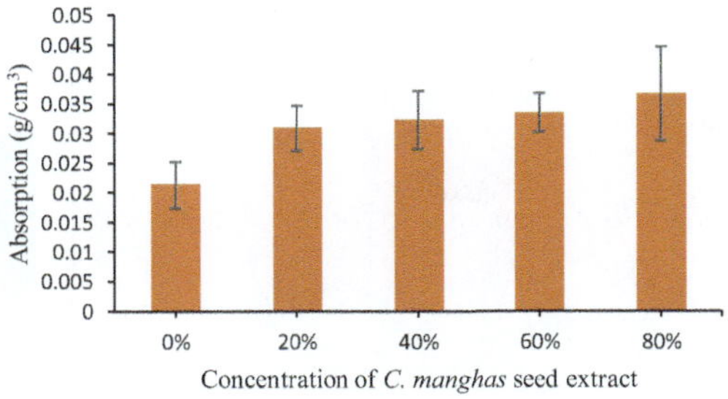

Fig. 3 Correlation between concentration of *C. manghas* seed extract toward absorption

Retention value is an important indicator in wood preservation. This value determines the amount of *C. manghas* seed extract bounded in the preserved wood. It correlates in increasing treated wood durability. For the concern, this factor is also affected by moisture content and density of wood [12, 26].

3.3 Absorption of C. manghas Seed Extract

The results showed that the concentration of *C. manghas* seed extract had a significant effect on the absorption value (Fig. 3). This analysis showed that concentrations of 20%–80% had no significant difference to the absorption result. The 20% concentration is most optimal based on the absorption value. In addition, the concentration of *C. manghas* seed extract is also directly proportional to its absorption in the wood.

The absorption value in this study shows the absorption of *C. manghas* seed extract. The higher the absorption value correlates with the more *C. manghas* seed extract are absorbed in the wood. This condition is able to provide a better protective effect on wood against wood-destroying organisms [15].

3.4 Wood Weight Loss

The results showed that the differences concentration of *C. manghas* extract had a significant effect on the wood weight loss due to the attack of *S. commune* (Fig. 4). This result phenomenon is similar to the previous study [28]. The result also shows that between the 20%, 40%, and 60% concentrations have not a significantly different effect on the wood weight loss, but these three concentrations markedly different from the 80% *C. manghas* seed extract concentration. Besides that, all the concentrations have a significantly different effect to control (0%). The concentration of 80% provides the lowest wood weight loss compared to other concentrations based on statistical tests.

Based on Indonesian National Standard [22] the wood weight loss of *A. scholaris* at the concentrations of 0% and 20% is in the IV (not resistant) wood class of

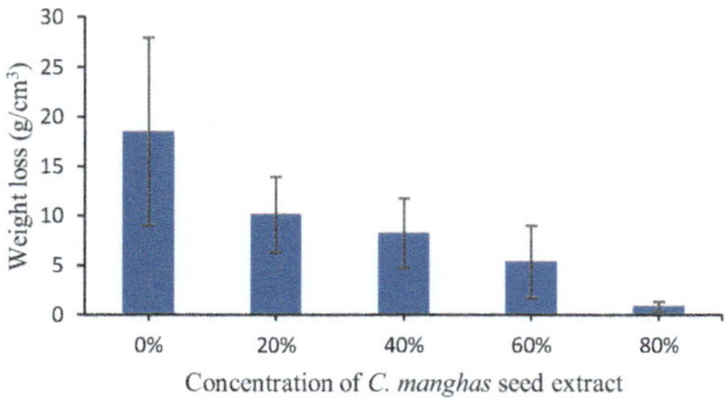

Fig. 4 Correlation between concentration of *C. manghas* seed extract toward weight loss

durability. The concentrations of 40% and 60% are in the III (moderate resistant) wood class of durability. In contrast to other concentrations, the 80% concentration of *C. manghas* can increase the wood durability into class I (extremely resistant). In line with statistical tests, a concentration of 80% is the best concentration. This condition also indicates that *C. manghas* seed extract can be used to reduce the attack of *S. commune*.

The wood weight loss due to *S. commune* attack decreases as the concentration of *C. manghas* seed extract increases (Fig. 5). This indicates that the greater the concentration, the more toxic the treated *A. scholaris* wood. The *C. manghas* seed extract is classified as a toxic material to the fungus and bacteria because it contains several chemical substances, such as cerberin, saponins, polyphenols, alkaloids, tannins, and flavonoids. Cerberin is a group of alkaloids or glycosides that are thought to play a role in insect mortality [24]. In addition, secondary metabolite compounds namely flavonoids, terpenoids, and alkaloids are also able to protect against pest and disease attacks. Including saponins and tannins are also classified as toxic compounds [24].

Fig. 5 Treated and control wood after 12 weeks attacked by *S. commune*

3.5 C. manghas *Seed Extract Compound Content*

Component of *C. manghas* seed extract compound is one of the parameters tested. This aims to ensure the type of active compound or substance that acts as an antidote to the attack of *S. commune*. GC-MS analysis showed that the *C. manghas* seed ethanol extract contained 35 types of compounds (Table 1). The five dominant types of compounds are: Octadecane (CAS) n-Octadecane, Nonadecane (CAS) n-Nonadecane, Heneicosane (CAS) n-Heneicosane, Tetratetracontane (CAS) n-Tetratetracontane, and Tetracontane, 3, 5, 24-trimethyl—(CAS). These compounds were bolded in the Table 1. Several of these compounds have an antibacterial and antifungal effect, such as Octadecane, Nonadecane, and Heneicosane [10, 20, 21].

4 Conclusion

Concentration of *C. manghas* seed ethanol extract had a significant effect on retention, absorption, and weight loss of *A. scholaris* wood. The giving of *C. manghas* seed ethanol extract concentrations both of 40% and 80% was able to increase the durability of *A. scholaris* wood. These concentrations increase the durable class into III and I, respectively. *C. manghas* seed ethanol extract contained 35 types of compounds. The five dominant types of compounds are: Octadecane (CAS) n-Octadecane, Nonadecane (CAS) n-Nonadecane, Heneicosane (CAS) n-Heneicosane, Tetratetracontane (CAS) n-Tetratetracontane, and Tetracontane, 3, 5, 24-trimethyl—(CAS).

Table 1 GC-MS result of *C. manghas* seed extract

Compound name	Area (%)	Chemical formula
Octadecane (CAS) n-Octadecane	1.39	$C_{14} H_{30}$
Tritetracontane (CAS) n-Triatetracontane	1.38	$C_{43} H_{88}$
Pentatriacontane (CAS) n-Pentatriacontane	1.95	$C_{35}H_{72}$
Eicosane, 2, 4-dimethyl	1.02	$C_{22}H_{46}$
Nonadecane (CAS) n-Nonadecane	1.14	$C_{19}H_{40}$
1-Heptanal, 3, 5, 5-Triethyl	1.61	$C_{13}H_{26}O$
Nonadecane (CAS) n-Nonadecane	0.98	$C_{19}H_{40}$
Pentadecane, 8-hexyl-(CAS) 8-n-Hexylpentadecane	1.81	$C_{21}H_{44}$
Pentacosane (CAS) n-Pentacosane	1.72	$C_{25}H_{52}$
Tetratetracontane (CAS) n-Tetratetracontane	3.58	$C_{44}H_{90}$
Docosane, 11-decyl-(CAS) 11-n-Decyldocosane	1.53	$C_{32}H_{66}$
Cyclopropane, 1-(1, 2-dimethylpropyl)-1-methyl-2-nonyl-(CAS)	3.26	$C_{18}H_{36}$
Tetrapentacontan, 1, 54-Dibromo	0.97	$C_{54}H_{10}BR_2$
Tetracosane, 11-decyl-(CAS) 11-n-Decyltetracosane	4.23	$C_{34}H_{70}$
Nonane, 2-methyl-(CAS) 2-Methylnonane	2.83	$C_{10}H_{22}$
Docosane (CAS) n-Docosane	3.79	$C_{22}H_{46}$
Hentriacontane (CAS) Untriacontane	2.53	$C_{31}H_{62}$
Heneicosane (CAS) n-Heneicosane	5.81	$C_{21}H_{44}$
Nonadecane (CAS) n-Nonadecane	5.98	$C_{19}H_{40}$
Tetracontane, 3, 5, 24-trimethyl-(CAS)	3.02	$C_{40}H_{82}$
Docosane (CAS) n-Docosane	2.22	$C_{22}H_{46}$
Tetrapentacontan, 1, 54-Dibromo	1.55	$C_{54}H_{10}BR_2$
Pentatriacontane (CAS) n-Pentatriacontane	4.05	$C_{35}H_{72}$
Tetracontane, 3, 5, 24-trimethyl-(CAS)	5.08	$C_{40}H_{82}$
Benzoic acid, 5-hydroxy-2-methoxy	2.06	$C_7H_6O_2$
Octadecane, 4-methyl-(CAS)	3.06	$C_{19}H_{40}$
Nonahexacontanoic acid (CAS) n-Nonahexacontanoic acid	2.00	$C_{69}H_{138}O_2$
Tetradecane (CAS) n-Tetradecane	5.28	$C_{14}H_{30}$
Octadecane (CAS) n-Octadecane	6.41	$C_{14}H_{30}$
Pentatriacontane (CAS) n-Pentatriacontane	2.91	$C_{35}H_{72}$
Docosane (CAS) n-Docosane	3.04	$C_{22}H_{46}$
Tetratetracontane (CAS) n-Tetratetracontane	5.58	$C_{44}H_{90}$
2-Butoxysulfonylhexadecane	2.69	$C_{12}H_{26}O_2$
Tetracontan	2.15	$C_{40}H_{82}$
Undecane, 3, 8-dimethyl-(CAS)	1.4	$C_{13}H_{28}$

Acknowledgements We thank the Japan Society for the Promotion of Science (JSPS) for partial funding of the publication fee.

References

1. Abanikanda JO, Adetogun AC, Mukhtar RB (2020) Evaluation of honey bee propolis as wood preservative using weight loss. Sci World J 15(1):45–47
2. Adetogun AC, Adegeye AO, Omole AO (2009) Evaluation of cashewnut shell liquid (CNSL) as a wood preservative using weight loss. Agric J 4(1):32–35
3. Afrouzi YM, Marzbani P, Omidvar A (2015) The effect of moisture content on the retention and distribution of nano-titanium dioxide in the wood. Ciencie y Tecnologia 17(2):385–390
4. Al Edrus SSAO, Tahir PM, Guan CT, Hua LS, Halip JA, Ilyas RA, Sapuan SM (2019) Degradation of medium density fibreboard and particleboard mechanical performance after exposed to different environmental condition. Int J Recent Technol Eng 8(24):528–532
5. Altgen M, Kyyrö S, Paajanen O, Rautkari L (2020) Resistance of thermally modified and pressurized hot water extracted Scots pine sapwood against decay by the brown-rot fungus *Rhodonia placenta*. Eur J Wood Wood Prod 78:161–171
6. Arinana DF (2009) Kualitas kayu pulai (*Alstonia scholaris*) terdensifikasi (sifat fisis, mekanis dan keawetan). Jurnal Ilmu dan Teknologi Hasil Hutan 2(2):78–88
7. Artama IPAB, Adnyana PB, Santiasa IMP (2016) Variasi konsentrasi ekstrak kasar daun bintaro (*Cerbera odollam* Gaertn.) mengakibatkan perbedaan mortalitas kutu putih (*Paracoccus marginatus*). Jurnal Penelitian Pendidikan Biologi UNDIKSHA 4(2)
8. Azwanida NN (2015) A review on the extraction methods use in medicinal plants, principle, strength and limitation. Med Aromat Plants 4(3)
9. Faisal M, Utari S, Hayvia Z, Maulana I (2019) A preliminary study of the utilization of Cu(II) modified liquid smoke to inhibit the activity of white-rot fungi (*Schizophyllum commune* FR) in a pinewood in-vitro. Int J GEOMATE 17(61):56–61
10. Ghavam M, Afzali A, Manconi M et al (2021) Variability in chemical composition and antimicrobial activity of essential oil of *Rosa damascene* Herrm. From mountainous regions of Iran. Chem Biol Technol Agric 8:22
11. Hunt GM, Garrat AG (1986) Pengawetan Kayu. Edisi 1 Cetakan 1: Penerjemah Mohamad Yusuf. Akademika Pressindo, Jakarta
12. Irbe I, Zommere Z, Kurnosova N (2020) Alternative wood preservation method: double treatment and its effectiveness against wood decay fungi. Eur J Wood Wood Prod 78:1233–1247
13. Jusoh I, Kamdem DP, Abu Samat SN (2005) Anatomical characteristics of tension and normal wood in rubber wood (*Hevea braziliensis*). In: 6th Pacific regional wood anatomy conference, Kyoto University, Japan, 1–5 Dec 2005
14. Khabibi J, Irawan B (2016) Proximate analysis of lignocellulosic material as alternative bioenergy resources. J Lignocellulose Technol 01:37–43
15. Kusumaningsih KR (2017) Sifat penyerapan bahan pengawet pada beberapa jenis kayu bangunan. CV Akademia Pressindo, Jakarta
16. Lie SK, Vestøl GI, Høibø O, Gobakken LR (2019) Surface mould growth on wood: a comparison of laboratory screening tests and outdoor performance. Eur J Wood Wood Prod 77:1137–1150
17. Mehramiz S, Oladi R, Efhamisisi D, Pourtahmasi K (2021) Natural durability of the Iranian domestic bamboo (*Phyllostachys vivax*) against fungal decay and its chemical protection with propiconazole. Eur J Wood Wood Prod 79:453–464
18. Okanlawon FB, Adegoke OA, Olatunji OA, Abiola JK (2020) Antifungal effectiveness of pyrolytic oil obtained from *Triplochiton scleroxylon* (De. Wild) sawdust on selected wood species. J Adv Microbiol 20(8):8–13

19. Owoyemi JM, Eniabiire OM, Emmanuel UO, Fuwape JA (2020) Efficacy of commonly used wood preservatives against subterranean termites in Akungba-Akoko, Ondo state, Nigeria. J Res Forest Wildl Environ 12(1):111–121
20. Rhetso T, Shubharani R, Roop MS et al (2020) Chemical constituents, antioxidant, and antimicrobial activity of *Allium chinense* G. Don. Futur J Pharm Sci 6:102
21. Saputri DD, Bintang M, Pasaribu FH (2015) Isolation and characterization of endophytic bacteria from tembelekan (*Lantana camara* L.) as antibacterial compounds producer. Curr Biochem 2(2):77–89
22. SNI 01-7207-2006 (2007) Uji ketahanan kayu terhadap organisme perusak kayu. BSN, Indonesia, Jakarta
23. TAPPI T 264 om-88 (1988) Preparation of wood for chemical analysis. TAPPI, United States of America, Norcross
24. Utami S (2010) Aktivitas insektisida bintaro (*Cerbera odollam* Gaertn.) terhadap hama *Eurema* spp. Pada Skala Laboratorium. Jurnal Penelitian Hutan Tanaman 7(4):211–220
25. Utami, S (2011) Bioaktivitas insektisida nabati bintaro (*Cerbera odollam* Gaertn.) sebagai pengendali hama *Pteroma plagiophleps* Hampson dan *Spodoptera litura*. Institut Pertanian Bogor, Bogor
26. Wang W, Zhu Y, Cao J (2013) Evaluation of copper leaching in thermally modified southern yellow pine wood impregnated with ACQ-D. BioResources 8(3):4687–4701
27. Xie Y, Ge S, Jiang S, Liu Z, Chen L, Wang L, Chen J, Qin L, Peng W (2018) Study on biomolecules in extractives of *Camellia oleifera* fruit shell by GC-MS. Saudi J Biol Sci 25:234–236
28. Yildiz ŰC, Killiç C, Gürgen A, Yildiz S (2020) Possibility of using lichen and mistletoe extracts as potential natural wood preservative. Ciencia Y Technologia 22(2):179–188

Characteristics and Potential Utilization of Yopo Wood (*Piptadenia peregrina* Benth.)

Yusup Amin, Danang S. Adi, Teguh Darmawan, Dimas Triwibowo, Wahyu Dwianto, and Ratih Damayanti

Abstract Yopo wood (*Piptadenia peregrina* Benth.) is one of lesser-known wood species with an *Mean Annual Increment* (MAI) of 2.77 cm/year. This research investigated the physical and mechanical properties, anatomical features, and chemical components of Yopo wood. The branches of Yopo wood by 10 cm in diameters obtained from the Purwodadi Botanical Garden in Pasuruan (East Java) were used in this study. Analysis of physical and mechanical properties were referred to BS 373-1957. Yopo wood powder (60 mesh) were used for chemical composition analysis referred to Mokushitsu Kagaku Jiken Manual. Determination of anatomical properties were referred to Standard of International Association of Wood Anatomist (IAWA). The results showed that based on the physical–mechanical properties, anatomical features and chemical components, Yopo wood could be classified into the strength class III–II; this wood has the potential and allows it to be developed and used as raw materials for construction, furniture, or as raw material for bioethanol.

Keywords Yopo wood · Physical and mechanical properties · Anatomical features · Chemical components · Potential utilization

1 Introduction

The increasing population growth and the rising advancement of wood industries lead to the increasing number of global wood consumptions. Based on Food and Agriculture Organization (FAO) data [1], the global production of log-wood was increased approximately 15% over the period of 2000–2018, i.e. 3.5 billion m^3 in 2000 and

Y. Amin (✉) · D. S. Adi · T. Darmawan · D. Triwibowo · W. Dwianto
Research Center for Biomass and Bioproducts, National Research and Innovation Agency (BRIN), Cibinong, Indonesia
e-mail: yusu007@brin.go.id

R. Damayanti
Directorate of Environment, Maritime, Natural Resources, and Nuclear Policy, National Research and Innovation Agency (BRIN), Jakarta, Indonesia

M. A. Rahandi Lubis (ed.), *Proceedings of the 10th International Symposium for Sustainable Humanosphere*, Lecture Notes in Civil Engineering 255,
https://doi.org/10.1007/978-981-19-1422-5_17

slightly more than 4 billion m^3 in 2018. The global production of sawn-wood and wood-based panels in 2018 were 28% and 127% increase from the 2000 productions. In Indonesia, the productions of log-wood and sawn-timber were increased 27.5% and 14.8% over period of 2016–2018 [2]. Unfortunately, the communities and industries are still utilizing commercial woods, whereas the availability of those woods currently is reducing due to high degradation and deforestation of natural forests. Quite the reverse, the lesser-known wood species (LKWS), which are plentiful but have not yet been much explored, seem to have a good potential for being used as an alternative wood resource in the future [3].

The increasing of forest degradation and deforestation in Indonesia has encourage to the use of LKWS, which are abundantly and recently has not much been utilized [4]. By promoting the utilization of LKWS, the utilization of wood resources can be efficient [5]. Yopo wood (*Piptadenia peregrina* Benth.) is one of LKWS with an *Mean Annual Increment* (MAI) of 2.77 cm/year [6] and grow up to 20 m tall [7], which has rarely been studied. This species originates from South America. In native habitat this species is known as jopo or cohoba [8]. This wood has a wide geographical distribution and with good properties for various uses. In Indonesia, it can be found in the Purwodadi Botanical Garden, East Java. Recently, the new paradigm of the Botanical Garden will essentially transform scarce wood species into the breeding center of potential species [4], which is expected to support the development of this species in the future through community forest or industrial plantation forest. The objective of this research was to investigate the basic characteristics of wood: physical and mechanical properties, anatomical features, and chemical components of Yopo wood. It was expected that the research would provide information on its potential utilization.

2 Materials and Methods

The specimens used were the branch of Yopo wood by 10 cm in diameters obtained from Purwodadi Botanical Garden, Pasuruan region, East Java, Indonesia. The physical–mechanical properties and chemical composition analysis was done at the Research Center for Biomass and Bioproducts, National Research and Innovation Agency (BRIN), Cibinong, while the observation of anatomical properties was done at the Plant Anatomy Laboratory, Standardization Instrument of Environment and Forestry Agency, Ministry of Environment and Forestry, Bogor.

The technical procedures of this research were referred to BS 373-1957 [9] for physical–mechanical properties; Mokushitsu Kagaku Jiken Manual [10] for chemical composition analysis; standard of International Association of Wood Anatomist—IAWA [11, 12] for anatomical properties. Parameter of physical properties tested was density, while parameters of mechanical properties included Modulus of Elasticity (MOE) and Modulus of Rupture (MOR). The testing of mechanical properties was carried out by the Universal Testing Machine (UTM) Shimadzu with a maximum load capacity of 50 kN, loading speed 5 mm/min.

Yopo wood powder (60 mesh) were used for chemical composition analysis referred to Mokushitsu Kagaku Jiken Manual [10]. The chemical content analyzed includes the component of α-cellulose, hemicellulose, lignin, and extractives soluble in alcohol-benzene.

The parameter observation of wood anatomical properties includes macroscopic and microscopic characteristics. The macroscopic observations that were carried out included the color, texture, grain orientation, gloss, appearance of the wood surface, and odor [12]. The microscopic characteristics were observed from wood specimens in cross-section, radial, and tangential sections (15–25 μm in thickness) which cutting by microtome [3], and then compared with IAWA standard [11].

3 Results and Discussion

3.1 Physical–Mechanical Properties

The physical and mechanical properties of wood are important properties that must be recognized before the wood is used [13]. Wood density is one of the main parameters in determining the use of wood. It can affect wood-hygroscopicity, strength, shrinkage, acoustics, and electricity as well as other properties related to subsequent woodworking [14]. Table 1 shows the physical and mechanical properties of Yopo wood (density, MOR, and MOR). The density of Yopo wood (0.61–0.73 g/cm^3) was higher than others LKWS, i.e. sengon "0.33 g/cm^3" [15]; *Schizolobium amazonicum* "0.47 g/cm^3" [5]; gmelina "0.42–0.61 g/cm^3" and *Acacia mangium* "0.43–0.66 g/cm^3" [12].

Wood density is very influential on the mechanical strength of wood (MOE and MOR). Mechanical properties of wood are properties related to the wood's ability to withstand loads (external forces) acting on it. The mechanical properties of wood are usually the most important characteristics of wood products to be used for structural building materials. The structural use can be defined as any use where mechanical properties are the first criterion for material selection [16].

The strength of wood (MOE and MOR) is directly proportional to the density value. But in generally the level of wood strength is identical to MOR value. Wood with a high MOR value tends to have high strength as well. It is necessary to

Table 1 Physical–mechanical properties of Yopo wood

Physical–mechanical properties	
- Density (g/cm^3)	0.61–0.73
- MOE (kg/cm^2)	76,151.20–92,120.80
- MOR (kg/cm^2)	657.74–842.27
Strength-class*	(III-II)

* Martawijaya et al. [15]

Table 2 Chemical components of Yopo wood

Chemical components (%)	
- α-Cellulose	42.33–54.87 (49.81)
- Hemicellulose	25.38–37.74 (31.11)
- Lignin	23.81–26.15 (25.13)
- Extractive	3.03–7.42 (5.46)

consider ratio the strength to weight of wood (strength to weight ratio), because the higher value ratio (MOR/weight), the wood getting better and more suitable for construction use. Base on the result, the strength to weight ratio of Yopo wood was higher than 1000, so that it is possible to use as a lightweight construction material. Based on MOR value, Yopo wood could be categorized into the strength class III-II [15]. It was higher than both of sengon and *S. amazonicum* (IV-III), but lower than *Acacia mangium* (II). Yopo wood has potential to be used for construction material or furniture.

3.2 Chemical Components

Table 2 shows the chemical compositions of Yopo wood.

The cellulose content of Yopo wood almost similar to *S. amazonicum* i.e. about 80% [4], but lower than *A. mangium* [16]. Yopo wood has low content of lignin (23.81–26.16%), lower than *S. amazonicum* i.e. aboat 80% [4] and *A. mangium* [17]; but almost similar with lignin content of sengon [15]. Nevertheless, Yopo wood has more of extractives (3.03–7.42%). It's higher than *S. amazonicum, A. mangium* dan sengon. Based on these chemical components, it is possible to use Yopo wood as raw material for bioethanol productions.

3.3 Anatomical Features

Macroscopic features:

The color of Yopo wood is cream and slightly brownish. Fiber direction is blends with decorative pattern (cross section). Medium to weight in hardness with smooth touch impression (shiny), even though the wood texture is medium to rough. The surface of the slashed wood is slightly shiny. In the transverse plane there are interesting patterns due to the presence of a clear growth ring, but the presence of these patterns is less visible in the radial and tangential planes. It has no special odor in Yopo wood (Fig. 1).

Microscopic features:

Fig. 1 Cross section of Yopo wood (macroscopy)

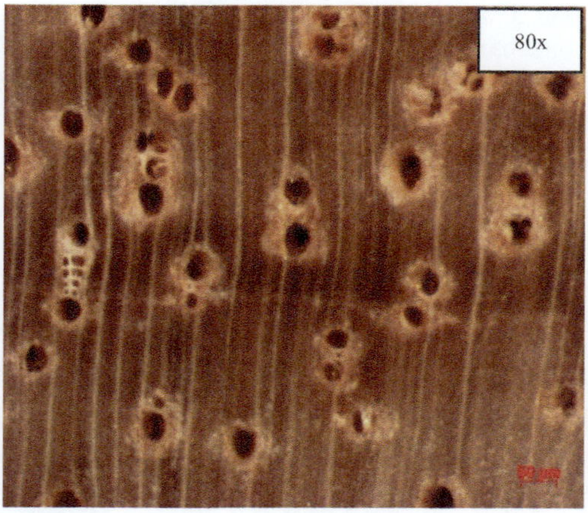

The growth ring of Yopo wood was district. The vessels cell was diffuse (diagonal pattern), with average of diameter 146–160 μm and frequency about 6–9 per mm^2 (Fig. 2). The perforation plates are simple with axial parenchyma were paratracheal, aliform lozenge until confluent. Some prismatic crystals were found inside the axial parenchyma chamber (Fig. 3); the number of cells per parenchyma strand were 5–8 units. The type of insulated fibers was found, the cell walls were very thin (early wood), and thin until thick (late wood). Rays: homocellular (lying cell), composed by 1–3 series (Fig. 2).

Fig. 2 Radial section (microscopy)

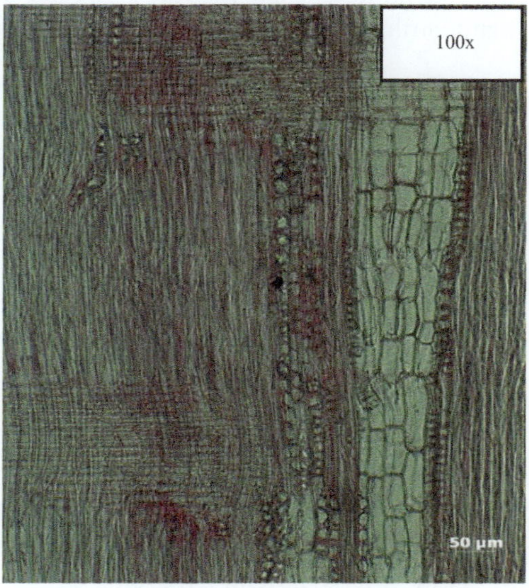

Fig. 3 Tangential section (microscopy)

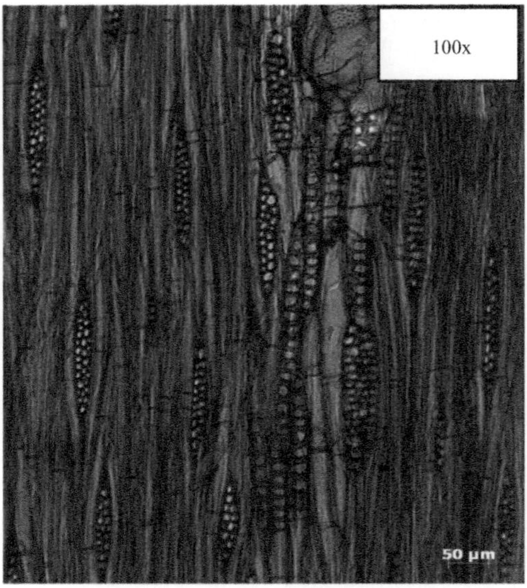

4 Conclusions

Based on the physical–mechanical properties (density and MOR) and anatomical features, Yopo wood (*Piptadenia peregrina* Bent.) could be classified into the strength class III-II; this wood is suitable and potentially use as raw material for light construction and furniture. Yopo wood has high contents of holocellulose and relatively low of lignin, so this wood has potential to be used as raw material for bioethanol production.

Acknowledgements Authors would like to thank to the JSPS Alumni Association of Indonesia (JAAI) program for financial support of this research, as well.

References

1. FAO (2020) World Food and Agriculture—statistical yearbook 2020, Rome. https://doi.org/10.4060/cb1329en
2. https://www.bps.go.id/indicator/60/167/1/produksi-kayu-hutan.html. Accessed 03 Sept 2021
3. Amin Y, Wahyudi I (2014) Anatomical characteristics of Wangkal wood (*Albizia procera* Roxb. Bent). In: Proceedings the 3rd International Symposium for Sustainable Humanosphere. Bengkulu, Indonesia, 17–18 September 2013, pp 305–310. ISSN 2088-9127
4. Amin Y, Adi DS, Wahyuni I, Kusumah SS, Damayanti R (2013) Anatomical characteristics and chemical properties of the branch-wood of *Schizolobium amazonicum* Ducke species and its potential uses. J For Res 10(2):123–129. ISSN 0216-0919. Terakreditasi LIPI, Nomor 625/E/2013 (538/AU3/P2MI-LIPI/06/2013
5. Amin Y, Wahyuni I, Darmawan T, Dwianto W, Sunarko (2011) Physical and mechanical properties of *Schizolobium amazonicum* Ducke wood branches. J Trop Wood Sci Technol 9(2):182–187. ISSN 1693-3834. Accredited LIPI, Number 185/AU1/P2MBI/08/2009

6. Budiharta S, Mukaromah L (2006) Average annual increment of legume species (*Fabaceae*) collection of Purwodadi Botanical Garden. In: Proceedings of One-Day Seminar "Conservation and Utilization of Plant Diversity in Dry Areas II" UPT Plant Conservation Center of Purwodadi Botanical Garden, Pasuruan–East Java
7. http://en.wikipedia.org/wiki/Anadenanthera_peregrina#Related_species. Accessed 02 Jul 2010
8. Souza CO, Silva JGM, Arantes MDC, Vidaurre GB, Dias Júnior AF, Oliveira MP (2020) Pyrolysis of *Anadenanthera peregrina* wood grown in different spacings from a forest plantation in Brazil aiming at the energy production. Environ Dev Sustain 22:5153–5168. https://doi.org/10.1007/s10668-019-00418-0
9. British Standard Institution (1957) Methods of testing small clear specimens of timber. BS 373:1957. Decorporated by Royal Charter. British Standard House, London
10. Mokushitsu Kagaku Jiken Manual (2000) Japan Wood Research Society Publisher
11. Wheeler EA, Baas P, Gasson E (1989) IAWA list of microscopic features for hardwood identification. IAWA Bull 10(3):219–332
12. Mandang YI, Pandit IKN (2002) Guidelines for identification of wood types in the field. Prosea Foundation, Bogor and Forestry Human Resource Employee Training Center, Bogor, 194 p
13. Kasmudjo (1998) Information technology in determining the use of wood for industrial raw materials. Duta Rimba-Perum Perhutani, Jakarta
14. Hadjib N, Hadi YS, Setyaningsih D (2007) Physical and mechanical properties of ten provenances of mangium wood (*Acacia mangium* Wild.) from Parung Panjang, West Java. J Sci Technol Wood Trop 5(1):7–11
15. Martawijaya A, Kartasujana I, Kadir K, Prawira SA (2005) Indonesian wood atlas volume II revised edition. Department of Forestry, Forestry Research and Development Agency, Bogor
16. Haygreen JG, Bowyer JL (1996) Forest products and wood science: an introduction. Translated by SA Hadikusumo. Gadjah Mada University Press, Yogyakarta
17. Yahya R (2004) Chemical composition of branchwood including bark of *Acacia mangium* Willd as raw material for pulp and paper manufacture. Prosiding Seminar Nasional Mapeki VII, Makassar, 5–6 Agustus 2004

Bioactivities and Phytochemicals of Some Ferns from East Kalimantan, Indonesia

Ashlikhatul Mahmudah, Irawan Wijaya Kusuma, Harlinda Kuspradini, Widya Fatriasari, and Enih Rosamah

Abstract Fern belongs to Pteridophyte is a group of plants that can be found in almost every region in Indonesia. This study aimed to analyze the bioactivities and phytochemicals of some fern species i.e., *Pteridium caudatum, Acrostichum aureum, Lygodium circinnatum, Lycopodium cernuum, Ceratopteris thalictroides, Acrostichum speciosum, Blechnum orientale,* and *Salvinia molesta.* This study includes qualitative phytochemical analysis to evaluate the presence of alkaloid, triterpenoid, steroid, flavonoid, tannin, coumarin, carotenoid and carbohydrate. Antimicrobial assay was conducted using an agar-well diffusion against *Propionibacterium acnes, Escherichia coli, Streptococcus sobrinus,* and a yeast-like fungus, *Candida albicans.* Toxicity of the fern extracts was evaluated by a Brine Shrimp Lethality Test using gallic acid as a positive control. The results of the phytochemical showed the appearance of coumarin, tannin, carbohydrate, steroid, and carotenoid. The results displayed that *Pteridium caudatum, Acrostichum aureum, Lygodium circinnatum, Ceratopteris thalictroides,* and *Blechnum orientale,* possessed potential growth inhibition on microorganism tested. Three fern species, *Pteridium caudatum, Blechnum orientale,* and *Salvinia molesta* were considered non-toxic based on the toxicity assay, with LC_{50} values of 2672 ppm, 1410 ppm, and 1670 ppm, respectively.

Keywords Antimicrobial activity · Biological activity · Ferns · Phytochemical · Toxicity

---

A. Mahmudah · I. W. Kusuma · H. Kuspradini · E. Rosamah (✉)
Faculty of Forestry, Mulawarman University, Samarinda, Indonesia
e-mail: enihros@fahutan.unmul.ac.id

W. Fatriasari
Research Center for Biomaterials, Indonesian Institute of Science, Cibinong, Indonesia

M. A. Rahandi Lubis (ed.), *Proceedings of the 10th International Symposium for Sustainable Humanosphere*, Lecture Notes in Civil Engineering 255,
https://doi.org/10.1007/978-981-19-1422-5_18

1 Introduction

Ferns are a group of plants that can be found in almost every area in Indonesia, including humid tropical area such as swamps, mangroves, secondary and primary forests. Ferns are classified as a plant division with kormas and are divided into three major parts: roots, stems, and leaves [1]. Many ferns have very fast growth rate in favorable conditions which are spread in every region with typical functions, efficacy, and benefits.

Ferns are not only used by people as a food ingredient, but also in many purposes including ornamental, and healthcare treatment [2]. *Blechnum orientale* leaves are applied externally to treat cut wounds, boils, and stomach ulcers in Malaysia. The plant leaves are also reported to be used to treat stomach pain and bladder [3]. *Acrostichum aureum* is included in ferns that contain phenolic compounds with antioxidant and total phenolic content, the antioxidant in the ethanol extract of *Acrostichum aureum* reaches 41.95 g/ml [4]. *Lycopodium cernuum* is also believed to be cure gout, menstrual and neurological disorders, fever, and as an anesthetic [2]. Young leaves on *Acrostichum speciosum* can inhibit the growth of *Staphylococcus aureus* that causes ulcers and acne [5].

Therefore, intensive research on the bioactivity and content of compounds as basic materials for the utilization and development of their derivative products is important. This study aimed to analyze the content of phytochemicals, antimicrobial activity, and toxicity of some ferns from East Kalimantan, Indonesia.

2 Materials and Methods

2.1 Sample Preparation

The ferns taken from East Kalimantan have been identified by Arbain (Institute of Agriculture, East Kutai, East Kalimantan). Eight ferns used were *Pteridium caudatum, Acrostichum aureum, Lygodium circinnatum, Lycopodium cernuum, Ceratopteris thalictroides, Acrostichum speciosum, Blechnum orientale*, and *Salvinia molesta.* Samples of ferns were ground and following with maceration using methanol solvent to obtain fern extracts. Methanol maceration was conducted for 2×24 h followed with filtration and solvent removal using a rotary evaporator at temperature of 30–40 °C. At the end of the process, crude extracts obtained were weighed and kept in the refrigerated until further investigation.

Table 1 Analysis phytochemicals

No.	Component	Method	Presence indicator
1	Alkaloids [6]	Dragendorff test	Orange or red
2	Steroid or Triterpenoid [6]	Salkowski test	Red or purple indicates triterpenoids and green or blue indicates steroids
3	Flavonoids [7]	Sodium hydroxide test	Colorless
4	Coumarin [8]	Sodium hydroxide test	Yellow
5	Carbohydrate [6]	Molisch	Purple ring
6	Tannin [7]	Lead sub-acetate test	Yellow precipitate
7	Carotenoids [8]	Sulphuric acid test	Blue surface

2.2 Phytochemicals

Phytochemical analysis was done to evaluate the presence or absence of alkaloids, steroid or triterpenoids, flavonoids, coumarin, carbohydrate, tannin, and carotenoids using a qualitative method with reference methods as follow (Table 1).

2.3 Antimicrobial Activity Assay

Culture Media

Glucose and nutrient broth were added with distilled water and then heated until dissolves. Then the media were sterilized using an autoclave along with Petri dishes and tools for 15 min at 121 °C.

Preparation of Microbial Suspension

Three bacteria and a yeast-like fungus were used as test microorganisms, including *Propionibacterium acnes*, *Escherichia coli*, *Streptococcus sobrinus*, and *Candida albicans*. Test microbial strains left on an agar medium at 37 °C for 24 h. The microbial suspension density was determined by using a UV/Vis spectrophotometer type UV-1280 at a wavelength of 600 nm [9].

Antibacterial Activity Assay

Sterile hot media was added to the petri dish of 20 ml and allowed to harden. The median was added with 100 ml of a bacterial suspension, spread with a cotton swab, then set aside to dry for 15 min.

The dried media is cut with a cork borer to make a 7-mm well. Each well was poured with 20 ml on concentration series or solvent only as a control. The culture media were then incubated at 37 °C for 18–24 h. The diameter of clear zone around the hole indicating the area inhibitory to bacteria or fungi were measured in millimeter.

2.4 Toxicity

Preparation of Larvae *Artemia salina*

A certain amount of *Artemia salina* eggs were put into a tank filled with saline. After 48 h, *Artemia salina* eggs hatched into *Artemia salina* larvae called nauplii that further used for toxicity assay.

Toxicity Assay

Positive control or sample solution consisting of 5 ml of seawater. The solvent was removed from the sample or positive control vial. Vials containing sea water were then filled with 10 marine shrimp larvae while negative control contained distilled and sea water only. Vials were kept in a fairly light. After 24 h, the number of dead shrimp larvae for each concentration was calculated and recorded [10].

3 Results and Discussion

3.1 Phytochemicals

The phytochemical analysis was conducted with a qualitative method. The color reaction was compared to the reference using papaya extract (alkaloid), spirulina extract (steroids/triterpenoids), apples extract (flavonoids), parsley extract (coumarin), glucose (carbohydrates), guava extract (tannins), and carrots extract (carotenoids). The results of phytochemical analysis were presented in Table 2.

The phytochemical analysis results showed the occurrence of coumarin, tannins, carbohydrates and some samples contain steroids. Alkaloids, flavonoids and

Table 2 Results of phytochemicals analysis of 8 fern extracts

Samples	Alk	Flv	Kum	Tan	Ste	Ter	Kart	Carbs
Pteridium caudatum	–	–	++	++	+	–	–	++
Acrostichum aureum	–	–	++	++	+	–	+	++
Lygodium circinnatum	–	–	++	++	+	–	–	++
Lycopodium cernuum	–	–	++	++	–	–	–	++
Ceratopteris thalictroides	–	–	++	++	+	–	–	+
Acrostichum speciosum	–	–	++	++	+	–	–	++
Blechnum orientale	–	–	++	++	–	–	–	++
Salvinia molesta	–	–	++	++	+	–	–	++

* *Alk* alkaloid, *Flv* flavonoids, *Kum* coumarin, *Tan* tannins, *Ste* steroids, *Ter* triterpenoids, *Kart* carotenoids, *Carbs* carbohydrates

** Test results using references; (+ +): the high content, (+): low content and (–): none carotenoids were not seen in our test, except a species, *Acrostichum aureum* that indicated the presence of carotenoids.

3.2 Antimicrobials

The antimicrobial activity was determined by the percentage of inhibition relative to the positive control. Their antimicrobial activity was indicated by the formation of inhibition zone [11].

A standard classification of response inhibition on bacterial growth was presented in Table 3.

With the existence of the above parameters, the antimicrobial activity of the fern extracts can be determined whether they are belonged to weak, moderate, strong or very strong activity.

The results of antimicrobial activities of 8 fern extracts were presented in Table 4.

The antimicrobial activity of the ferns tested were evaluated against *Candida albicans, Propionibacterium acnes, Escherichia coli* and *Streptococcus sobrinus. Acrostichum aureum* inhibited all microorganism tested at low concentration of 62.5 g/well with moderate to weak results. *Lycopodium cernuum* inhibited *Escherichia coli* at the concentration of 62.5 g/well.

The phytochemicals analysis showed the presence of one of interesting secondary metabolites, tannins. Tannins is a large class of polyphenols compounds. Regarding antimicrobial activity, some hydrolyzed tannins were proven to be more reactive and have a strong inhibitory effect than the condensed tannins. Tannins work by depositing proteins and may damage cell membranes resulted in fungal growth inhibition [13].

Each bacteria strain has a different sensitivity to the sample, including the case where bacteria may form a resistance to antibiotic which is a natural mechanism in maintaining its survive [14]. Our results showed that the one or more phytochemicals in some fern extracts may possess potential as antimicrobial agents.

Table 3 Classification of bacterial growth inhibition [12]

Inhibition zone diameter (mm)	Classification
≥ 20	Very strong
16–20	Strong
10–15	Moderate
≤ 10	Weak

Table 4 Results of antimicrobial assay

No.	Microbe	Con (ug/well)	*Candida albicans*			*Propionibacterium acnes*			*Escherichia coli*			*Streptococcus sobrinus*		
			IZ (mm)	Class	RIA (%)	IZ (mm)	Class	RIA (%)	IZ (mm)	Class	RIA (%)	IZ (mm)	Class	RIA (%)
1	*Pteridium caudatum*	Control (+)	30.1 ± 0.1925	Very strong	100	29.1 ± 0.7698	Very strong	100	27.1 ± 1.3472	Very strong	100	29.7 ± 0.3333	Very strong	100
		500	13.3 ± 0.8819	Moderate	44.28	13.1 ± 0.5092	Moderate	45.04	12.7 ± 0.6667	Moderate	46.72	11.7 ± 0.6667	Moderate	39.33
		250	11.7 ± 0	Moderate	38.75	12.0 ± 0.5774	Moderate	41.22	12.0 ± 0	Moderate	44.26	10.4 ± 0.5092	Moderate	35.21
		125	10.2 ± 0.3849	Moderate	33.95	11.1 ± 0.1925	Moderate	38.17	10.3 ± 0.5774	Moderate	38.11	9.8 ± 0.6939	Weak	32.96
		62.5	–	–	–	–	–	–	10.4 ± 0.3849	Moderate	38.52	–	–	–
		31.25	–	–	–	–	–	–	9.6 ± 0.8389	Weak	35.25	–	–	–
2	*Acrostichum aureum*	Control (+)	30.6 ± 0.5092	Very strong	100	30.0 ± 0	Very strong	100	27.3 ± 0	Very strong	100	29.8 ± 0.6939	Very strong	100
		500	15.7 ± 0.3333	Moderate	51.27	14.9 ± 0.8389	Moderate	49.63	13.2 ± 1.9532	Moderate	48.37	13.4 ± 1.2620	Moderate	45.15
		250	14.6 ± 0.3849	Moderate	47.64	13.7 ± 0.5774	Moderate	45.56	12.2 ± 1.2620	Moderate	44.72	13.1 ± 1.8359	Moderate	44.03
		125	14.0 ± 0.3333	Moderate	45.82	13.2 ± 0.3849	Moderate	44.07	11.6 ± 0.3849	Moderate	42.28	12.7 ± 1.1547	Moderate	42.54
		62.5	13.2 ± 0.5092	Moderate	43.27	11.9 ± 0.1925	Moderate	39.63	10.3 ± 0.3333	Moderate	37.80	11.3 ± 1.4530	Moderate	38.06
		31.25	10.6 ± 0.6939	Moderate	34.55	10.2 ± 0.3849	Moderate	34.07	–	–	–	–	–	–
3	*Lygodium circinnatum*	Control (+)	29.7 ± 0.3333	Very strong	100	29.9 ± 0.1925	Very strong	100	27.1 ± 0.6939	Very strong	100	29.2 ± 0.1925	Very strong	100
		500	–	–	–	12.2 ± 0.1925	Moderate	40.89	15.0 ± 0.8819	Moderate	55.33	10.8 ± 0.5092	Moderate	36.88
		250	–	–	–	11.0 ± 0.3333	Moderate	36.80	13.6 ± 1.8359	Moderate	50.00	9.9 ± 0.1925	Weak	33.84
		125	–	–	–	10.1 ± 0.3849	Moderate	33.83	12.6 ± 0.3849	Moderate	46.31	–	–	–
		62.5	–	–	–	–	–	–	10.3 ± 0.3333	Moderate	38.11	–	–	–
		31.25	–	–	–	–	–	–	9.8 ± 0.5092	Weak	36.07	–	–	–

(continued)

Table 4 (continued)

No.	Microbe	Con (ug/well)	*Candida albicans*			*Propionibacterium acnes*			*Escherichia coli*			*Streptococcus sobrinus*		
			IZ (mm)	Class	RIA (%)	IZ (mm)	Class	RIA (%)	IZ (mm)	Class	RIA (%)	IZ (mm)	Class	RIA (%)
4	*Lycopodium cernuum*	Control (+)	30.1 ± 0.3849	Very strong	100	29.8 ± 0.3849	Very strong	100	27.2 ± 0.8389	Very strong	100	29.3 ± 0.3333	Very strong	100
		500	–	–	–	–	–	–	14.9 ± 0.3849	Moderate	54.69	–	–	–
		250	–	–	–	–	–	–	11.7 ± 0.5774	Moderate	42.86	–	–	–
		125	–	–	–	–	–	–	11.0 ± 0.6667	Moderate	40.41	–	–	–
		62.5	–	–	–	–	–	–	9.8 ± 0.1925	Weak	35.92	–	–	–
		31.25	–	–	–	–	–	–	9.8 ± 0.2357	Weak	36.12	–	–	–
5	*Ceratopteris thalictroides*	Control (+)	30.6 ± 0.3849	Very strong	100	29.8 ± 0.3849	Very strong	100	27.3 ± 0.5774	Very strong	100	29.4 ± 0.1925	Very strong	100
		500	12.4 ± 0.1925	Moderate	40.73	13.4 ± 0.5092	Moderate	45.15	14.2 ± 1.6777	Moderate	52.03	11.4 ± 0.1925	Moderate	38.87
		250	11.0 ± 0.8819	Moderate	36.00	12.1 ± 0.3849	Moderate	40.67	12.3 ± 1.2019	Moderate	45.12	10.7 ± 0.6667	Moderate	36.23
		125	10.2 ± 1.0184	Moderate	33.45	10.9 ± 0.1925	Moderate	36.57	10.2 ± 0.5092	Moderate	37.40	9.7 ± 0.3333	Weak	32.83
		62.5	–	–	–	–	–	–	–	–	–	–	–	–
		31.25	–	–	–	–	–	–	–	–	–	–	–	–
6	*Acrostichum speciosum*	Control (+)	30.3 ± 0.3333	Very strong	100	29.8 ± 0.1925	Very strong	100	27.3 ± 0.3333	Very strong	100	29.6 ± 0.5092	Very strong	100
		500	–	–	–	–	–	–	–	–	–	–	–	–
		250	–	–	–	–	–	–	–	–	–	–	–	–
		125	–	–	–	–	–	–	–	–	–	–	–	–

(continued)

Table 4 (continued)

No.	Microbe	Con (ug/well)	*Candida albicans*			*Propionibacterium acnes*			*Escherichia coli*			*Streptococcus sobrinus*		
			IZ (mm)	Class	RIA (%)	IZ (mm)	Class	RIA (%)	IZ (mm)	Class	RIA (%)	IZ (mm)	Class	RIA (%)
		62.5	–	–	–	–	–	–	–	–	–	–	–	–
		31.25	–	–	–	–	–	–	–	–	–	–	–	–
7	*Blechnum orientale*	Control (+)	30.3 ± 0.3333	Very strong	100	29.8 ± 0.3849	Very strong	100	26.9 ± 0.1925	Very strong	100	29.7 ± 0.3333	Very strong	100
		500	15.2 ± 0.5092	Moderate	50.18	14.2 ± 0.3849	Moderate	47.76	12.7 ± 0.3333	Moderate	47.11	12.1 ± 0.1925	Moderate	40.82
		250	14.1 ± 0.7698	Moderate	46.52	12.4 ± 0.7698	Moderate	41.79	10.8 ± 0.1925	Moderate	40.08	10.4 ± 0.7698	Moderate	35.21
		125	–	–	–	–	–	–	10.3 ± 5.9753	Moderate	38.43	–	–	–
		62.5	–	–	–	–	–	–	–	–	–	–	–	–
		31.25	–	–	–	–	–	–	–	–	–	–	–	–
8	*Salvinia molesta*	Control (+)	30.3 ± 0.3333	Very strong	100	29.8 ± 0.3849	Very strong	100	26.0 ± 0.8819	Very strong	100	29.6 ± 0.1925	Very strong	100
		500	–	–	–	–	–	–	–	–	–	–	–	–
		250	–	–	–	–	–	–	–	–	–	–	–	–
		125	–	–	–	–	–	–	–	–	–	–	–	–
		62.5	–	–	–	–	–	–	–	–	–	–	–	–
		31.25	–	–	–	–	–	–	–	–	–	–	–	–

**Con* Concentration, *IZ* Inhibition zone, *RIA* relative inhibitory activity

3.3 Toxicity

Acute toxicity assay using brine shrimp lethality test is a simple assay to show the toxicity of the extracts. The toxicological properties of a substance are determined by the dose or the concentration of toxins [15]. The results on toxicity assay of 8 fern extracts were presented in Table 5.

Classification of the toxicity assay was presented in Table 6 as a basis to determine the toxicity of 8 fern extracts tested.

Lethal concentration is defined as how many samples may cause lethality to 50% of shrimp nauplii. Compounds with $LC_{50} < 1000$ ppm can be considered as toxic [10].

The results of the 8 ferns tested were categorized according to their LC_{50} value. Four ferns were determined to have moderate toxicity, i.e., *Lycopodium cernuum, Lygodium circinnatum, Lycopodium cernuum, Lygodium circinnatum,* and *Ceratopteris thalictroides. Acrostichum aureum* was classified as less toxic, while three ferns i.e., *Pteridium caudatum, Blechnum orientale, Salvinia molesta* were belong to non-toxic extracts.

4 Conclusion

Phytochemical analysis of 8 fern extracts showed the occurrence of coumarin, tannins, carbohydrates, steroids, and carotenoids. Antimicrobial activity assays displayed that 4 fern species inhibited *Candida albicans* and *Propionibacterium acnes*, 6 species inhibited *Escherichia coli* and 5 species inhibited *Streptococcus sobrinus*. Brine shrimp lethality test of 8 fern extracts showed that three samples were non-toxic, a samples was less toxic and four samples were moderately toxic.

Table 5 Results toxicity testing

No.	Samples	Concentration (ppm)	Nuplii the dead	%Dead	LC_{50}	Toxicity category
1	Control (+) Gallic acid	1000	10 ± 0	100	113.930	Moderate toxic
		500	10 ± 0	100		
		250	2.3 ± 0.577	23.3		
		100	1.7 ± 0.577	16.7		
		50	0	0		
2	*Acrostichum aureum*	1000	7 ± 1	70	514.909	Less toxic
		500	4.5 ± 0.707	45.0		
		250	3.3 ± 0.577	33.3		
		100	2.7 ± 1.155	26.7		
		50	0.3 ± 0.577	3.3		
3	*Lygodium circinnatum*	1000	9.3 ± 0.577	93.3	156.375	Moderate toxic
		500	9 ± 0	90		
		250	7.3 ± 0.577	73.3		
		100	6.3 ± 0.577	63.3		
		50	0.3 ± 0.577	3.3		
4	*Acrostichum speciosum*	1000	6.7 ± 1.528	66.7	761.473	Less toxic
		500	3 ± 1.732	30		
		250	1.7 ± 2.082	16.7		
		100	1 ± 1.732	10		
		50	10 ± 0	100		
1	Control (+) Gallic acid	1000	9.7 ± 0.577	96.7	296.560	Moderate toxic
		500	10 ± 0	100		
		250	1.7 ± 1.528	16.7		
		100	1 ± 1	10		
		50	0	0		
2	*Pteridium caudatum*	1000	2.7 ± 1.528	26.7	2672.833	Non-toxic
		500	1.3 ± 1.155	13.3		
		250	1 ± 1	10		
		100	0	0		
		50	0	0		
3	*Lycopodium cernuum*	1000	8.3 ± 0.577	83.3	176.037	Moderate toxic
		500	8.7 ± 1.155	86.7		
		250	6.3 ± 2.082	63.3		
		100	5 ± 1	50		
		50	1.7 ± 0.577	16.7		
4	*Ceratopteris thalictroides*	1000	3.3 ± 2.082	33.3	478.091	Moderate toxic

(continued)

Table 5 (continued)

No.	Samples	Concentration (ppm)	Nuplii the dead	%Dead	LC_{50}	Toxicity category
		500	5 ± 3.464	50		
		250	2.7 ± 2.517	26.7		
		100	1.7 ± 1.528	16.7		
		50	1 ± 1	10		
5	*Blechnum orientale*	1000	3.3 ± 0.577	33.3	1410.169	Non-toxic
		500	1.7 ± 1.528	16.7		
		250	0.7 ± 0.577	6.7		
		100	0	0		
		50	1.3 ± 1.528	13.3		
6	*Salvinia molesta*	1000	3 ± 1	30	1670.408	Non-toxic
		500	0.3 ± 0.577	3.3		
		250	2.7 ± 2.309	26.7		
		100	0	0		
		50	0	0		

The results of toxicity testing on determining the sample to toxic or no categorized based on the LC_{50} values

Table 6 Classification of toxicity based on the LC_{50} value [16]

Class	LC_{50} values
Very toxic	0–100 ppm
Moderate toxic	100–500 ppm
Less toxic	500–1000 ppm
Non-toxic	> 1000 ppm

Acknowledgements Research grant from the Ministry of Culture, Education, Research and Technology under contract number 069/SP2H/LT/DRPM/2019 was gratefully acknowledged. The authors thank the late Dr. Sugiarto (Institute of Agriculture, Sangatta, East Kalimantan, Indonesia) for collecting the fern samples.

References

1. Arini DID, Julian K (2012) Biodiversity of plants fern (Pteridophyte) in the nature mount threshold North Sulawesi, vol 2(1). Manado Forestry Research Institute
2. Grob BL (2007) *Lycopodium* (spp) L. ground pine. Kasey Hartz Natural Area Reference Sheet, Muskegon Community College
3. Dash GK, Abdullah MS, Yahaya R, Majeed S (2012) *Blechnum orientale* Linn: an important medicinal edible fern. Pharmacogn Phytochem Res 7(4):723–726

4. Khan SA, Md. Aslam H, Sandesh P, Md. Asadujjaman, Arif H (2015) Assessment of antioxidant and Analgesic activity of *Acrostichum aureum Linn.* (Family Pteridaceae). Pharmacology 1:166–217
5. Ernianingsih SW, Mukarlina R (2014) Ethnopharmacology of mangrove plants *Achantus ilicifolius* L., *Acrostichum speciosum* L. and *Xylocarpus rumphii* Mabb. in Sungai Tekong Village, Sungai Kakap District, Kubu Raya Regency. Protobiont 3(2):252–258
6. Harborne JB (1987) Phytochemical methods applied methods by Kosasih Padmawinata phytochemicals and Iwang Soediro. ITB Publishers, Bandung
7. Trease GE, Evans WC (2002) Pharmacognosy, 15th ed. Saunders Publishers, London, pp 42–44, 221–229, 246–249, 304–306, 331–332, 391–393
8. Senthilmurugan G, Vasanthe B, Suresh K (2013) Screening and antibacterial activity analysis of some important medicinal plants. Int J Innov Appl Stud 2(2):146–152
9. Pujol I, Capilla J, Ferna'ndez B, Torres, Ortoneda M, Guarro J (2002) Use of the sensitritre colorimetric microdillution panel for antifungal susceptibility testing of dermatophytes. J Clin Microbiol 30(7):2618–2621
10. Meyer BN, Ferrigni NR, Putnam JE, Jacobsen LB, Nichols DE, McLaughlin JL (1982) Brine shrimp: a convenient general bioassay for active plant constituents. Planta Med 45:31–34
11. Muharni F, Sofa F (2017) Antibacterial activity test ethanol extract plant medicine parts Musi Musi Banyuasin, South Sumatra. Pharm Indonesia 7(2):127–135
12. Greenwood D (1995) Antibiotics, susceptibility (sensitivity) test antimicrobial and chemotherapy. The McGraw-Hill Companies, New York, NY
13. Sudira IW, Merdana I (2011) Inhibition test of Kedondong leaf extract (*Lannea grandis* Engl.) on the growth of *Erwinia carotovora*. Udayana Vet J Bull 3(1):45–50
14. Hidayat S, Napitupulu RM (2015) Medicinal plant guidance (in Indonesia). Agriflo, Jakarta
15. Wirasuta IMAG, Rasmaya N (2007) General toxicology. Textbook of pharmacy. Faculty of Mathematics and Natural Sciences, University of Udayana, Bandung
16. Clarkson C, Maharaj VJ, Crouch NR, Grace OM, Pillay P, Matsabisa MG, Bhagwandin N, Smith PJ, Folb PI (2004) In vitro activity of antiplasmodial medicinal plants native to or naturalized in South Africa. J Ethnopharm 92:177–191

Anatomical and Fiber Quality Characterization of the Least-Known Wood Species of Icacinaceae, Cardiopteraceae and Stemonuraceae

Rais Anggara, Listya Mustika Dewi, Ratih Damayanti, and Hadisunarso

Abstract The least known wood is categorized as wood species that have not been used and included into commercial timber classification. The anatomical and fiber quality characterization of the least known wood is important to complete the database for wood identification and assess the quality of fiber for pulp and paper material. In this study, the anatomical properties and fiber quality of wood were observed. The wood samples were provided by Xylarium Bogoriense which belongs to three families namely Icacinaceae (*Apodytes* sp.), Cardiopteraceae (*Citronella* spp., *Gonocaryum* spp.), and Stemonuraceae (*Gomphandra* sp., *Medusanthera papuana*, and *Urandra* sp.). The observation on general characteristics of macroscopic features was conducted using magnification lens. The permanent slides of thin sliced wood samples were obtained following Sass. The anatomical characterization was described according to IAWA Committee. Fiber maceration was undertaken to measure fiber dimensions and determine fiber quality for pulp and paper materials. The results showed that in Icacinaceae family, the diagnostic characteristics of *Apodytes* sp. were thick-walled fiber cells, ray composition procumbent, square and upright cells mixed throughout the ray. In Cardiopteraceae family, *Citronella* spp., wood had two distinct sizes of rays, axial parenchyma diffuses, diffuse in-aggregate, and scanty paratracheal. Two different wood species were identified under *Gonocaryum* genera. Lastly, in Stemonuraceae, *Gomphandra* sp. and *Medusanthera papuana* were almost similar macroscopically and microscopically, except in the type of crystals. The diagnostic features of *Urandra* sp. were axial parenchyma diffuse,

Supplementary Information The online version contains supplementary material available at https://doi.org/10.1007/978-981-19-1422-5_19.

R. Anggara · Hadisunarso
Biology Department, Faculty of Mathematics and Natural Sciences, IPB University, Kampus IPB Darmaga, Bogor, Indonesia

L. M. Dewi (✉) · R. Damayanti
National Research and Innovation Agency (BRIN), Bogor 16911, Indonesia
e-mail: list008@brin.go.id

R. Damayanti
e-mail: ratih.damayanti@brin.go.id

M. A. Rahandi Lubis (ed.), *Proceedings of the 10th International Symposium for Sustainable Humanosphere*, Lecture Notes in Civil Engineering 255,
https://doi.org/10.1007/978-981-19-1422-5_19

uniseriate and all square rays cells. The fiber quality of all samples were categorised into class III for pulp and paper raw material.

Keywords Commercial timber classification · Fiber quality · Pulp and paper · The least-known wood species · Wood anatomy · Wood utilization

1 Introduction

Indonesia is one of mega biodiversity countries which has more than 4000 tropical wood species [1]. Wood is a complex material which each of those species has different characteristics. Wood characteristics are important aspect to determine the appropriate utilization of wood species effectively and efficiently. In South-East Asia, timber classification has been differentiated into major commercial timber, minor commercial timber, and lesser-known wood species [2–4]. Each of these classifications describe 51, 62, and 309 genera of wood, respectively. In Indonesia, there are many wood species have not been studied, categorized and commercialized, therefore they are grouped as the least-known wood species. Based on Xylarium Bogoriense database, there are 577 species, 281 genus, and 77 families of wood species that are categorized as the least-known wood species [5]. Some unstudied wood species that distribute in many regions in Indonesia can be potentially utilized as bioproducts.

Six wood species from three families namely Icacinaceae (*Apodytes* sp.), Cardiopteraceae (*Citronella* spp., *Gonocaryum* spp.), and Stemonuraceae (*Gomphandra* sp., *Medusanthera papuana*, and *Urandra* sp.) are categorized as the least-known timber. Previous study on Icacinaceae family was conducted by Dewi et al. [6] to investigate the anatomical and fiber quality of *Platea* spp. which showed high fiber quality for pulp and paper raw material. Therefore, further studies are required to scrutinize other wood species belongs to Icacinaceae and also other unstudied families namely Cardiopteraceae, and Stemonuraceae to complete the database for utilization purposes.

The appropriate utilization of wood species must be aligned with its inherent characteristics. Among the basic properties of wood, a comprehensive understanding of its anatomical structure serves as a critical foundation, as the physical and mechanical properties of wood are linked to structure of the constituent cells [7, 8]. Not only to complete the database for wood identification, wood anatomy characterization plays vital role to predict the overall performance of wood as well as for determining potential utilization purposes.

Meanwhile, assessing fiber quality is also essential to determine the fiber quality for pulp and paper raw materials. Fiber length is the main indicator to assess the feasibility of a wood species as raw material for high-quality pulp and paper [9]. Several studies have proven that the quality of pulp and paper can be predicted through the amount of fiber derived value without doing the cooking process. The derived values consist of Runkle Ratio (RR), Muhlstep Ratio (MR), Felting Point (FP),

Flexibility Ratio (FR) and Stiffness Coefficient (CR). Therefore, this study aimed to observe the anatomical properties and assess the fiber quality of the least-known wood species.

2 Materials and Methods

Six wood species were selected for this study which belong to three families namely Icacinaceae (*Apodytes* sp.), Cardiopteraceae (*Citronella* spp., *Gonocaryum* spp.), and Stemonuraceae (*Gomphandra* sp., *Medusanthera papuana*, and *Urandra* sp.) (Table 1).

The observation on general characteristics of macroscopic features was conducted using magnification lens [11]. The permanent slides of thin sliced wood samples (1 cm^3) were obtained following Sass [12]. After softening for about a week, the samples

Table 1 Selected the least-known wood species belongs to Icacinaceae, Cardiopteraceae, and Stemonuraceae

No.	Species	Families	ID	Local names	Origin	Specific gravity [10]
1	*Apodytes* sp.	Icacinaceae	8563	Kakamba	Sumba	0.77
2	*Citronella* sp.	Cardiopteraceae	5463	Tuntun surimanu	Simalur, Sumatera	
			17155	Lenguru tanru, tambulu	–	–
			27394	–	Maluku	–
			20799	–	Malili, Sulawesi	–
			12535	Sabui	Martapura, Kalimantan	–
			27249	–	Maluku	–
3	*Gonocaryum* sp.		6342	Karo	Buton, Sulawesi	0.82
			24072	Gomamatari	Tobelo, Maluku	–
			29646	–	–	–
			29834	–	–	–
4	*Gomphandra* sp.	Stemonuraceae	29676	–	–	–
5	*Madhusanthera papuana*		19054	Tambai:Tenggehu	–	0.49
6	*Urandra* sp.		29704	Karappi	Serui, Maluku	0.86

were then sliced in 20–30 μm thick using sliding microtome (American Optical AO 860) for cross, radial and tangential sections; stained in 2.5% safranin, and mounted by using Entellan. The anatomical characterization was described according to IAWA Committee [13].

Fiber maceration was undertaken to measure fiber dimensions [14] and determine fiber quality for pulp and paper material [15, 16]. Matchsticks size samples were prepared and placed into a glass tube, then mixture of 50:50 (v/v) of glacial acetic acid and hydrogen peroxide was added until all the matchsticks were completely immersed. Test tubes were then placed in a water bath heated to 60 °C until the samples were softened. The tubes were shaken vigorously to separate the fibres and then stained with 2.5% safranin. In addition, the separated (macerated) wood cells (fibres, vessels and parenchyma) were mounted on object-glass by using entellan, and ready for observation. The number of fibres and vessel elements used for the measurement were 30–50 individual cells per segment. Anatomical structure observation and cell dimension measurement were carried out by using ZEISS Axio Imager Microscope.

3 Results and Discussions

3.1 General Characteristics

The tangential surfaces of studied wood samples are presented in Fig. 1. Generally, all wood samples were not significantly different in terms of wood color, texture, and grain orientation.

3.2 Anatomical Characteristics

The result showed that the anatomical characteristics of *Apodytes* sp., *Citronella* spp., *Gomphandra* sp., *Gonocaryum* sp., *Medusanthera papuana*, and *Urandra* sp. can be differentiated from their cell types such as rays, parenchyma, vessel, and fiber both quantitatively and qualitatively. Tables 2 and 3 presents the anatomical characteristics of each wood specimens according to IAWA committee [13]. The microscopic images of transverse, radial and tangential sections of all species are presented in Appendix.

Apodytes sp. (Icacinaceae) can be clearly identified from the axial parenchyma types namely diffuse (76), diffuse in-aggregate (77) and scanty paratracheal (78). The other diagnostic characteristics are the thick-walled fiber cells (70) and the ray composition with procumbent, square and upright cells mixed throughout the ray (109).

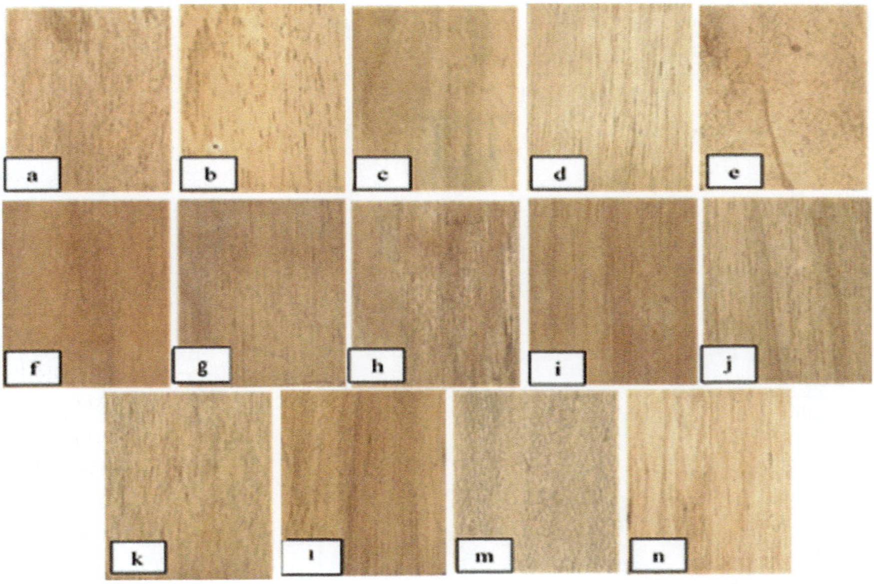

Fig. 1 Tangential surface of Icacinaceae, Cardiopteraceae, and Stemonuraceae wood samples: **a** *Apodytes* sp. ID 8563, **b** *Citronella* sp. ID 5463, **c** *Citronella* sp. ID 17155, **d** *Citronella* sp. ID 27394, **e** *Citronella* sp. ID 20799, **f** *Citronella* sp. ID 12535, **g**. *Citronella* sp. ID 27249, **h** *Gomphandra* sp. 29676, **i** *Gonocaryum* sp. ID 6342, **j** *Gonocaryum* sp. ID 29649, **k** *Gonocaryum* sp. ID 29384, **l** *Gonocaryum* sp. ID 24072, **m** *Medusanthera* papuana No. 19054, **n** *Urandra* ID 29704

In Cardiopteraceae family, six observed specimens of *Citronella* spp. show almost similar characteristics except for sample 27394 which has thin-walled fibers, while the others have very thick wall fibers. The main characteristics of *Citronella* spp. are the rays with two distinct sizes (103), axial parenchyma diffuse (76), diffuse in-aggregate (77) and scanty paratracheal (78). Interestingly, four specimens of *Gonocaryum* spp. showed different features that can be clearly seen transversely. It can be differentiated from the vessel frequency, ray width, and axial parenchyma. The differences indicate that that these specimens might be categorized as different species under *Gonocaryum genera*, except for *Gonocaryum* sp. ID 6342 which has identical characteristics with *Gonocaryum* sp. ID 24072.

In family Stemonuraceae, *Gomphandra* sp. has key specific characteristics such as axial parenchyma diffuse (76), diffuse-in-aggregates (77), narrow bands or lines up to three cells wide (86), scalariform (88), very thick-walled fibers (70), very wide rays (99), ray height > 1 mm (102), rays in two distinct size (103) and the presence of crystals sand (152, 153). *Gomphandra* sp. looks very similar to *Medusanthera papuana* if observed in transverse section, but in radial section we can found the differences in crystal types. For *Urandra* sp., the diagnostic features are axial parenchyma diffuse (76), the uniseriate rays which can be observed in tangential surface (96), and all ray cells square in radial section (105).

Table 2 Anatomical characteristics of *Apodytes* sp. and *Citronella* spp.

Species	*Apodytes*	*Citronella*					
ID	8563	5463	12,535	17,175	20,799	27,249	27,394
Anatomical characteristics							
Growth rings	(2)	(2)	(2)	(2)	(2)	(2)	(2)
Vessels	(5, 9, 14)	(5, 9, 12, 14)	(5, 9, 12, 14)	(5, 9, 12, 14)	(5, 9, 12, 14)	(5, 9, 12, 14)	(5, 9, 12, 14)
Diameter (μm)	75 (41)	81 (41)	96 (41)	86 (41)	99 (41)	119 (42)	165 (42)
Frequency (/mm^2)	16–21 (47, 48)	11–13 (47)	10–12 (47)	13–14 (47)	12–13 (47)	10.1–12 (47)	1–2 (46)
Length (μm)	291 (52)	167 (52)	134 (52)	196 (52)	156 (52)	203 (52)	324 (52)
Pits	(21, 30)	(21, 22, 30)	(21, 22, 30)	(21, 22, 30)	(21, 22, 30)	(21, 22, 30)	(21, 22, 30)
Fiber	(62, 63, 66, 70 73)	(62, 63, 66, 70, 72)	(62, 63, 66, 70, 72)	(62, 63, 66, 70, 72)	(62, 63, 66, 70, 72)	(62, 63, 66, 70, 72)	(62, 63, 66, 69, 72)
Parenchyma	(76, 77, 78, 93, 94)	(76, 77, 78, 92, 93)	(76, 77, 78, 92, 93)	(76, 77, 78)	(76, 77, 78)	(76, 77, 78)	(76, 77, 78)
Rays	(97, 107, 108, 109)	(98. 102, 103, 108, 110)	(98. 102, 103, 108, 110)	(98. 102, 103, 108, 110)	(98. 102, 103, 108, 110)	(98. 102, 103, 108, 110)	(98. 102, 103, 108, 110)
Frequency (/mm)	8–10 (115)	1–2 (114)	2–3 (114)	2–3 (114)	2–3 (114)	2–3 (114)	4–5 (115)
Mineral inclusions	(136)	(136, 138, 139)	(136, 138, 139)	(136, 138, 139)	(136, 138, 139)	(136, 138, 139)	(136, 138, 139)

Note Number in parentheses is the code of anatomical characteristics according to IAWA Committee[13]

3.3 Fiber Quality

The quality of the fiber was determined based on the dimension value of the fiber and its derivative value which consists of Runkel ratio, Felting Power, Flexibility Ratio, Coefficient of Rigidity, and Muhlsteph Ratio. The results of fiber quality measurements for the fourteen wood samples are presented in Table 4.

The results showed that the fiber quality of all wood samples belong to class III with a value of < 225. *Apodytes* sp. and *Urandra* sp. had the lowest fiber quality values (125) compared to the other five observed genera, as they had the shortest fiber length values (1413 μm). While *Gomphandra* sp. and *Medusanthera* sp. had

Table 3 Anatomical characteristics of *Gomphandra* sp., *Gonocaryum* sp., *Medusanthera papuana*, and *Urandra* sp.

Species	*Gonocaryum*				*Gomphandra*	*Medusanthera*	*Urandra*
ID	6342	29646	29834	24072	29676	19054	29704
Anatomical characteristics							
Growth rings	(2)	(2)	(2)	(2)	(2)	(2)	(2)
Vessels	(5, 9, 12, 13, 30)	(5, 9, 13, 30, 58)	(5, 9, 13, 30, 58)	(5, 9, 13, 30, 58)	(5. 9. 7. 13, 21, 30, 58)	(5. 9 12, 13. 13. 41. 47.54. 58)	(5, 10, 13, 36, 37, 58)
Diameter (μm)	80 (41)	173 (42)	1112 (42)	95 (41)	78 (41)	73 (41)	94 (41)
Frequency (/mm^2)	15–17 (47)	3–5 (46)	17 (47)	11 (47)	8 (47)	8 (47)	4 (46)
Length(μm)	177 (52)	177 (52)	1553–1645 (54)	2353–2505 (54)	2565–2679 (54)	2929–3075 (54)	1358–1467 (54)
Pits	(22, 30	(22, 24, 30	(22. 30)	(21, 22, 30)	(21, 31)	(20, 21, 30, 32, 34)	(22, 31)
Fiber	(62, 63, 65, 66, 70, 73)	(61, 63, (65, 69, 73)	(61, 63, 65, 66, 69, 73)	(62, 63, 66, 70)	(62. 63. 64. 65. 66. 70. 73)	(62, 63, 66, 70, 73)	(61, 63, 64, 66, 70, 73)
Parenchyma	(76, 77, 78, 92, 93)	(76, 77)	(75, 77, 78)	(76, 77, 78)	(76, 77, 86, 88	(77, 78, 86)	(76)
Rays	(98, 102, 103, 108, 110)	(97, 105, 106)	(97, 105)	(98, 101, 103, 105, 115)	(98. 99, 102, 103 101, 105)	(93, 98, 99, 102, 108)	(96, 105)
Frequency (/mm)	2–3 (114)	4–5 (115)	4–5 (115)	4–6 (115)	4–6 (115)	4–6 (115)	4–6 (115)
Oil and mucilage cells			–	–	–		
Mineral inclusions	(136, 137, 138)		(136, 137, 138)	–	(152, 153)	(136, 137, 138, 145, 151)	–

Note Number in parentheses is the code of anatomical characteristics according to IAWA Committee [13]

Table 4 Fiber dimensions and fiber derivative values of *Apodytes*, *Citronella*, *Gonocaryum*, *Gomphandra*, *Medusanthera*, and *Urandra* species

Fiber dimensions and derivative values	L		RR		FP		FR		CR		MR		Quality class
Wood ID	N_{hit}	N_k	N_{hit}	N_k	N_{hit}	N_k	N_{hit}	N_k	N_{hit}	N_k	N_{hit}	N_k	
***Apodytes* sp.**													
8563	1412.75	50	6.37	0	37.67	25	0.24	25	0.76	25	94.28	0	125/III
***Citronella* sp.**													
12535	3052.53	100	8.15	0	36.71	25	0.2	25	0.8	25	96.12	0	175/III
27394	2955.64	100	10	0	49.55	25	0.17	25	0.83	25	97.22	0	175/III
5463	2961.74	100	10.87	0	45.03	25	0.16	25	0.84	25	97.59	0	175/III
17155	2221.29	100	11.42	0	46.76	25	0.15	25	0.85	25	97.78	0	175/III
20799	2602.44	100	10.51	0	32.96	25	0.16	25	0.84	25	97.44	0	175/III
27249	2602.44	100	11.26	0	45.48	25	0.15	25	0.85	25	97.73	0	175/III
***Gonocaryum* sp.**													
24702	1554.63	100	3.69	0	35.22	25	0.35	25	0.65	25	87.66	0	175/III
29646	2313.18	50	3.08	0	20.44	25	0.39	25	0.61	25	84.51	0	125/III
6342	1599.06	100	2.93	0	43.75	25	0.41	25	0.59	25	83.55	0	175/III
29384	1599.06	50	4.01	0	72.58	50	0.33	25	0.67	25	88.93	0	150/III
***Gomphandra* sp.**													
29676	2633.88	100	7.26	0	66.13	50	0.22	25	0.78	25	95.33	0	200/III
***Medusanthera* sp.**													
19054	3002	100	9.46	0	79.4	50	0.17	25	0.83	25	96.96	0	200/III
***Urandra* sp.**													
29704	1412.75	50	8.55	0	16.18	25	0.19	25	0.81	25	96.41	0	125/III

L Fiber length, *FP* felting power, *RR* Runkel ratio, *FR* flexibility ratio, *CR* coefficient of rigidity, N_k: valued criteria, N_{hit} = mean value (μm)

the highest quality value among the fourteen samples observed, with a value of 200 because the fiber length values of both species were highest among fourteen samples.

Fiber length affects the physical properties of paper such as strength and stiffness. Long fibers allow for wider bonds between fibers, but the longer the fibers, the coarser the paper will be. Longer wood fibers will produce paper sheets that have better strength properties because they have a wider bonding area during milling and better stress transfer properties. On the other hand, shorter wood fibers are able to produce paper sheets that are smoother and more uniform [15].

Six samples of Citronella and four samples of Gonocarium had fiber quality values that tended to be the same which given value of 175. This was caused by the low Felting Power value of these species. The Felting Power affects the tear strength of the paper. Thin-walled fibers will tend to provide low tear strength. A good bond between fibers can be obtained from longer fibers because they play a role in increasing the tear strength of the paper. This is because the tearing force will be divided into a long area. The diameter of the fiber indicates the degree of slenderness. The higher the value of Felting Power, the fiber tends to be more flexible [17].

The Runkle Ratio value of the fourteen samples were zero. This is due to the very narrow lumen diameter of all fiber samples so that the calculated Runkle Ratio value calculated was high. A high Runkle Ratio value indicates that the fiber has a thin wall but a small lumen diameter so that the pulp produced is more difficult to grind and has a narrower bonding area between the fibers so that it is expected to produce pulp sheets with lower tear, tensile, and folding strengths [6].

The analysis of the measurement of the Muhlstep Ratio value of all observed samples resulted in a high value (above 80) which scored as zero. This is due to the large fiber diameter and small lumen which affect very high Muhlstep Ratio results. High Muhlstep Ratio indicates the small surface area of fiber so that the bonding and contact area between the fibers decreases. This causes the resulting paper sheets tend to have low tensile and crack resistance [16].

Overall, all observed wood samples belonged to class III which means that the wood fiber can produce sheets of moderate tear, break and tensile strength because the fiber pulp sheet is easy to flatten and produces a poor weave [15]. The priority for pulp and paper production can be given to wood species with the total highest scoring namely *Gomphandra* sp. and *Medusanthera* sp. While for the other species such can be considered for other purposes.

4 Conclusions

The anatomical properties of selected species can be used for key identification database. All wood samples could be microscopically differentiated from their anatomical characteristics especially from the type of rays, parenchyma, vessel, fiber, and mineral inclusion. The diagnostic characteristics of *Apodytes* sp. (Icacinaceae) were thick-walled fiber cells, the ray composition was procumbent, square and upright cells mixed throughout the ray. In Cardiopteraceae family, *Gonocaryum* sp. ID 6342 had identical characteristics with Gonocaryum sp. ID 24072, while the others were different, thus two species under *Gonocaryum* genera were identified. *Citronella* spp., wood had two distinct sizes of rays, axial parenchyma diffuses, diffuse in-aggregate, and scanty paratracheal. *Gomphandra* sp. and *Medusanthera papuana* (Stemonuraceae) had almost similar macroscopic and microscopic characteristics, except in the type of crystals. Urandra was the most different than the other wood species in Stemonuraceae family with axial parenchyma diffuse, uniseriate and all square rays cells. The fiber quality of all samples were categorised into class III for pulp and paper raw material. Further studies are required to investigate other potential utilizations.

Acknowledgements Authors would like to thank Forest Products Research and Development Center for providing wood samples and laboratory facilities, and special thanks to technicians in the Lignocellulose Anatomy Laboratory for their support during the research: Tutiana, Romi, and Ida. Thanks are also dedicated to Japan Society for the Promotion of Science (JSPS) and Japan ASEAN Science, Technology and Innovation Platform (JASTIP) for providing funding in publication process. All authors are equally contributed to the research and paper writing.

References

1. Kartasujana I (1993) 4000 trees species in Indonesia and its index by region. Forestry Research and Development Center, Forestry Department, Jakarta
2. Soerianegara I, Lemmens R (1994) Plant resources of South-East Asia, vol 5(1). In: Timber trees: major commercial species. Backhuys Publisher, Leiden
3. Lemmens RHMJ, Soerianegara I, Wong WC (1995) Plant resources of South-East Asia, vol 5(2). In: Timber trees: minor commercial timbers. PROSEA, Bogor
4. Sosef MSM, Hong LT, Prawirohatmodjo S (1998) Plant resources of South-East Asia: Timber Trees, vol 5(3): lesser-known timbers. Backhuys, Leiden
5. Mandang YI (2013) Xylarium Bogoriense dan perannya dalam penelitian anatomi dan pengenalan aneka jenis kayu di Indonesia
6. Dewi LM, Supartini S (2011) Anatomical properties of Shorea Mujongensis Ps Ashton, a critically endangered species of Dipterocarps from Kalimantan. Indones J For Res 8(2):91–100. https://doi.org/10.20886/ijfr.2011.8.2.91-100
7. Zobel BJ, Van Buijtenen JP (1989) Wood variation: its causes and control. Springer-Verlag, Heidelberg. https://doi.org/10.1007/978-3-642-74069-5
8. Bowyer J, Shmulsky R, Haygreen J (2003) Forest products and wood science an introduction, 4th ed., Iowa, Iowa State Press (Blackwell Publishing Company)
9. Marbun SD, Wahyudi I, Suryana J, Nawawi DS (2019) Anatomical structures and fiber quality of four lesser-used wood species grown in Indonesia. J Korean Wood Sci Technol 47(5):617–632
10. Oey DS (1990) Berat jenis dari jenis-jenis kayu Indonesia dan pengertian beratnya kayu untuk keperluan praktek. Pengumuman 3
11. Damayanti R, Mu'min, Andianto, Herlina N (2018) Identifikasi jenis kayu secara makroskopis (Macroscopic wood identification), p 31
12. Sass J (1961) Botanical microtechnique, 3rd edn. The Iowa State University Press, Iowa
13. Comittee IAWA (1989) IAWA list of microscopic features for hardwood identification. IAWA Bull 10(3):219–332
14. Damayanti R, Rulliaty S (2010) Anatomical properties and fiber quality of five potential commercial wood species from Cianjur. West Java. Indones J For Res 7(1):53–69
15. Rachman A, Siagian R (1976) Dimensi Serat Jenis Kayu Indonesia. Laporan No. 75, Bogor, Indonesia
16. Dewi LM, Wahyudi I, Mandang Y (2010) Wood anatomy and fiber quality of Platea spp. (Icacinaceae). In: Proceedings of the First International Symposium of Indonesian Wood Research Society, pp 27–38
17. Sofyan K, Nawawi D, Priadi T (1993) Sifat Pulp Jenis-Jenis Kayu Cepat Tumbuh. Faculty of Forestry, IPB, Bogor

Low-Formaldehyde Emission Particleboard Resistance Against Three Different Wood-Decay Fungi Species

Yuliati Indrayani, Evi Septiani, Alkhadi, and Wiwik Winarsih

Abstract In the manufacture of composites, synthetic adhesives are generally used, which cause formaldehyde emissions and impact human health and the environment. Therefore, efforts are needed to reduce formaldehyde emission, one of which is using activated charcoal. The use of bagasse-activated charcoal in the manufacture of particleboard mixture of bagasse and sawdust to minimize formaldehyde emission has been carried out. This research aims to evaluate the resistance of low-emission particleboard against three different wood-decay fungi species. Activated charcoal is made by carbonizing bagasse at 300 °C for 2.5 h followed by carbon activation using 0.1 M HCl solution for 24 h. The particleboard used in this study has to vary formaldehyde emission, such as 5.03, 3.95, 3.61, 3.38, 3.31, 3.28, 3.2, 3.19, 3.13, 3.05 mg/L. Particleboard was exposed to two white-rot wood-decay species white-rot *Coriolus versicolor* (COV) and *Schizophyllum commune* (SC), and brown-rot *Tyromyces palustris* (TYP), for three months testing period. The results showed that particleboard with low formaldehyde emission (high activated charcoal content) was more resistant to all wood-decay species.

Keywords Activated charcoal · Bagasse · Particleboard · Sawdust · Wood-decay fungi

1 Introduction

Particleboard is a composite product made of lignocellulosic materials which are bonded using adhesives. In general, synthetic adhesives based on formaldehyde are used, such as urea–formaldehyde (UF), melamine–formaldehyde (MF), melamine–urea–formaldehyde (MUF), phenol–formaldehyde (PF), and resorcinol–formaldehyde (RF). According to Kumar and Pizzi [10], 95% of formaldehyde-based adhesives produce wood-based composites. Among synthetic adhesives used worldwide,

Y. Indrayani (✉) · E. Septiani · Alkhadi · W. Winarsih
Tanjungpura University, Pontianak, West Kalimantan, Indonesia
e-mail: mandaupermai@yahoo.com

M. A. Rahandi Lubis (ed.), *Proceedings of the 10th International Symposium for Sustainable Humanosphere*, Lecture Notes in Civil Engineering 255,
https://doi.org/10.1007/978-981-19-1422-5_20

UF adhesive is the most dominant, accounting for around 85% of all uses, followed by MF at 10% and PF at 5% [1, 3, 10].

Synthetic adhesives are widely used because they have some advantages, such as high reactivity, cost efficiency, adhesive strength, and moisture resistance [8, 9, 11, 20]. On the other hand, the use of synthetic adhesives causes formaldehyde emission that impacts humans and the environment, especially when used indoors. According to WHO [18], a formaldehyde emission level of 0.1 mg for 30 min of usage can lead to health problems. An effort that can be made to reduce the formaldehyde emission in composite products is adding activated charcoal. Activated charcoal is charcoal that has been chemically or physically activated to have a high absorption capacity with varying carbon content [12].

Previously, research on the use of activated charcoal in composite products only discussed its influence on formaldehyde emission, physical properties, and mechanical properties of composites [13, 15, 17]. Meanwhile, the effect of adding activated charcoal to the composite on biological properties has not yet been reported. Biological degradation can include the termite attack or wood decay fungus. On the other hand, previous studies used wood as a raw material in producing activated charcoal [12, 13, 15, 17]. Meanwhile, there has been no mention of the use of bagasse in the production of activated charcoal on bagasse particleboard. Bagasse was selected for this study because it was readily available, inexpensive, and did not pose a health risk. Later, it may be used to reinforce composites as a sugarcane fiber, which can help to solve environmental issues. Therefore, the aim of this study was to assess the resistance of particleboard made from a mixture of bagasse and sawdust with varying amounts of formaldehyde emission to wood decay fungus.

2 Materials and Methods

2.1 Manufacturing of Particleboard

The particleboard manufacture was done by mixing activated charcoal, sawdust, bagasse, and UF adhesive (Solid Content 52%) with a concentration of 12%, 0.1% catalyst, and 1% paraffin. UF, catalyst, and paraffin were weighed based on the dry weight of the particles. The levels of activated charcoal added to the particleboard were 2%, 4%, and 6%. The target density of particleboard is 0.7 g/cm^3 with a size of 30 × 30 × 10 cm. The board was pre-pressed for ± 1 min, then hot-pressed at a temperature of 140 °C for 10 min with a pressure of 35 kg/cm^2 [7]. After that, the particleboard was conditioned for ± 7 days at room temperature before testing the wood decay fungus.

2.2 Wood Decay Bioassay

The testing method of wood decay fungi was based on SNI 01–7207-2006 with modification.

Media Preparation

The media of the test used was 39 g of PDA (potato dextrose agar) dissolved in 1 L of distilled water. 100 ml of PDA solution was put into each bottle and then sterilized using an autoclave at a temperature of 121 °C with a pressure of 1.5 atmospheres in 30 min. Furthermore, above the PDA media, wood decay fungus was inoculated and stored in the incubation room until the mycelium growth and thick on the surface of the PDA media. The wood decay fungi used were brown-rot *Tyromyces palustris* (TYP), and white-rot *Coriolus versicolor* (COV), and *Shyzophylum commune* (SC) (Fig. 1). The three types of fungi were chosen because they were classifies as wood decay with a high level attack on wood and cause weight loss up to 65% on *T. palustris* [4], up to 70% on *S. commune* [6], and reach 53% on *C. versicolor* [19].

Wood Decay Bioassay

The test sample, which had been known its initial weight of oven-dry, was put in a bottle containing the wood decay culture. Each bottle was placed with one test sample, and it was done for three species of wood decay in separate bottles. The bottle containing the test sample was incubated for 12 weeks at a temperature of 22.7 °C and 50% humidity. Three replications was run for each species of wood-decay fungi. At the end of the test, the test sample was carefully removed from the bottle, cleaned from the attached mycelium of fungi, dried in the oven, and then weighed to determine the weight loss. The weight loss of the test sample was obtained by calculating the percentage of weight before and after testing divided by the initial weight of the test sample with the following formula:

$$Weight\ Loss = \frac{(W1 - W2)}{W1} \times 100\%$$

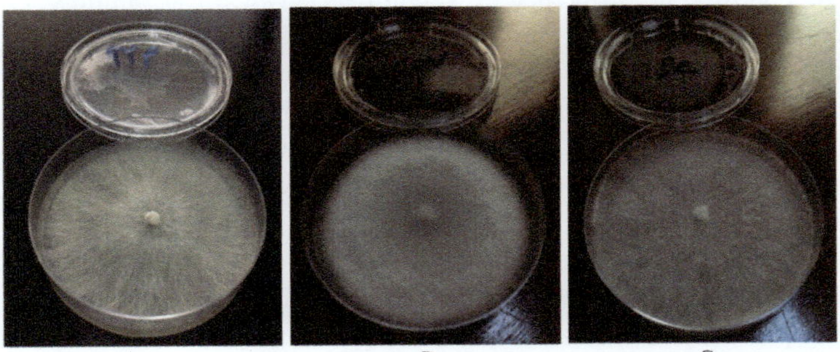

A B C

Fig. 1 Wood-decay fungi: **a** *T. palustris*, **b** *C. versicolor*, **c** *S. commune* where;

$W1$ = Filter paper weight before testing (g).

$W2$ = Filter paper weight after testing (g).

2.3 Data Analysis

The data analysis in this research was done using Microsoft Excel 2013 to calculate the weight loss of the test samples. While for statistical tools, we use analysis of variance (ANOVA). Further analysis to compare between treatments uses Tukey's test according to Gaspersz [5].

3 Results and Discussion

The particleboard used in this research has varying levels of formaldehyde emission, as presented in Table 1.

Table 1 shows that more activated charcoal added in the particleboard manufacture caused lower the formaldehyde emission level. The results of this study are in line with Tristaya et al. [17], who investigated the formaldehyde emission of particleboard of rubberwood (*Hevea brasiliensis* Muell. Arg.). It can be explained because the activated charcoal in particleboard can absorb free formaldehyde of adhesives so that formaldehyde emission became lower [15]. Furthermore, Solovyov et al. [16] reported that the element carbon (C) in the activated charcoal may absorb anion, cation, and molecules in the form of organic and inorganic compounds, either as a solution or as a gas. In addition, the high content of activated charcoal could increase the pattern of amorphous and macropore structures [12].

The value of weight loss of particleboard after being tested on SC, COV, and TYP fungi are presented in Figs. 2, 3 and 4.

Figure 2 showed no difference in weight loss between test samples A1 and A2, A3, and A4, but there was a significant difference with test samples A5, A6, A7, A8, and A9 (Tukey's test: $P < 0.01$). The weight loss of test samples A2 was not significantly different from that of A3, A4, and A6, but it was significantly different from that of A5, A7, A8, and A9 (Tukey's test: $P < 0.01$). The sample test A3 was not significantly different from A4, but significantly different from A5, A6, A7, A8, and A9 on Tukey's test: $P < 0.01$. Meanwhile, the A4 test samples were not significantly different from A5, A6, A7, and A8, but they were significantly different from A9 (Tukey's test: $P < 0.01$). The weight loss of test sample A5 was not different from that of test samples A6, A7, and A8, but it has significant differences from test sample A9 (Tukey's test: $P < 0.05$). The sample test A6 was not different from A7 and A8, but it was different from A9 (Tukey's test: $P < 0.05$). The A7 test sample was not different from the A8 test sample, but it has significantly different from the A9 test

Table 1 Particleboard with vary formaldehyde emission for test sample

No.	Sample	Average emission formaldehyde content (mg/L)
1	A1	0
2	A2	5.03
3	A3	3.95
4	A4	3.38
5	A5	3.31
6	A6	3.2
7	A7	3.19
8	A8	3.61
9	A9	3.13
10	A10	3.28
11	A11	3.05

Note A1: Solid wood control
A2: The control of particleboard without activated charcoal
A3: 2% activated charcoal with 75% particle composition of sawdust: 25% bagasse
A4: 2% activated charcoal with 50% particle composition of sawdust: 50% bagasse
A5: 2% activated charcoal with 25% particle composition of sawdust: 75% bagasse
A6: 4% activated charcoal with 75% particle composition of sawdust: 25% bagasse
A8: 4% activated charcoal with 50% particle composition of sawdust: 50% bagasse
A7: 4% activated charcoal with 25% particle composition of sawdust: 75% bagasse
A9: 6% activated charcoal with 75% particle composition of sawdust: 25% bagasse
A10: 6% activated charcoal with 50% particle composition of sawdust: 50% bagasse
A11: 6% activated charcoal with 25% particle composition of sawdust: 75% bagasse

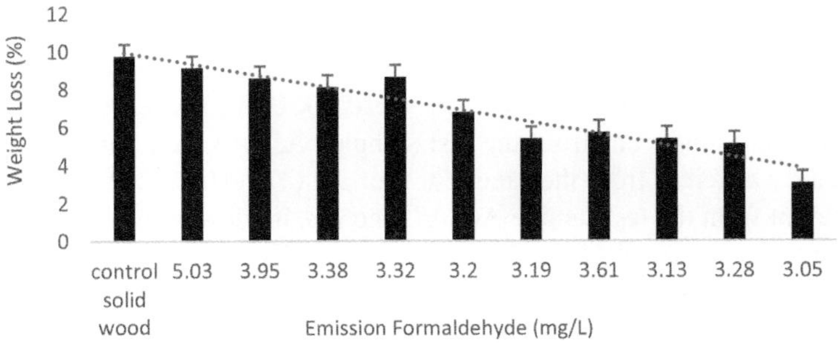

Fig. 2 Weight loss of the test sample after 12 weeks exposed to *S. commune*

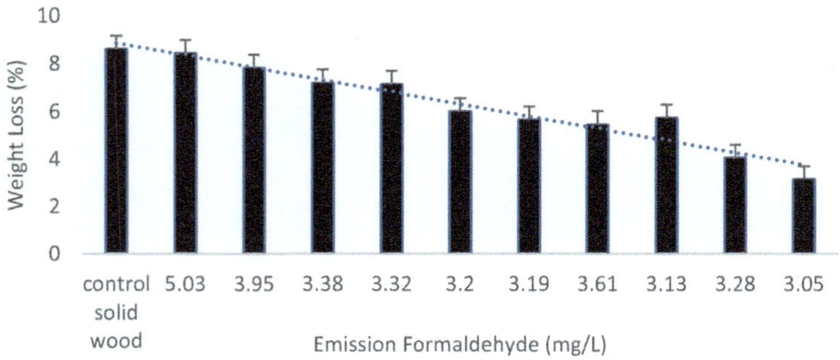

Fig. 3 Weight loss of the test sample after 12 weeks exposed to *C. versicolor*

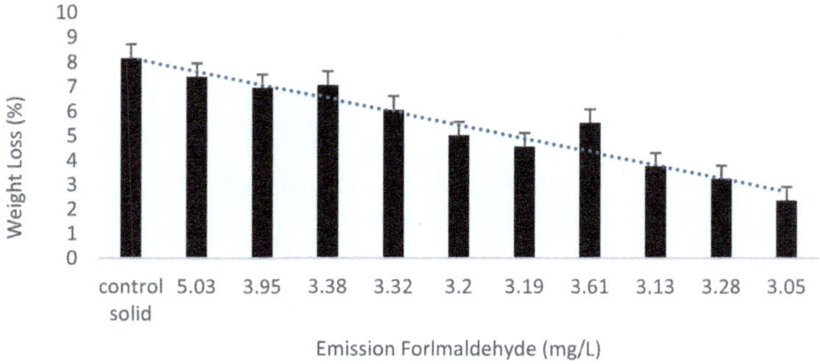

Fig. 4 The weight loss of the test sample after 12 weeks exposed to *T. palustris*

sample (Tukey's test: $P < 0.05$). The test examples A8 and A9 were not significantly different.

Figure 3 showed that the test sample A1 was not different from the test sample A2, A3, A4, A5, A6, and A7, but it was significantly different from A8 and A9 in Tukey's test: $P < 0.01$. The weight loss of test samples A2 was not different from that of A3, A4, A5, A6, and A7, but it was significantly different from those of test samples A8 and A9 (Tukey's test: $P < 0.01$). Meanwhile, there was no difference between test samples A3, A4, A5, A6, and A7, but it has significant differences from the test samples A8 and A9 (Tukey's test: $P < 0.01$).The test samples A4 were not significantly different from the test samples A5, A6, A7, and A8, but it was significantly different from the sample test of A9 (Tukey's test: $P < 0.01$). A5 was not different from the test sample A6, A7, and A8, but it was significantly different from the test sample A9 (Tukey's test: $P < 0.05$). The sample test of A6 was not different from the A7 and A8, but it was different from A9 (Tukey's test: $P < 0.05$). Similarly, the test sample of A7 was not different from A8, but it had a significant difference from A9 in Tukey's test: $P < 0.05$. Meanwhile, the test sample of A8 was not different from A9.

Figure 4 showed no difference between the test samples A1 and A2, A3, A4, and A6, but they were significantly different from A5, A7, A8, and A9 (Tukey's test: $P < 0.01$). The sample test of A2 was not different from A3, A4, and A6, but it had significantly different from that of A5, A7, A8, and A9 (Tukey's test: $P < 0.01$). The weight loss in test samples of A3 was not different from A4, A5, and A6, but it was different from that of A7, A8, and A9 (Tukey's test: $P < 0.01$). Meanwhile, the test sample of A4 was not different from A5, A6, A7, and A8, but it was different from A9 (Tukey's test: $P < 0.01$). The sample test of A5 was not different from A6, A7, and A8, but it was different from A9 (Tukey's test: $P < 0.01$). The sample test of A6 was not different from A7, but it was different from A8 and A9 (Tukey's test: $P < 0.05$). The test sample of A7 had no difference from A8 and A9, and the test sample of A8, and there was no difference from A9.

The average weight loss of samples after exposure to wood decay varied from 3.04 to 8.64% for SC, 3.18 to 8.47% for COV, and 2.35 to 7.4% for TYP (Figs. 2, 3 and 4). In general, it can be said that the lower the formaldehyde emission level, the lower the sample weight loss (Figs. 2, 3 and 4). The sample with low formaldehyde emission had a higher content of activated charcoal so that the sample contained more pure activated carbon; consequently, wood rot fungus did not like the sample. This study's characteristics of activated charcoal of bagasse were 10.54% moisture content, 6.12% ash content, and 66.68% pure activated carbon. It is known that white-rot fungus degraded the lignin chemical components, while the brown wood-rot fungus degraded cellulose and a small amount of lignin chemical components to obtain food. White-rot fungus degrade the chemical component of lignin so that the remaining chemical component in the wood is cellulose, causing the wood attacked by this fungus has a white color. On the other hand, brown-rot fungus degrade cellulose chemical components, as a results, the wood attacked by this fungus has a brown color because the remaining chemical component in the wood is lignin. The activation process of charcoal into activated charcoal could increase the adsorption ability of activated carbon itself [17]. Adsorption properties could attract other substances around it to interact and bound to these substances, and this adsorption process could occur between solid and solid or between solid and liquid. It might cause a bond between cellulose and lignin on particleboard with activated charcoal, as a result, wood rot fungus did not recognize the cellulose and lignin as a food source. Furthermore Pari et al. [13] explain that the element carbon (C) in activated charcoal is able to absorb organic compounds including cellulose and lignin, then the carbon is covalently bonded in a hexagonal lattice.

The weight loss of the test sample was also influenced by the composition of sawdust and bagasse in particleboard manufacture. Figures 2, 3, and 4 show that the higher the sawdust content, the higher the weight loss of the test sample. Particleboard with a composition of 75% sawdust and 25% bagasse experienced a higher weight loss than other compositions in testing all species of wood rot fungus (Figs. 2, 3 and 4). It was understandable because the cellulose and lignin content in wood was higher than the cellulose and lignin content in bagasse. As previously explained, for their survival, wood rot fungus degraded lignin (for white-rot fungus) and cellulose (for brown-rot fungus). The cellulose content in wood was 38–49% and lignin 23–30% [14], while the cellulose content in bagasse was 42–46% and ligni18–20% [2].

In this study, two species of white-rot fungus were used, SC and COV, and one species of brown-rot fungus, TYP. The test results on those three types of wood decay fungus in this study showed that the weight loss of the test samples after being exposed to TYP was smaller than SC and COV. It was understandable because TYP only degraded lignin while SC and COV degraded cellulose and in a small amount of lignin. Among the white-rot fungus, SC caused higher weight loss when compared to COV. During the test, the mycelium of the SC was thicker than COV may explain this phenomenon.

In conclusion, this study showed that the activated charcoal of bagasse could be used as an antifungal material for wood-decay fungus based on the weight loss of the test sample. The results of this study indicated that a high level of activated charcoal could resist the attack of wood-decay fungus, and it was an environmentally friendly material that was not harmful to humans and the environment. The activated charcoal level of sugarcane bagasse was as much as 6% on the particleboard it was able to repress wood-decay fungus by 53% for SC, 48.76% for COV, and 57.97% for TYP. Further research is needed to examine low formaldehyde emission particleboard to termite resistance to develop wood-decay fungus and termite resistant particleboard technology.

References

1. Byung-Dae P, Jae-Woo K (2008) Dynamic mechanical analysis of urea-formaldehyde resin adhesives with different formaldehyde-to-urea molar ratios. J Appl Polym Sci 108(3):2045–2051
2. Carvalho W, Canilha L, Castro PF, Bardosa LDFO (2009) Chemical composition of the sugarcane bagasse. In: 31st Symposium on Biotechnology for fuels and chemicals, pp.3–32. San Fransisco, CA
3. Costa N, Pereira J, Ferra J, Cruz P, Martins J, Magalhães F, Mendes A, Carvalho LH (2013) Scavengers for achieving zero formaldehyde emission of wood-based panels. Wood Sci Technol 47:1261–1272
4. Darma IGKT, Hadi YS, Atmojo AT (2002) Decay resistance of wood polymer composite (WPC) against brown rot fungi tyromyces palustris. J Trop For Manage VIII(1):31–38 (Indonesian)
5. Gaspersz V (1991) Metode Perancangan Percobaan. CV, Armico. Bandung
6. Herliyana EN, Maryam LF, Hadi YS (2011) Schizophyllum commune Fr. as Indonesian National Standard Wood Resistance test fungi on four kinds of community wood: sengon, rubber, tusam, and mangium. Trop Silvicult J 2(3):176–180 (Indonesian)
7. Iskandar MI, Supriadi A (2013) The effect of adhesive levels on properties of sugarcane bagasse particleboard. J For Prod Res 31(1):19–26
8. Jin Y, Cheng X, Zheng Z (2010) Preparation and characterization of phenol-formaldehyde adhesives modified with enzymatic hydrolysis lignin. Biores Technol 101:2046–2048
9. Jivkov V, Simeonova R, Marinova A, Gradeva G (2013) Study on the gluing abilities of solid surface composites with different wood based materials and foamed PVC. In: 24th International Scientific Conference Wood is Good–User Oriented Material, Technology and Design Proceedings, pp 49–55. Zagreb, Croatia

10. Kumar RN, Pizzi A (2019) Environmental aspects of adhesives—emission of formaldehyde. In: Adhesives for wood and lignocellulosic materials. Wiley-Scrivener Publishing, Hoboken, NJ, pp 293–312
11. Mantanis GI, Athanassiadou ET, Barbu MC, Wijnendaele K (2018) Adhesive systems used in the European particleboard, MDF and OSB industries. Wood Mat Sci Eng 13:104–116
12. Pari G, Sofyan K, Syafii W, Buchar (2004) Active charcoal as a formaldehyde capture material in plywood. J Tek Ind Pert 14(1):17–23
13. Pari G, Santoso A, Hendra D (2006) Manufacturing and application of activated charcoal as reductor of plywood formaldehyde emission. For Prod Res J 24(5):425–436 (Indonesian)
14. Rowell RM, Pettersen R, Tshabalala MA (2012) Handbook of wood chemistry and wood composites, 2nd edn. CRC Press, London, New York
15. Santoso A, Pari G (2012) The effect of active charcoal in raw material mixture on particleboard properties. For Prod Res J 30(3):235–242 (Indonesian)
16. Solovyov LA, Shmakov AN, Zaikovski VI, Joo SH, Ryoo R (2002) Detailed structure of the hexagonally packed mesostructured carbon material CMK-3. Carbon 40:2477–2481
17. Trisatya DR, Prastiwi DA, Santoso A (2018) Adsorptivity of activated charcoal on formaldehyde emission of rubberwood (*Hevea brasiliensis* Muell. Arg.) particleboard. J ITEKIMA 3(1):48–59 (Indonesian)
18. World Health Organization (2010) WHO guidelines for indoor air quality: selected pollutants. WHO Regional Office of Europe
19. Yamamoto K (1988) Decay resistance of extractives from chengal (*Neobalanocarpus heimii*) 1(1):51–55
20. Zhang W, Ma Y, Wang C, Li S, Zhang M, Chu F (2013) Preparation and properties of lignin-phenolformaldehyde resins based on different biorefinery residues of agricultural biomass. Ind Crops Prod 43:326–333

Discourse on Exploring of Natural Agarwood: A Case Study of Papua-Indonesia

Herman Hidayat and Robert Siburian

Abstract Agarwood is a non-timber forest product (NTFP) which is a domestic and global commodity, because economically it has a high selling price of non-timber forest products. Indonesia became a supplier of agarwood products in the late 1990s and the early 2000s. The paper examines four critical questions as mentioned below: First is to highlight the significance of agarwood as trading commodity. Second is to examine agarwood known by the Papua people as commodity. Third is to know agarwood traders in Papua from province level, district and sub-district level. The conceptual and theoretical framework is used 'political ecology' which highlights the role of stakeholders to analysis the significance of agarwood as trading commodity whether domestic and international as well. Besides, to know trading network of agarwood from province level, subdistrict and village. The findings of this fieldwork are that Ministry of Environment and Forestry still ignoring on non-timber forest products, including agarwood; Indonesia receives foreign exchange earnings from agarwood trading reached US$ 1.6 billion in 2017 and IDR 28 billion as PNBP (non-taxes as government revenue); to know effectively economic and social improvement among Papua people due to natural agarwood.

Keywords Agarwood · Trading · Political ecology · Stakeholders · Commodity · Local people

Herman Hidayat and Robert Siburian contributed equally as the main contributor of this paper. All authors read and approved the final paper.

H. Hidayat (✉) · R. Siburian
The Research Center for Society and Culture, National Research and Innovation Agency (BRIN), Jakarta, Indonesia
e-mail: hermanhidayat246@gmail.com

M. A. Rahandi Lubis (ed.), *Proceedings of the 10th International Symposium for Sustainable Humanosphere*, Lecture Notes in Civil Engineering 255,
https://doi.org/10.1007/978-981-19-1422-5_21

Fig. 1 *Kemedangan* level of agarwood (Nature Conservation Agency of Papua Province, April 2018)

1 Introduction

1.1 The Significance of Agarwood in Trading

Agarwood or familiar so-called 'kayu gaharu' in the Indonesian language is one of the non-timber forest products (NTFP) that has become a commodity both domestically and globally with high economic value, compared to other NTFP, including rattan, agarwood, resin, honey, fruits, and so on [5: 22]. Agarwood that is relatively widely traded is the form of chips. Other forms that are gaining popularity for trading are in the form of blocks, logs, flakes, perfume and powder [17: 150]. Agarwood oil as an extract from agarwood is also widely traded, with prices ranging from US$ 50,000 to US$ 80,000/litre [20]. In the domestic context, the price of agarwood commodity is also relatively high. In 2006, for example, the price of agarwood in the market in Samarinda city (*the capital of East Kalimantan province*) reached IDR 30,000,000/kg for the *super-super-king* level (best quality), while the average price for the super level reached IDR 10,000,000/kg, and the lowest quality or low level which is called *kemedangan* (Fig. 1) around IDR 25,000-IDR 50,000/kg [28: 9]. Meanwhile, the results of field research on the island of Lombok in May 2018 showed that the price of super-king level for natural agarwood reached IDR 70,000,000/kg (Fig. 2).[1] Therefore, efforts to trade agarwood, including its derivative products (perfume, powder, flakes) are also relatively high in price.

Indonesia is one of the agarwood-producing countries in the world. Other countries in Southeast Asia (ASEAN) that produce agarwood, include Malaysia, Vietnam, Cambodia, Thailand, Laos, and Papua New Guinea [11]. Agarwood is one of the local and global commodity products. Southeast Asia, with Indonesia as one of the largest agarwood producers, has been providing agarwood products that are traded both locally and internationally. This is reinforced by the opinion of Soeharto et al. [25: 1] that the global trade related to 'agarwood' has been carried out since two thousand

[1] Interview with informan (Z) on May 10, 2018. As natural agarwood trader in Mataram (the capital of Nusa Tenggara Barat Province), he had agarwood *super-king* level with the price reaches US$ 5000/kg. US$1 conversion to become Indonesian currency reaches IDR 14.000.

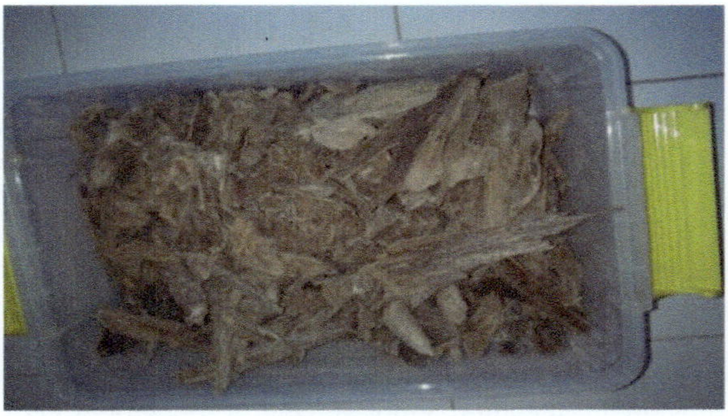

Fig. 2 Super king level of agarwood (Team Research Document, May, 2018)

years ago. In the context of this agarwood trading, Indonesia and Malaysia have played a role as the main agarwood suppliers in international trade in the early 1990s [17: 163]. This is related to the main suppliers before the 1990s recorded including India, Bangladesh, Thailand, Cambodia, Vietnam and China. Indonesia appears in the international agarwood trade after agarwood producing tree in former supply countries it was exploited.

According to a report from CITES (*Convention on International Trade of Endangered Species*) [3], it is shown that Indonesia and Malaysia are the largest agarwood producing countries compared to other ASEAN countries in the international market. In the 1995–1997 period for instance, the CITES report showed that Indonesia's total exports reached 923 tones, almost three times that of Malaysia, which exported 341 tones. Data from the Central Statistics Agency [1, 2] regarding the export of agarwood originating from Indonesia for the last ten years, namely from 2008 to 2018, is interesting to study further because the volume of agarwood exported by Indonesia shows a decreasing trend (Fig. 3). The largest decrease was seen especially in the last two years (2017–2018) with an export volume of less than 900,000 kg of agarwood per year, but the export value in year similar was increased (Fig. 4). This is due to the increasing price of agarwood in the international market. Meanwhile, in previous years the export volume of agarwood was always more than 1.4 million kg and the share had reached more than 4.5 million kg in 2011–2014.

Access to international markets largely determines agarwood products in Indonesia, as a global commodity. Conditions that occur in other countries, especially countries that are consumers of agarwood from Indonesia, cannot be separated

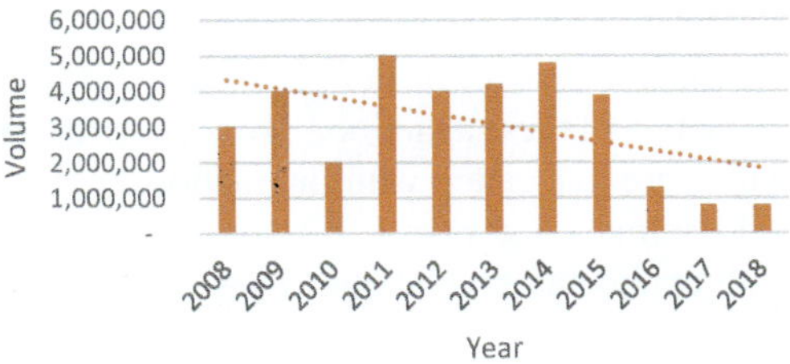

Fig. 3 Export volume of agarwood Indonesia (Kg) Year 2008–2018 (Conservation Office on Natural Resources (BKSDA), Abepura, 2018)

Fig. 4 Export value of agarwood Indonesia (US$) Year 2008–2018 (Conservation Office on Natural Resources (BKSDA), Abepura, 2018)

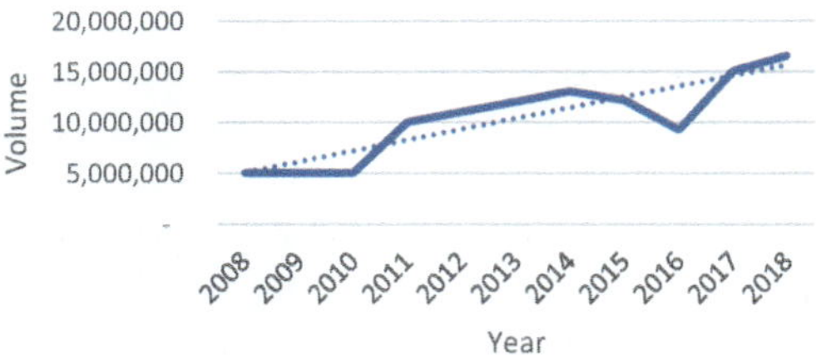

from the influence of globalization. This is because globalization is related to the movement of capital, labor and cultural practices between nations [18], determining the ups and downs of demand for agarwood from Indonesia. This is reinforced by research findings [16], that globalization has resulted in wider social relations that transcend national boundaries. Therefore, the local transformation that occurs in a place is part of the influence of globalization which is the development of social connections that transcend time and space. As an example of what is experienced by agarwood traders in Lombok, when there is economic and social instability in the world, due to the bombing of the twin towers in the United States known as the "nine-eleven" (September 9, 2001) incident, agarwood trade transactions in Lombok are almost non-existent. Likewise, when Saddam Hussein's regime from Iraq invaded Kuwait in 2003, it also resulted in a decrease demand for agarwood from Lombok and other islands in Indonesia (Interview Y, May 11, 2018). In this case, globalization has increased global connections and inter-regional dependence, and the capacity possessed by a country in other countries.

Indonesia cannot supply all the world's consumer demand, this is related to the increasingly limited natural agarwood production from forest areas, also because each agarwood importing country has a more specific choice of agarwood products [12: 11]. As an illustration, consumers who come from the Middle East (Arab society) and Indians prefer to choose the type of agarwood *Aquilaria malaccensis* as an ingredient in perfume and religious traditions,Chinese people prefer *Aquilaria sinensis* because it is intended for medicinal purposes; South Korean society recognizes only *Aquilaria malaccensis* as true agarwood; and Japanese consumers prefer *Aquilaria crassna*, which is used for meditation.

With the excessive exploitation of agarwood from natural forests without considering the carrying capacity of agarwood-producing trees and their regeneration, coupled with the high conversion of forest functions to non-forest such as agricultural, forestry industry and mining sector, this has accelerated the extinction of the agarwood-producing trees. As stated by Mohamed and Lee [17], over exploitation of the wild mother free from which agarwood is derived of the young tree. To prevent the extinction of these agarwood-producing trees in natural forests, the parties to CITES, which are associated with international agreements that focus on the protection of wild animals and plants that are traded internationally that are not in accordance with applicable regulations. Thus, it is possible for wild animals and plants to be threatened (*extinct*), agarwood, including its derivatives such as extracts, finished products packaged and ready for retail trade, powder, and woodchips, are included in the CITES Appendix II list. The categories of wild animals and plants included in Appendix II are all species which although not necessarily now threatened with extinction may become so unless trade in specimens of such species is subject to strict regulation in order to avoid utilization incompatible with their survival. Agarwood species obtained from the species *Aquilaria malaccensis*, entered the CITES list in Appendix II in 1995, while all *Aquilaria* and *Gyrinops* species were included in the Appendix II list in 2004 [21: 2]. With the inclusion of agarwood in the list at CITES, the agarwood trade is regulated by a quota system where the amount of circulation in each country is evaluated annually.

The government also receives *non-tax state revenue* (PNBP) from the yield of agarwood trade transactions from year to year. We examine the PNBP that the government has received from the 2013–2017 period has decreased. This is related, because the target quota given by the Conservation Office on Natural Resources (BKSDA) to agarwood entrepreneurs at the provincial level has not met the target, because the yield of natural agarwood is decreasing (Fig. 5).

In addition, the rational reason expressed by the informant (H) of Provincial Agency on Planning and Development Agency (Interview, 19 July 2018) was that the central government was more oriented towards developing a capitalist-based macro-scale economy to exploit natural resources. On the other hand, attention to NTFP, including *natural agarwood* products, has been neglected, even though the NTFP sector is a large economic potential involving the participation of local communities, immigrants and agarwood commodity traders whose economic value is considered high, contributing up to IDR 28 billion as PNBP. The economic activities of NTFPs can be seen from collecting natural agarwood from villages, sub-districts, districts, provinces and even sending natural agarwood products to major central market centers such as Surabaya, Jakarta, and international markets (Singapore, Hong Kong, Sanghai, Taiwan, Saudi Arabia, and Abu Dabi).

According to Barden et al. [3] states that from 1995 to 2018 it is seen that the largest agarwood export destinations in international trade are Middle Eastern countries (Saudi Arabia, Arab Emirates, Kuwait, etc.), Hong Kong and Taiwan (Table 1). Meanwhile, Singapore plays an important role as an importer of agarwood products from South and Southeast Asian countries, especially Indonesia in the form of agarwood and oil products. As an illustration, in year 1995–1997, around 70% of the total 1113 tons of *Aquilaria malaccensis* agarwood (nearly 800 tones) were exported by Indonesia and Malaysia, then re-exported from Singapore. In addition, according to a report by the Taiwan Customs Office, Indonesia is the largest supplier of agarwood *Aquilaria spp* in Taiwan, which was recorded in 1993–1998 with a total of more than 2800 tons, more than five times the volume exported by Vietnam as the country of largest agarwood second in Taiwan at that time.

On the other hand, capitalist economic practice is only centered on the economic elite in the center of Jakarta and the provinces, including the forestry sector (*Hak Pengusahaan Hutan*/HPH and *Hutan Tanaman Industri*/HTI) and the industries of sawmills, plywood and pulp-paper; agricultural sector (oil palm); mining sector (coal, nickel, oil and gas). It must be admitted that the foreign exchange income obtained

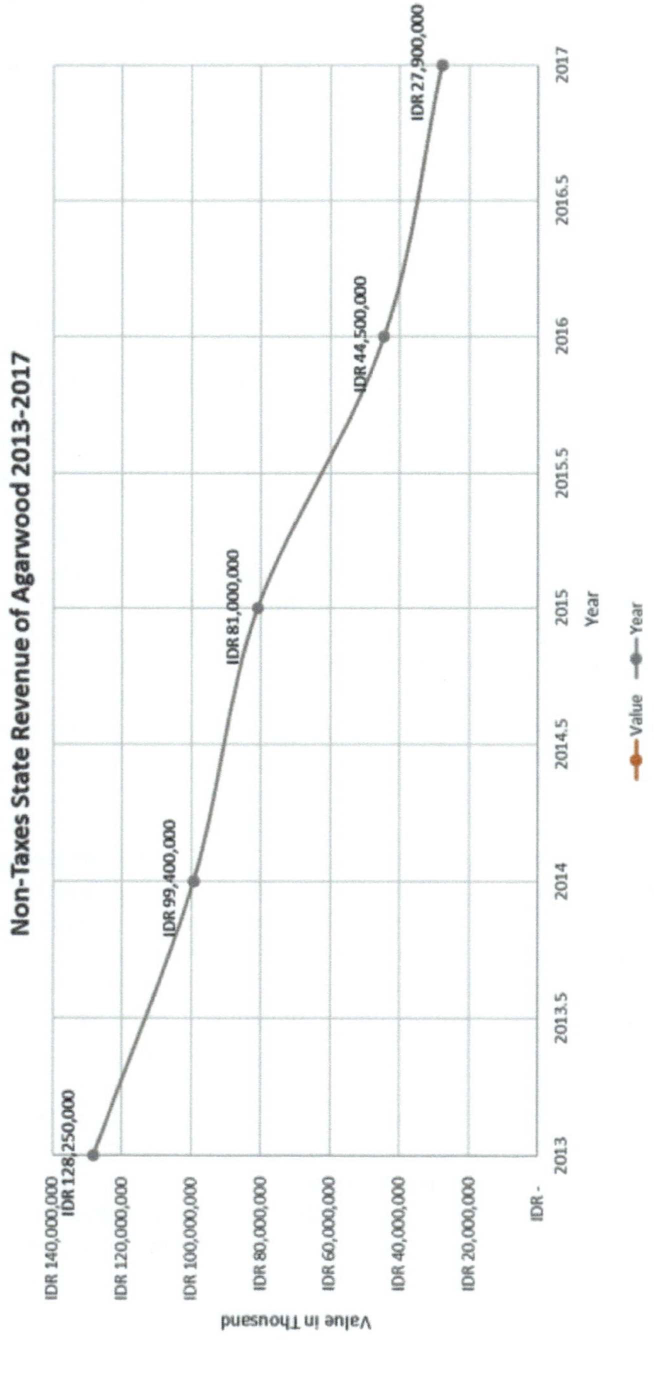

Fig. 5 Non-tax state revenue (PNBP) in 2013–2017 (East Kalimantan BKSD Office 2017)

Table 1 Estimation of export value of Indonesia agarwood in 2018

Country	Export volume per Kg (US$)	Export volume per Kg (IDR)
Singapore	9.60	137,523
Saudi Arabia	27.16	388,910
United Arab Emirates	15.47	221,502
Taiwan	1.85	106,893
China	1.04	315,774
Korea Selatan	105.00	525,870

Note In 2018 (US$ 1 = Rp. 14,320). Export volume per kg from US$ is conversed to become Indonesian currency (Rupiah) based on current curs in 2018
Source Badan Pusat Statistik/BPS Report [2]

from various capitalist economic activities is classified as large, can create employment opportunities and *regional revenue* (PAD) for the province and taxes for the central government. As an illustration, it can be seen in the distribution of production forests in East Kalimantan, that 83% of the area of 13 million hectares of production forest in East Kalimantan[2] has been granted for HPH and HTI licenses, business use rights (*Hak Guna Usaha*/HGU) for oil palm, mining, and mix use in the plantation and mining sectors. The forest potential is controlled by large investors who get permits from the central government, so that the remaining forest land for customary communities and *other use areas* (*Areal Penggunaan Lain*/APL) is very limited for use by the community. The challenge going forward is how to invite the participation of local communities to manage and develop NTFPs at the level of *social forestry* (*Perhutanan Sosial*/PS) with a *partnership* forest scheme between the community, the provincial Forestry Service through the Forest Management Unit (*Kesatuan Pengelolaan Hutan*/KPH) and the Director General of Social Forestry and Protection Forest to develop *agro forestry* including community-based agarwood.

With the existence of CITES which has been adopted by the Indonesian government in Government Regulation (*Peraturan Pemerintah*/PP) number 8 of 1999, the use of agarwood must follow the stages and regulations such as determining quotas, taking from nature or cultivated products (*captive breeding*), transportation for domestic circulation and transportation for marketing abroad [23: 13]. In the context of local communities also have their own rules regarding harvesting of agarwood from the forest. As an illustration, the *Dayak Kenyah* tribe in Metulang village, Kayan District, Malinau Regency, for example, according to Subiakto and Bakrie's research [26: 306] states that they have their own system for agarwood producing trees. Agarwood trees can only be felled if *aing* (black and fragrant contents) is found in the tree. Otherwise, the tree is allowed to continue to grow and may not be cut down until *aing* is there. Therefore, the purpose of cutting down trees is to get *aing*, so cutting down agarwood-producing trees without getting *aing* is pointless and

[2] Production forests in East Kalimantan are used for the allocation of HPH, HTI, and plantations (oil palm) around 10.6 million ha (source: Directorate of Public Forest Production, 1997) and Hidayat, H, *Politik Lingkungan* (Political Ecology) [9: 43].

consumes energy. Another illustration, the Ngata Toro indigenous people in Central Sulawesi also have customary sanctions in the management of natural resources in the form of non-timber forest products, including rattan, *panakangi*, agarwood, resin, and fruits. If those who manage the forest resources are not based on the provisions of customary law, they will be subject to customary sanctions in the form of *Tolu Ongu, Tolu Mpulu, Tolu Tingkau* (3 buffalo or cows, 30 panes, and 30 pieces of Mbesa cloth) [15: 198]. As a result, the role of agarwood is increasingly important, not only in domestic and international economic trade, but also for the benefit of sustainable ecological development that relies on local plant species, one of which is an agarwood-producing tree. Due to the fact that agarwood-producing trees are increasingly difficult to obtain in Indonesia's natural forest areas, the cultivation of agarwood-producing trees is one solution so that agarwood production and export of agarwood to other countries continues, so that economic benefits can be obtained by the community. At the same time, Indonesia's natural forests are not getting damaged due to over exploitation, so that the agarwood-producing trees in natural forests are maintained. It's just that, until now, the quality of cultivated agarwood (*inoculation)* has not been able to reach the highest quality of natural agarwood, including the super level (double-super) and super quality [19: 42]. The obstacle to developing agarwood production for cultivated agarwood is because the quality of agarwood produced is still low, far below the quality of agarwood obtained from natural forests [10: 235]. Therefore, the demand for cultivated commodities is not yet high. In addition, there are not many *inoculants* to produce cultivated agarwood that are available for sale. *Inoculants* traded in the market are relatively expensive. However, agarwood tree cultivation is absolutely necessary for the future.

Natural agarwood from Indonesian forests has a long history and is known globally. Indonesian natural agarwood is recorded as the main commodity bartered between the Chinese dynasty and the kingdoms in Indonesia in ancient times, namely since the era of the Silk Road [27]. Although Indonesian natural agarwood entered global trade in the 1990s, natural agarwood production continued to decline, as seen from the quota given by CITES for the distribution of Indonesian agarwood never fully realized. For example, the 2015 quota of 746,500 kg was realized only 156,485 kg,the 2016 quota of 3,593,800 kg realized 1,523,747 kg; and the quota in 2017 of 1,064,600 kg which was realized was only 352,265 kg (Ministry of Environment and Forestry RI, 2019).

Observing from above description, threfore due to the increasing scarcity of agarwood production from natural forests while the global demand for agarwood is still high, so that the demand for agarwood is higher than what natural forests can provide [7], it is hoped that efforts to meet the demand for agarwood are still high. Therefore, apart from still very limited quantities of agarwood being produced from Indonesia's natural forests, cultivated agarwood is also being developed on community plantation lands. The paper would like to focus on Papua province in eastern Indonesia (Fig. 6) with dominant natural agarwood from viewpoint of history of agarwood an initial came to Papua.

Fig. 6 Map of Indonesia where Papua Province is located. *Source* Google Search (2020)

2 Methodology

This paper utilizes 'political ecology' as the theoretical framework of analysis which emphasises the role of stakeholders' movement (the role of central and local government as well, NGOs, private sector/agarwood traders, local community, etc.) [4]. The reason why Papua province is selected to be field work of agarwood based two reasons. First is the potential of natural agarwood being large and the quality of Papua's agarwood is classified as old and very good. Second agarwood is very significant to the local government's income due to it provides billions of rupiah (IDR 1.8 billion) in the form of the non-tax state revenue (Pendapatan Nilai Bukan Pajak/PNBP) of Papua Province.

Meanwhile, the method to collect data during field work in April–May 2019 used literature reviews; participation observation; and interview with several informants (17 persons) as follows: forestry agency officers in province level, trading agency officers in province level and district, the Regional Development Planning and Development Agency (*Badan Perencanaan dan Pembangunan Daerah*/Bappeda) in province and district, Natural Resource Conservation Agency office (*Balai Konservasi Sumber Daya Alam*/BKSDA), agarwood traders, NGO/Indonesia Farmers of Agarwood Association, local community and also use *Forum Group Discussion* (FGD) with invites agarwood traders, local people, NGOs, head of customary rights forest, government officers, etc.

The paper focuses on discussion to adopt some issues to be examined for instance agarwood initially known by the Papua people, the profile of agarwood traders, stakeholder actor's discussion on agarwood for instance provincial forestry agency officer, the regional development planning agency (Bappeda) officer, BKSDA, University of Cendrawasih (UNCEN) officer, and NGO/Indonesia Farmers of Agarwood Association (Asosiasi Petani Gaharu Indonesia/PEGATRI).

3 Agarwood Known by the Papua People

Agarwood for the people of Papua is a new commodity even though the plant has been around for a long time in Papua's natural forests. However, as a commodity that has high economic value and is traded domestically and globally as well, its existence was only known around the 1990s. The introduction of agarwood by the state of Papua New Guinea which borders the land of Papua Province (Indonesia) is even more recent because agarwood was found and harvested in that country only in 1998 [8].

After agarwood was introduced to Papuans, over time more and more local people were involved in finding agarwood in natural forests to sell to agarwood entrepreneurs. The interest of the Papuan people in looking for agarwood is due to the high selling price and also the absence of other alternative jobs. Their active involvement in the search for agarwood is not based on their understanding of the characteristics of agarwood-producing trees containing agarwood, so that the selection of trees that produce agarwood is not carried out. Therefore, whenever they meet a tree that produces agarwood, that tree is cut down. This reality makes the stands of agarwood-producing trees that grow in natural forests quickly run out.

The search for agarwood in Papua reached its peak around 1995–2000s as many businessmen applied for permits related to agarwood to be processed at the Papua Provincial Forestry Service. Those who applied for the permit did not only come from Papua Province but also from Jakarta, South Sulawesi and Sumatra Island. This phenomenon resulted in the occurrence of 'agarwood fever' in Papua Province and was able to make agarwood *a prima donna.* As known that natural agarwood trees was found in 28 districts of forest areas in Papua and West Papua province (Table 2 and Fig. 7).

In the era of 'agarwood fever', many parties left their daily activities, such as the sub-district head, sector police chief, and military district command, to go into the forest to look for agarwood. According to Lianawati [13], many school children in Asmat Regency are involved in looking for agarwood with their parents in the forest by skipping school for months. Leaving work or skipping school to look for agarwood

Table 2 The spread of natural agarwood trees in 28 districts of Papua and West Papua Province

Province	The spread of districts	Amount
Papua	Asmat, Mappi, Mimika, Merauke, Nabire, Puncak Jaya, Tolikara, Pegunungan Bintang, Yahukimo, Waropen, Yapen, Biak, Membramo Raya, Membramo Tengah, Panial, Kabupaten Jayapura, Boven Diegul, Supiori, Jayawijaya, dan Keerom	20 districts
West Papua	Fak Fak, Kaimana, Teluk Wondama, Teluk Bintuni, Manokwari, Sorong Selatan, Sorong district, Raja Ampat district	8 districts
Total		28 districts

Source BKSDA Office of Papua Province, Abepura, 2019

Fig. 7 The map of spread of agarwood in Papua and West Papua Province (BKSDA Office of Papua Province, Abepura, 2019)

in the forest because they are tempted by the yields that will be obtained from the agarwood, because based on their observations, agarwood seekers who are successful in obtaining agarwood from natural forests can easily improve their economic and welfare. To bring agarwood seekers to natural forests, there are businessmen who transport them by helicopter, including bringing agarwood that was harvested from natural forests.

The 'agarwood fever' which occurs in Papua with a relatively high selling price, results in agarwood not only being traded for a certain amount of money but also bartered for foodstuffs [22] and sexual services from female commercial sex workers, as happened in Merauke Regency [14] and the interior of Asmat Regency and Mappi Regency in 1998 (Wijaya n.d.).

The rampant uncontrolled felling of agarwood-producing trees in natural forests has resulted in 'agarwood fever' which never happened that did not last long. Even if there are agarwood-producing trees until 2019, the places where they grow were farther away in the middle of the forest. The explanation of the agarwood entrepreneur in Jayapura Regency states that in 1994, it was very easy for agarwood seekers to get agarwood because that period was the beginning of agarwood being taken from the forest so that the position where agarwood was grown was not so far from the settlements of agarwood seekers and the population was still large. In 1994, the cost to enter the forest to buy food items that were brought in was IDR 150 thousand for a period of 15 days to 20 days. Food ingredients that were brought into the forest include: rice, super noodles, cigarettes, coffee, sugar, milk, soy sauce, sauces, and candles. The results obtained by the agarwood seekers after selling the agarwood ranged from IDR 750 thousand to IDR 1 million. Then in 2000, the cost increased to IDR 500 thousand for a period of 15 days to 3 weeks living in the forest. The results obtained after agarwood is sold are around IDR 1 million. Even so, agarwood seekers who search for agarwood in the forest do not always bring results because sometimes they also fail to get agarwood even though it has been about 15 days to 3 weeks looking for it in the forest.

The presence of agarwood of Papua in the agarwood trade in Indonesia after the existence of natural agarwood, especially from Sumatra and Kalimantan, has become thinner due to excessive exploitation or conversion of forest to non-forest functions. With the increasingly scarce condition of agarwood on the two islands, agarwood entrepreneurs are expanding their search for agarwood to new places, one of which is Papua. Therefore, the parties who introduced agarwood as a commodity with high selling value to Papuans were agarwood entrepreneurs from Sumatra. Before being introduced by these agarwood entrepreneurs from Sumatra, the Papuan people were not familiar with the types of trees that produce agarwood which have high selling value. However, estimates are based on science (*scientific*) that the population of agarwood-producing trees in Papua's natural forests is only 433 trees per hectare ([24] cited by [6]). This means that the population of agarwood-producing trees in Papua is not that high, referring to the estimate referred to, so that if the exploitation of the agarwood producing trees is not carried out, the trees in question will soon become extinct from the land of Papua.

3.1 The Profile of Agarwood Trader in Papua

Junaedi was born in 1966 and originally come from Payakumbuh district, West Sumatra province. He started looking for agarwood to Manokwari and continued to Jayapura in 1994. He founded *Papua Intan Lestari* (PIL) company, which is based in Sentani City, Jayapura Regency. Junaedi sees that the potential for agarwood in the forests of Papua is huge. Agarwood is seen as a non-timber forest product whose results have not been optimized for the community. Junaedi through Papua Intan Lestari company obtained location permits for agarwood searches from various districts including Merauke, Mappi, Asmat, Mambramo Raya, Intan Jaya, Yapen, Waropen, etc. A work system for obtaining agarwood products, the Papua Intan Lestari company employs a group of seekers of 5–7 people to enter the forest area. One group, consisting of the chairperson and members, usually asks for initial funds in advance for operations (down payment is around IDR 10–12 million). So if the members are 5–6 people, each member gets IDR 2 million. If the group obtains natural agarwood with various level such as *kemedangan*, it is usually sold agarwood products at market prices to entrepreneurs who have provided initial capital (Interview with informant Y, 8 April 2019). The golden day of agarwood products from various districts including Mappi, Merauke, Asmat, Membramo Raya occurred in 1997-early 2000s. Rupiah currency at that time reached IDR 18,500 from IDR 3500/per US dollar. This condition impact toward agarwood traders in province level and big traders in Jakarta and Surabaya obtained a big profit.

Besides, another way to obtain agarwood also shows in the (Fig. 8) the flow of capital in Papua Province. Usually, when local entrepreneurs lack the capital to buy agarwood from agarwood seekers through *plasma* (agarwood collector in sub-district), these Papuan businessmen sometimes contact exporters in Jakarta to ask for a certain amount of capital. The capital sent by the exporters is then forwarded

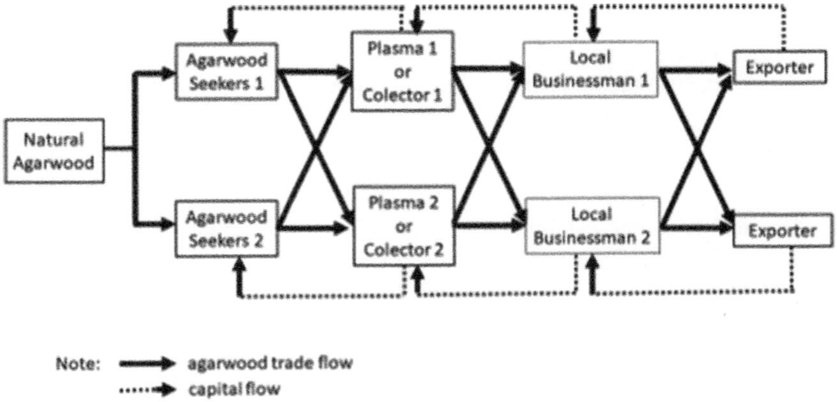

Fig. 8 Natural agarwood trade flow and capital in Papua Province

to some *plasma* to be given to agarwood seekers who will enter the forest to buy food while in the forest and family living expenses while they are left to look for agarwood. Capital requested from exporting entrepreneurs is usually returned after the agarwood has been collected and then taken to Jakarta and after buying and selling transactions between local entrepreneurs and exporters, the money received by local entrepreneurs is first given the capital previously given.

The implication of obtaining agarwood at the village and sub-district levels is that economic growth increases and the purchasing power of local people increases. Although, the economic changes of the local community are still felt to be underdeveloped, because the money obtained from selling agarwood is used by the majority for consumption and purchases of primary needs. The price of this level of *agarwood* (ordinary) or so called "*kemedangan*" is around IDR 25,000–70,000/kg. Usually, this level of company is exported to buyers of Saudi Arabia, Kuwait, United Arab Emirates, Qatar, etc. The price of 1 kg agarwood with super level reached IDR 15–20 million in 2014. The most expensive price was the super super-king level which reached IDR 50 million/kg in the provincial level. This level of super king agarwood is very popular for carving wood products and is installed for home and building decoration by consumers in China, including Sanghai, Taiwan and Hong Kong. Agarwood seekers, if they succeed in obtaining only the super-king level, are said to gain income. And the price on the road reaches IDR 75 million/kg in Jakarta can be sold by big businessmen for export consumption abroad (Hong Kong and Taiwan). Papua Intan Lestari's partner for agarwood sales in Jakarta is Graha Mitra Utama in Cengkareng, Jakarta.

As an illustration, the Papua Natural Resources Conservation Agency (BKSDA) office, which is headquartered in Abepura-Jayapura, has given permission to collect agarwood quota to Papua Intan Lestari company amounting to 2 tons of agarwood in 2018, and 1 ton in 2019. The reason why the quota of agarwood for Papua Intan Lestari company was reduced? According to Junaedi, the reduction in the company's quota was based on two rational reasons. First, because many agarwood seekers in the forests of Papua take a break to do their activities, it is difficult to obtain natural agarwood products. Second, the demand from Chinese consumers has greatly decreased, because the entry permit for natural agarwood products from Indonesia is very limited by the Chinese government.

There are several difficulties faced by agarwood entrepreneurs at the provincial level:

(1) Political and social stability, in some regions still politically volatile, namely threats from the Free Papua Organization (OPM), so that the agarwood search process has stopped, due to security and safety factors;
(2) Natural agarwood is increasingly difficult to obtain, due to the expansion of oil palm plantations, industrial forest plantations (HTI), HPH (forest management rights) by private parties, so that many production forest areas are converted into plantation areas;
(3) The high cost of transporting agarwood by plane to Jakarta. Because previously the cost per kg only reached IDR 17,000/per kg. The price currently reaches IDR 50,000/per kg since the end of 2018/2019;
(4) Entrepreneurs ask the BKSDA office to make permits easier for transportation, so they don't wait long for the permit process.

4 Stakeholder Actor's Discussion on Agarwood

4.1 Government Institutions Discuss on Agarwood Cultivation

The contribution of agarwood to the Non-Tax State Revenue (*Penerimaan Negara Bukan Pajak*/PNBP) of Papua Province reaches IDR 1.8 billion. PNBP from agarwood commodity can actually be increased even more, if agarwood is one of the leading commodities besides sago and is contained in the 2019–2024 Medium Term Development Plan (*Rencana Pembangunan Jangka Panjang Menengah*/RPJM). This paper focuses on the study of institutional structures from the perspective of the Provincial Forestry Service, BKSDA, the Bappeda, and University of Cendrawasih (UNCEN). How do the four institutions discuss the agarwood commodity as a superior commodity in the 2019–2024 provincial RPJM. Even though, in reality the large potential of agarwood in Papua is still facing bureaucratic challenges, so the regional government of Papua province still has not included the agarwood commodity in the RPJM year 2019–2024.

5 Provincial Forestry Service

The trade in natural agarwood recorded in the BKSDA revenue report is very significant in Papua. From 1994 to the late 2000s, the demand trend for agarwood for the commodity of agarwood was high. Most of the agarwood production is obtained from the districts of Asmat, Mappi and Merauke. Agarwood is important to the local government and the people of Papua, for two rational reasons. First, because it provides billions of rupiah (IDR 1.8 billion) in the form of the PNBP for Papua Province. The PNBP comes from the Forest Resources Provision (*Provisi Sumberdaya Hutan*/PSDH) and is recorded and given to the Provincial Forestry Service office. Second, the potential for natural agarwood is large and the quality of Papua's agarwood is classified as old and very good. So, if natural agarwood is increasingly difficult to find and obtain, of course, it correlates with falling PSDH income. There are two ways to prevent decreasing income from agarwood. First, the central government, particularly the Ministry of Environment and Forestry (*Kementerian Lingkungan Hidup dan Kehutanan*/KLHK) prohibits conversion of production forest land to carry out land clearing (forest clearing) for the expansion of oil palm plantations. Second, the local government issued a policy to encourage facilities to cultivate agarwood by involving the participation of local communities. For this cultivation, the Provincial Forestry Service provides facilities for the development of agarwood cultivation in various districts, including Keerom, Sarmi, Yalimo, Waropen, Yapen, Biak, etc. To plant agarwood, the Forestry Service pays for the agarwood seeds that the community collects. This is because the Provincial Forestry Service does not directly take care of agarwood technically. After all, those who take care of technically are in the Forestry Service Branch (*Cabang Dinas Kehutanan*/CDK) and Forest Management Units in each region. As an illustration, in 2016–2017, Kesatuan Pengelolaan Hutan/KPH (Forestry Management Unit) as the leading institution for managing social forestry in agroforestry schemes at the district and sub-district levels has developed 2000 agarwood seeds and distributed them for planting by local communities. The planting scheme allocated 10 ha of land which supported 20 people to participate in planting agarwood.

The problem that arises in Papua is that the large potential of agarwood is still not included in one of the leading commodities (NTFP) in Papua Province and has not been recorded in the 2019 RPJM. It seems that two reasons must be addressed. First, officials from the Provincial Forestry Service and Bappeda officials do not know for sure that the agarwood commodity has high economic value. Second, the implications of agarwood have not been recorded in the provincial RPJM. So as a consequence, the 2019 provincial regional revenue and expenditure budget (*Anggaran Pendapatan Belanja Daerah*/APBD) allocation cannot be given, among other things, for economic incentives and capacity building for human resources (HR), *inoculant* development, integrated agarwood nurseries and planting. This is related to the request of the farmer group, if the agarwood plant is 7–8 years old for harvesting, the farmers usually ask the Provincial Forestry Service to participate in providing sponsorship for inoculant to obtain agarwood sap that smells good and has economic value.

According to the staff of the Provincial Forestry Service (X), the institution has collaborated with the Forestry Research and Development Agency, Gunung Batu, Bogor, which is coordinated by Maman Nurjaman to provide training in understanding *inoculation techniques* and to make inoculant laboratories. Because if the inoculant must be sent from the Bogor Forestry Research and Development Agency, it is considered ineffective, it does not provide the knowledge transfer process to the Papuan people. The hope of the Forestry Service, with the inclusion of agarwood as one of the *leading commodities* that will be included in the Papua RPJM (2020–2024) by the Regional Planning and Development Agency (Bappeda) institution, the budget allocation will increase for the development of inoculants, socialization, and the distribution of agarwood planting is wider in the community, thus making it the agarwood province for Papua (Interview Y, 8 April, 2019). According to the staff of the Forestry Service (X), the institution also hopes that the provincial government will also be active in seeking partners from funding (fund raising) from *Corporate Social Responsibility* (CSR) from both Government Business Entrepreneurship (*Badan Usaha Milik Negara*/BUMN) and private companies. The central of BUMN includes *Pertamina, Bank Indonesia, Bank Rakyat Indonesia* (BRI), *Bank Mandiri, Bank Nasional Indonesia* (BNI). It is hoped that Local Government Business Entrepreneurship (*Badan Usaha Milik Daerah*/BUMD) will contribute from *Bank Papua.* From private companies, among others, Freeport Indonesia, HPH (*Hak Pengusahaan Hutan*/Logging Forest Concession) companies for instance Barito Pacific Timber, Sumalindo Lestari Jaya, Kayu Lapis, Kadeco, etc. Other challenges that must be realized in the future are coordination and synergy of programs to develop agarwood and *inoculant* engineering with various agencies including the Bappeda, the Agricultural Service, the Trade Service, the Industry Service, Research Institutions (regional and national), Universitas Cendrawasih/UNCEN, and developing a list of other commodities that have economic potential such as *sago, patchouli, sugar palm (aren), pineapple,* etc., to be developed in the RPJM year 2019–2024 and invite local community participation, so that there is an increase in income and economic growth and the welfare of the Papuan people.

6 Natural Resources Conservation Agency (BKSDA)

It is necessary to coordinate and synergize programs between the institutional structure of the Provincial Forestry Service and the BKSDA. This is related to the understanding of natural agarwood circulation data in Papua and the trade potential of agarwood from upstream (natural agarwood seekers) in forests, natural agarwood collectors at the village, sub-district, regency to provincial city levels downstream (traders/ large agarwood collectors). Apart from that, the trading of agarwood in the country and the export of agarwood commodities abroad. Thus, natural agarwood and domestic and foreign trading patterns as well as how much PNBP revenue for the state. All supporting elements regarding this important commodity of agarwood, the Forestry Service apparatus should raise the issue of agarwood as a "strategic" commodity such as sago and propose it to be a "superior" commodity and be recorded in the RPJM year 2019–2024 in a technical coordination and synergy meeting programs between other agencies, in particular, the Provincial Forestry Service and Bappeda to be decided. However, in reality, based on findings in the field, agarwood has not been proposed and recorded as a superior commodity in the RPJM year 2019–2024 which usually local governments have attention major in allocating funds and coaching training in human resources, inoculation research, intensification of results and markets.

The Natural Resources Conservation Agency (BKSDA) is an agency under the Directorate General of *Forest Protection and Nature Conservation* (*Perlindungan Hutan dan Konservasi Alam*/PHKA), *Ministry of Environment and Forestry*. The BKSDA office is usually located in the provincial capital. In the context of Papua, the Papua BKSDA office is located in the Abepura district on the middle road between Jayapura Regency and Jayapura city. According to the Decree of the Director General of PHKA P.25/IV.SET/2014, namely the need for regulations for registration of agarwood captive breeding/cultivation and the utilization of agarwood products. The BKSDA office is given the task and authority by the Director General of PHKA to provide quotas for permits for collecting, distributing, and transporting agarwood products out of town. In the context of granting quotas to agarwood traders/collectors an average of 400,000 kg (40 tons) per year. As an illustration, the accumulated types of quota fortunes given from 2010 to 8 April 2019 totalled 4,269,620 kg and the realization only reached 2,536,732 kg (Table 3).

Natural agarwood products originating from several districts in Papua are sent outside such as to Jakarta, Surabaya and Probolinggo. In Jayapura there are two (2) distribution and exploration permit holders such as Papua Intan Lestari and Indo

Table 3 The use of standing agarwood stands in Papua Province (2010–2019)

No	Year	The quota amount	Quota realization	Percentage (%)
1	2010	400,000	359,335	89.83
2	2011	400,000	244,694	61.17
3	2012	450,000	382,757	85.06
4	2013	450,000	179,840	39.96
5	2014	415,000	107,266	25.80
6	2015	401,800	188,605	46.94
7	2016	403,300	79,575	19.73
8	2017	403,300	403,300	100.00
9	2018	471,510	471,510	100.00
10	2019	473,910	120,005	25.32
Total		**4,269,620**	**2,536,887**	**59.42**

Source The Natural Resources Conservation Agency (BKSDA) Abepura Regency, Papua, 2019

Sejati company. Distribution and exploration permit, among others, in the districts of Membramo Raya, Yapen, Waropen, Jayapura, Intan Jaya and Merauke were granted to Papua Intan Lestari company which was held by Junaedi. In the context of obtaining natural agarwood, companies usually use the plasma system to seek natural agarwood in the forest. The plasma system is usually implemented, in which the head of the agarwood seeker is given an amount of money as a down payment (*persekot*) of IRD 10–15 million/group. One group usually has 5–6 members of agarwood seekers. This down payment money is usually used to buy logistical materials. Agarwood seekers (as *clients*), if they obtain natural agarwood, both the *kemedangan* and super level (there are sapwood) categories are given to the entrepreneurs/collectors (as *patrons)* according to the market price. The scheme for selling natural agarwood starts from obtaining agarwood seekers in production forests and protection forests, then transporting it and selling it to district-level entrepreneurs or provincial-level entrepreneurs/collectors who provide working capital to agarwood seekers. From provincial-level entrepreneurs, agarwood products of good quality and super quality are sent to national-level entrepreneurs in Jakarta and Surabaya. From the level of the national businessman, agarwood is wrapped and packaged according to the order and exported abroad, including in the Arab region, such as Qatar, Saudi Arabia, Kuwait, Dubai, United Arab Emirates, etc. Agarwood exports in Asia Pacific regions such as Singapore, Taiwan, Hong Kong, Sanghai, etc. The trade network from village, sub-district, district, provincial and central level collectors is difficult to control, so the agarwood commodity in Indonesia is still inaccurate in calculating the national level, how many national products are traded, both nationally and exported to international markets. The BKSDA office also provided a quota of 1 ton to Indo Sejati company in 2019 and 2 tons in 2018. The same thing was given 1 ton in 2019 and 2 tonnes in 2018 to Papua Intan Lestari company.

According to Juanedi (Interview, April 8, 2019), as Director of Papua Intan Lestari, the amount of natural agarwood is difficult to obtain today in various locations, both in production forests and protected forests, because stocks are running low. So, for agarwood products to be sustainable (*sustainability in product*), agarwood must be cultivated by farmers. Permit holders who obtain a quota, calculated per kg for natural agarwood, must pay in advance to the BKSDA office. On the other hand, the accumulated money that has been collected from distribution, collection and transport permits by the BKSDA office from 2010–2018 has reached IRD 1.8 billion. Three districts have the largest contribution as a contributor to PNBP, including Asmat Regency, amounting to IRD 1.3 billion, Mappi IRD 396,106,000, and Mimika of IRD 26,625,000 as seen in Table 4.

It is obvious that PNBP is registered amount in Table 4 is according to the valid regulation used over Indonesian territory, that PNBP is eventually returned by the central government to the local government in the formation of PSDH. The PSDH results were given to the central Ministry of Environment and Forestry for 20% and 80% for the regions. Its share of the 80% is 16% for the province; 32% for districts/cities as producers; and 32% for other districts in the province concerned.

The potential for natural agarwood is great, according to the staff of the Papua Province BKSDA (interview with Z, 8 April, 2019), among others, obtained in Asmat

Table 4 The utilization of non-tax value income (PNBP) in Papua Province year 2015–2018

No	District	2015	2016	2017	2018
1	Asmat	1,478,502,500	970,312,500	1,258,500,000	1,331,893,500
2	Mappi	468,750,000	545,062,500	507,000,000	396,106,500
3	Merauke	13,500,000	11,250,000	17,647,500	15,150,000
4	Membramo Raya	129,750,000	43,500,000	20,475,000	23,250,000
5	Puncak Jaya	0	0	1,500,000	225,000
6	Puncak	0	0	1,500,000	150,000
7	Nabire	9,000,000	12,750,000	6,000,000	7,500,000
8	Jayapura	6,000,000	23,250,000	13,125,000	8,850,000
9	Bouven Digul	3,750,000	3,750,000	0	0
10	Biak Numfor	1,125,000	3,750,000	0	0
11	Waropen	11, 25,000	9,750,000	6,000,000	5,520,000
12	Yapen	6,000,000	0	0	0
13	Mimika	5,250,000	4,500,000	15,075,000	26,625,000
14	Yahukimo	0	0	0	7,500,000
15	Keerom	0	0	0	1,200,000
16	Sarmi	42,000,000	24,750,000	11,325,000	825,000
17	Supiori	0	1,500,000	4,500,000	0
18	Pegunungan Bintang	0	0	0	600,000
19	Intan Jaya	0	0	0	105,000
Total		**2,175,252,500**	**1,654,125,000**	**1,862,647,500**	**1,825,500,000**

Source The Natural Resources Conservation Agency (BKSDA), Abepura Regency, 2019

Mappi District, Jayapura, Keerom, Membramo Raya, Intan Jaya, Yapen, Waropen, Nabire, Boven Diegul, Merauke in Papua. Agarwood is obtained from mud and has to be dug up. Many practices are found in Asmat district. Meanwhile, in Boven Diegul District, Mappi, Merauke, there are many *kemedangan* agarwood taken from standing agarwood. The Papua BKSDA office has not been able to map the number of quintals of natural agarwood potential in each district, because this data requires professional staff and large funds to conduct field surveys to assess the potential of natural agarwood in each district.

7 Conclusion

(1) Agarwood is a non-timber forest product that is a domestic and global commodity because economically it has a high selling price compared to other NTFP. Indonesia became a supplier of agarwood products in the late 1990s and the early 2000s. Agarwood producers can be found in various provinces in Indonesia, including Sumatra, Kalimantan, Java, West Nusa Tenggara, Sulawesi, Maluku and Papua.

(2) Agarwood products and their trade provide great economic, social, and cultural benefits not only to local communities as natural agarwood producers and seekers, but also Indonesian people from various tribes looking for agarwood to collectors at the subdistrict level, district-level traders, provinces and big traders in Jakarta and Surabaya. The largest exports of agarwood in international trade include the Middle East, Singapore, Taiwan, South Korea, and China (Hongkong). Indonesia got foreign exchange earnings from agarwood export reached IDR 28 billion in 2017 as example. However, along with the higher demand for agarwood products from consumers, including in the form of chips, blocks, logs, flakes, powders, and perfume; the demand has a correlation to encourage increased exploitation of this agarwood commodity. So, to prevent natural agarwood products from extinction, the United Nations in conjunction with CITES has reacted to this critical condition. In this regard, the Indonesian government has adopted Government Regulation number 8 of 1999, that the use of agarwood must follow the stages and regulations, namely determining quotas, taking from nature or cultivated products (*captive breeding*), transporting agarwood for domestic circulation and for foreign marketing. country.

(3) In the case of natural agarwood from Papua province, there are two types of natural agarwood obtained. First, natural agarwood is obtained from standing agarwood-producing trees and natural agarwood that are obtained from agarwood-producing trees that are buried in mud, among others in Merauke, Asmat, Boven Diegul and Mappi districts. Although agarwood obtained from the mud is not taken into account in international trade, this commodity contributes significantly to PNBP returned to the regions in the form of the provision of natural resources for the commodity-producing districts and provides alternative income for local people who seek it.

(4) Agarwood for the Papuan people was introduced by immigrants, especially agarwood traders from Sumatra, Sulawesi and West Nusa Tenggara who deliberately came to the Papua area to look for agarwood in the end1990s. Before the arrival of these seekers and traders of agarwood, the Papuan people were not familiar with the trees that produced agarwood, including its use as NTFP. As an illustration, at the time of Junaedi's arrival, for example, a agarwood entrepreneur from West Sumatra. According to Junaedi's testimony, that in the early days of the introduction of agarwood to the Papuan people, some indicated that the Papuan people were "exploited" by agarwood entrepreneurs from outside. This

is related, that entrepreneurs arbitrarily determine the price of agarwood to buy from local people, because Papuans were ignorant of agarwood. Agarwood obtained by Papuans both from the search for natural agarwood and agarwood obtained from the mud, is priced relatively low, is bartered with foodstuffs and relatively simple houses are built, and there is also bartering for sexual services by women in commercial sex work, and others.

Acknowledgements We are thankful to the Research Center Biology officer, Indonesian Institute of Sciences (LIPI), this article of 'Papua Agarwood' is originally based on funding research from National Research Priority Project (PRN) on CITES II plants and animals 2018–2019. The field-work was carried out in Papua and other provinces for instance West Nusa Tenggara/NTB, East Kalimantan, Riau and Bangka Belitung). Besides, we also thank to committee and reviewers of this symposium of the Sustainable Human sphere, JSPS Alumni (JAAI) for accepting and adopting the article to be published in the proceeding book publication.

References

1. Badan Pusat Statistik (BPS) (2018) Ekspor dan Impor (Export and import). https://www.bps.go.id/all_newtemplate.php
2. Badan Pusat Statistik (BPS) (2019) Statistik Perdagangan Luar Negeri Indonesia Ekspor 2018 Jilid I. BPS, Jakarta
3. Barden A et al (2000) Heart of the matter: agarwood use and trade and CITES implementation for Aquilaria malaccensis. In: TRAFFIC international. Cambridge, UK
4. Bryant R, Bailey S (1997) Third world political ecology. Routledge Press, London
5. de Beer H, McDermott MJ (1996) The economic value of non-timber forest products in Southeast Asia. Netherlands Committee for IUCN, Amsterdam
6. Donovan DG, Puri RK (2004) Learning from traditional knowledge of non-timber forest products: penan benalui and the autecology of Aquilaria in Indonesian Borneo, Dalam. Ecol Soc 9(3). http://www.ecologyandsociety.org/vol9/iss3/art3
7. Han T, Win PP (2016) Influence of inoculation techniques on agarwood chips formation in Aquilaria Malanccensis. Int J Curr Res Life Sci 05(07):604–607 https://www.Researchgate.Net/Publication/308168314_Influence-of_Inoculation_Techniqueson_Agarwood_Chips_Formationin_Aquilaria_Malanccens
8. Gunn BV, Stevens P, Singadan M, Sunari L, Chatterton P (2004) Eaglewood in Papua New Guinea. Workinng Paper No. 51. Resource Management in Asia-Pacific Program, Canberra
9. Hidayat H (2011) Politik Lingkungan: Pengelolaan Hutan Masa Orde Baru dan Reformasi (Political ecology: forest management in new order and reformation era). Obor Press, Jakarta
10. Jensen A (2003) Domestication of Aquilaria spp. and rural poverty-socio-economic and genetic aspect of planting boom in the "wood of the gods". In: NAFRI workshop proceedings shifting cultivation and poverty eradication in the uplands of the Laos PDR, pp 233–239
11. Jung D (2009) The value of agarwood: reflection upon its use and history in South Yemen. Dalam HeiDOK. http://www.ub.uni-heidelberg.de. Accessed 13 Mar 2018
12. Lee SY, Mohamed R (2016) The origin and domestication of Aquilaria, an important agarwood-producing genus. In: Mohamed R (ed) Agarwood: science behind the fragrance. Springer Press, Singapore, pp 1–20
13. Lianawati E (2020) Tertindas, Tinjauan Ekologi Politik Feminis (Oppressed, political ecological review on feminism). https://www.researchgate.net/publication/340870039. Accessed 21 Aug 2020

14. Liputan (Channel) 6 SCTV (2003) HIV di Suku Asmat (HIV in Asmat Tribe). https://www.liputan6.com/news/read/50900/hiv-di-suku-asmat. Accessed 9 June 2020
15. Mahfud R, Toheke RP (2009) The Indigenous Ngata Toro Community-Central Sulawesi: reforming adat to promote environmental, economic and cultural sustainability. In: Kleden EO, Chidley L, Indradi Y (eds) Forest for the future: Indigenous forest management in changing world. Aliansi Masyarakat Adat Nusantara (AMAN) and Down to Earth, Jakarta, pp 179–222
16. Majekodunmi A, Adejuwon KD (2012) Globalization and African political economy: the Nigerian experience. Int J Acad Res Bus Soc Sci 2(8):189–206
17. Mohamed R, Lee SY (2016) Keeping up appearances: agarwood grades and quality. In: Mohamed DR (ed) Agarwood: science behind the fragrance. Springer Press, Singapore, pp 149–167
18. Mondal SR (2007) Globalization and anthropology. Indian Anthropol 37(2):93–98
19. Muharromah (2015) Pengembangan Agarwood di Sumatra (Agarwood development in Sumatra). In: Sulistyo SA, Turjaman M (eds) Pengembangan Teknologi Produksi Agarwood Berbasis Pemberdayaan masyarakat (Technology development and agarwood production based on community empowerment). Pusat Penelitian dan Pengembangan Hutan dan Konservasi Alam, Bogor, pp 1–29
20. Persoon GA (2007) Agarwood: the life of a wounded tree. IIAS News Lett 45:24–25
21. Report of the Asian regional workshop on the management of wild and planted agarwood Taxa, Guwahati, Assam, India, 19–23 January 2015
22. Semiadi G, Wiriadinata H, Waluyo EB, Darnaedi D (2010) Rantai pasokan produk tumbuhan agarwood (*Aquilaria spp.*) asal Merauke, Papua (Supply chains agarwood plants production (Aquilaria spp) from Merauke, Papua). Buletin Plasma Nutfah 16(2):150–159
23. Siran SA (2010) Perkembangan Pemanfaatan Agarwood (Agarwood utilization and its development). In: Sulistyo A, Turjaman M (eds) Pengembangan Teknologi Produksi Agarwood Berbasis Pemberdayaan masyarakat (Technology development and agarwood production based on community empowerment). Pusat Penelitian dan Pengembangan Hutan dan Konservasi Alam, Bogor, pp 35–52
24. Soehartono TR (1997). Status and distribution of Aquilaria spp. in Indonesia, and the sustainability of the agarwood trade. Thesis submitted to the University of Edinburgh for the degree of Doctor of Philosophy for Institute of Ecology and Resource Management, The University of Edinburgh
25. Soeharto B, Suseno B, van Oordwijk M (2016). Agarwood (Agarwood) domestication: biotechnology, markets and agroforestry options. Working Paper number 27 published by the World Agroforestry Center
26. Subiakto WD, Bakrie I (2015) Peranan hukum adat dalam menjaga dan melestarikan hutan di desa metulang kecamatan kayan selatan kabupaten malinau propinsi kalimantan utara. Agrifor 14(2):293–314
27. Turjaman M, Hidayat A, dan Santoso E (2016) Development of agarwood induction technology using endophytic fungi. In Mohamed R (ed) Agarwood: science behind the fragrance. Springer Press, Singapore, pp 57–71
28. Turjamah (2015) Report of the Asian regional workshop on the management of wild and planted agarwood tax was held by ITTO, Wildlife Crime Control Bureau Government of India and CITES, in Guuwanhati, Assam, India on 19–23 January 2015

Enhancing Mitigation for Climate Change

A Preliminary Study of Temporary Shelter from Engineered Wood for Post-Disaster Mitigation

Agus Mudo Prasetyo, Agung Sumarno, and Luna Nurdianti Ngeljaratan

Abstract An intense climate-induced disaster such as flood and tropical cyclones has been the occasional season in Indonesia which affecting the vulnerable areas with poor housing construction. Moreover, in the area where there is a combined-risk of climate and natural hazards such as earthquakes, the vulnerability of the population increases with the urgent need for temporary shelter. The study focuses on the preliminary design of a temporary shelter in Lombok, NTB that experiences the tropical cyclone Seroja in 2021 as well as the Mw 6.9 earthquake in 2018. The wind load, as well as the seismic lateral load designs, follow the requirements in ASCE 7–05. The estimation of the hurricane load in the study area is computed from the impact of windborne debris using the missile criteria for hurricane shelters. Seismic analysis is conducted to determine the safety level of the structure under extreme earthquakes. Engineered wood in the form of glulam is designed as the load-bearing element and analyzed to ensure structural safety and stability. A sheathing from plywood is also designed considering the seismicity. Based on the stability and capacity checks on several structural components of the temporary shelter, the study concludes that the preliminary design shows adequacy in safety and stability against extreme cases.

Keywords Temporary shelter · Glulam · Timber · Cyclone · Seismic

1 Introduction

Tropical cyclone Seroja hit Lombok, Nusa Tenggara Barat (NTB), Indonesia with a maximum wind speed of 65–85 km/h that was further categorized by the Saffir Simpson Hurricane Wind Scale (SSHWS) as a level 1 hurricane. Most of East and West Nusa Tenggara experienced flash-flood, strong wind, and heavy rain that lasted for hours. The National Agency for Disaster Management (BNPB) of Indonesia

A. M. Prasetyo (✉) · A. Sumarno · L. N. Ngeljaratan
Research Center for Biomaterials, Indonesian Institute of Sciences, Cibinong, Indonesia
e-mail: agus@bu.lipi.go.id

M. A. Rahandi Lubis (ed.), *Proceedings of the 10th International Symposium for Sustainable Humanosphere*, Lecture Notes in Civil Engineering 255,
https://doi.org/10.1007/978-981-19-1422-5_24

reported that more than 500,000 people were affected by the tropical cyclone with more than 150 calamities, displaced population due to damaged houses. Besides hurricanes, Lombok, NTB also experienced several major earthquakes with the largest one happened in 2018 with a magnitude of Mw 6.9. Many casualties were reported with most of the deaths were caused by falling debris. These major events require strategic and efficient mitigation strategies and one of them is to provide the local people with a temporary shelter that accommodates them with greater protection than immediate shelters.

The presence of temporary shelter for a hurricane is still rare in Indonesia as it requires advanced design and analysis as well as durable materials that perform well during extreme events. Limited studies already investigated the utilization of engineered wood in the form of glue-laminated timber (glulam) or cross-laminated timber (CLT) as the building component to withstand a hurricane or any hydrometeorological hazards. For example, the study of [1] uses CLT as the main material, and glulam is used as supplemental materials. The CLT construction, Laminated veneer lumber (LVL), laminated strand lumber (LSL), and other commodity wood products [2–4] are investigated against impact and wind pressure testing to protect against tornadoes. As for glulam, it is widely used in seismic applications as an earthquake-proof shelter. For example, [5] considers glulam as the post that shows stable performance during design earthquake, [6] studies glulam as post-emergency housing systems, and [7] describes the ductile behavior of timber and engineered wood including glulam under dynamic excitation.

This study is focused on the preliminary design of temporary shelter for post-disaster mitigation that highlights the utilization of glulam as well as analysis of strong wind and seismic load that is implemented in the design of the structural component of the building. The general description of the shelter as well as the design references and design load are described first. Then, the results of wind and seismic analyses are explained with the design example of roof header and wall sheathing. The conclusion is drawn at the end of the study based on the finding and results.

2 Method

2.1 General Description of the Shelter and Scope of Work

A two-story wood-framed shelter is designed to be built in Lombok, NTB (with an elevation of 2 m), consists of roof joists, glulam roof ridge beams, plywood roof sheathing, plywood floor sheathing, floor joists, glulam floor girders, wood columns, and CLT shearwall. The ridge beams and floor girders, and the other two support only the floor girders. Stairs framing from the ground floor to the second floor is not shown and is not part of the design. The plan dimension of the building is 9 m × 15 m and floor to floor and floor to roof heights are both 3.5 m. The exterior

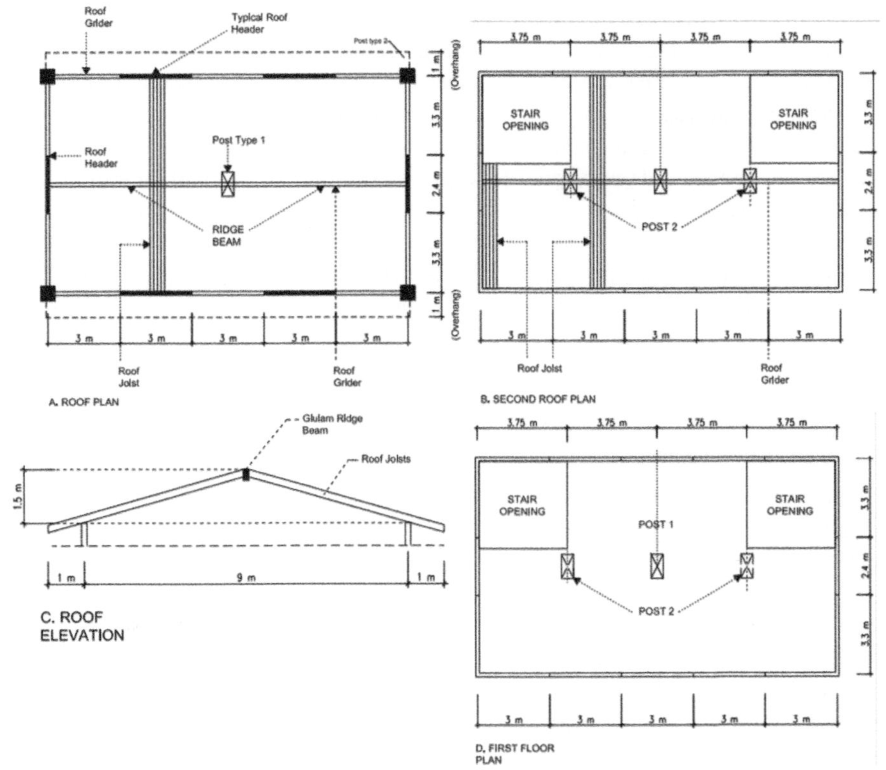

Fig. 1 Shelter plan

walls have 15 mm gypsum wallboard on the inside and 25 mm stucco on plywood sheathing on the outside. The ceiling is 15 mm gypsum wallboard and the second floor is finished with 20 mm thick hardwood flooring. The roofing is asphalt shingles, and the design also considers 7.3 kg/m^2 insulation and 4.8 kg/m^2 sprinklers in dead load of roof and floor. The building plans as well as the roof elevation are shown in Fig. 1. As there are several structural components of the shelter and each component should be specifically designed and checked for their capacity on the applied load, this preliminary study is only focused on the design of roof header and wall sheathing to highlight the implication of the design wind and seismic load on the structural components of the shelter.

2.2 Design References

The design of wood structural members was based on the dimension lumber/timber properties, requirements, and methods described in National Design Specification for Wood Construction [8]. The design of wind load, live load, as well as gravity and seismic load were based on the design requirements as per ASCE 7–10 [9]. The equivalent uniform weights of wood framing, as well as the weight of building materials used in the gravity load design, were chosen from Western Woods Use Book [10]. The designed response spectra were generated by Kementrian PUPR seismic design map [11] in which the location, the site latitude, and longitude were used as the input of the program. The design for roof, floor, and wall sheathings was based on Table Q225 D, Table 2304.7, Table 2306.3.1, and Table 2306.4.1 of the International Building Code [12]. Other supplemental guidelines and specifications are cited on the associated load or design components.

2.3 Design of Live Load and Gravity Load

According to ASCE 7–05, the live loads used in the design shall be the maximum loads expected by the intended use or occupancy but shall in no case be less than the minimum uniformly distributed unit loads required by Table 4–1 of ASCE 7–05. For hurricane shelter, the minimum uniformly distributed load at the floor (unreduced), L_0, was specified as 2.4 kN/m^2. As for the ordinary flat, pitched, and curved roofs, the roof live load (unreduced), L_r, was taken as 0.96 kN/m^2. The design of gravity load was based on the weight of building materials and equivalent weights of wood framing in Western Use Book with the estimated roof dead load, D_r of 1.05 kN/m^2, wall dead load, D_w of 0.96 kN/m^2, and floor dead load, D_f of 0.72 kN/m^2.

2.4 Design of Hurricane Load

Three major types of loads are associated with hurricanes, i.e. wind loads, missile impact, and storm surge. The design for storm surge is mostly focused on the structure built near the coastal region with pier support so it is not considered in this preliminary design. The wind load design is conducted using Method I—Simplified Procedure as described in ASCE 7–05 Sect. 6.4. To design the shelter using Simplified Procedure, the building in this project shall meet the requirements in ASCE 7–05 Sect. 6.4.1.1 for Main Wind-Force Resisting Systems (MWFRS) and Sect. 6.4.1.2 for Components and Cladding. Some of the requirements are the building has a short period that its fundamental frequency should be more than 1 Hz, no structural separations so windward and leeward wind loads are transmitted through the same vertical MWFRS, the mean roof height, *hmean*, is less than or equal to 18 m and does not exceed least horizontal dimension, and the building is symmetrical with roof $\theta \leq 45°$. The design wind pressure, p_s, is based on ASCE 7 as shown in Eq. (1) in which λ is the adjustment factor equal to 1.34 for mean roof height, h_{mean}, K_{zt} is a topographic factor, I is importance factor, and p_{s30} is the simplified design wind pressure for Exposure B, at h = 30 ft (9.1 m). The design wind pressure for component and cladding, p_{net} is given

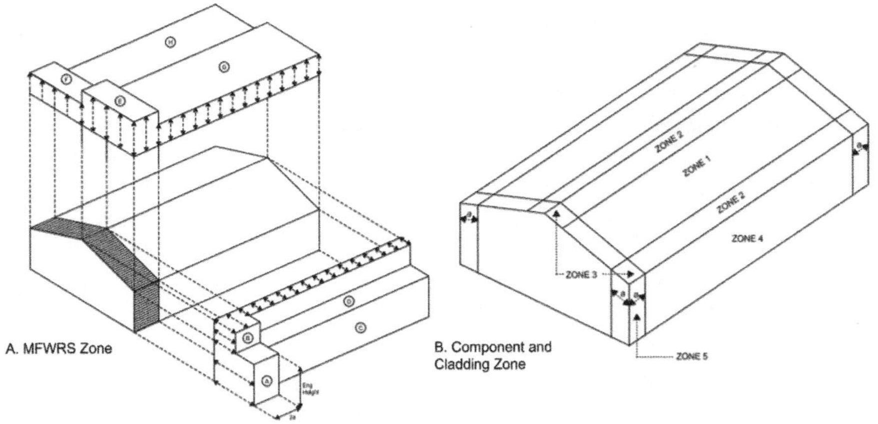

Fig. 2 MFWRS system and component and cladding system with the wind zone

by (2) that p_{net30} is the net design wind pressure on roof and wall also considering the adjustment factor, topographic factor, and importance factor (Fig. 2).

$$p_s = \lambda \times K_{zt} \times I \times p_{s30} \tag{1}$$

$$p_{net} = \lambda \times K_{zt} \times I \times p_{net30} \tag{2}$$

As per the requirement of ICC 500 [13], a shelter should be designed for the impact of windborne debris using the missile criteria for hurricane shelters. The debris impact test missile should be conducted for all components of the shelter envelope. A simplified procedure is adopted in this study [14] to estimate the design force as per (3):

$$F_{max} = VT(mK)\cos\theta w \tag{3}$$

where VT is total impact velocity, m is the missile mass, K is the stiffness of the building and θw is the angle between the impact velocity vector and the wall-normal.

2.5 Design of Seismic Load

The designed response spectrum of the building is generated by Cipta Karya seismic design map based on the location coordinates and the values used in the design are given in Table 1. The design of seismic load in this work is conducted using the response spectrum method of multi-degree of freedom (MDOF) structure as per (4):

$$M\ \ddot{x}(t) + C\dot{x}(t) + Kx(t) = f(t) \tag{4}$$

Table 1 Parameters for seismic load design

Parameters	Value
Soil class type	D (soft soil)
PGA (g)	0.46
S_{DS} (g)	0.76
S_{D1} (g)	0.54
Seismic design category	D
Importance factor, *I*	1.0

Equation (4) states that the sum of all forces acting on the mass should be equal to zero with $f(t)$ an externally applied force, $M\ \ddot{x}(t)$ the inertial force, $C\dot{x}(t)$ the (viscous) damping force, and $Kx(t)$ the restoring force. The variable $x(t)$ stands for the position of the mass M with respect to its equilibrium point, i.e. the position of the mass when $f(t) = 0$.

3 Result and Discussion

3.1 Design Wind Pressure and Hurricane Modelling

Separate provisions shall be considered in designing most of the structural members of the shelter such that the primary members are designed based on MWFRS and the secondary members use the Component and Cladding (CC) specifications. The MWFRS method presents the most crucial loads acting on the two major orthogonal directions of the building. The design provides support and stability for the structural members as the MWFRS system receives wind load on the building envelopes. As for the CC system, they also act as an envelope for the building, but they may transfer the wind load to the MWFRS components.

The minimum basic wind speed for the city of Lombok was estimated as 160 km/hr considering the extensive damages caused by tropical cyclone Seroja. The design wind pressure is computed using (1) with the results shown in Table 2. The load to the roof diaphragm is computed by first estimating the length of the end zone that presents the perimeter of the roof. The tributary height to the roof diaphragm is calculated as 3.2 m and the total wind forces to the roof diaphragm is greater at $p_s = 0$ in the roof zones EOH and GOH. The unit shear in roof diaphragm due to wind load is computed from the reaction forces due to wind load on shearwall as 2.64 kN/m. The load to floor diaphragm considers the tributary height to the 2nd-floor diaphragm as 3.35 m. Using a similar approach to the roof diaphragm, the unit shear in the floor diaphragm is calculated as 2.76 kN/m.

The wind design load for CC is shown in Table 3. The terrain category is assumed as Exposure C. The effective wind area, *A*, from ASCE 7–05 Sect. 6, *A* is the span length multiplied by effective width that is not less than 1/3 of the span length. To

Table 2 The design wind pressure, p_s, MWFRS

Horizontal pressure		Vertical pressure		Overhang	
Zone	p_s (kN/m^2)	Zone	p_s (kN/m^2)	Zone	p_s (kN/m^2)
A	1.37	E	– 1.23	EOH	– 1.71
B	– 0.38	F	– 0.84	GOH	– 1.34
C	0.91	G	– 0.85		
D	– 0.22	H	– 0.64		

Table 3 The design wind pressure, p_s, CC

Roof		Wall	
Zone	p_s (kN/m^2)	Zone	p_s (kN/m^2)
1	1.37	4	– 1.23
2	– 0.38	5	– 0.84
3	0.91		– 0.85

obtain maximum design wind pressure for components and cladding, A is assumed as 10ft^2 (0.93 m^2). The design values shown in Tables 2 and 3 are further used in designing the envelopes of the building and the stability check is required to assess the safety of each component following the specified maximum wind speed.

The missile criteria for hurricane shelter based on [15] is specified as 2 × 4 in wood stud with a weight of 4 kg. The missile speed is estimated as 40% of wind zone speed for the vertical surface and 10% of wind zone speed for the horizontal surface. The simplified method using (3) is used to compute the maximum impact of the windborne missile with the estimation values of VT as 64 km/h and 16 km/h for vertical and horizontal pressure, respectively. The missile mass m is estimated as 4 kg. The upper bound value of the stiffness, K_{max} and the lower bound, K_{min} are assumed as 7.3 kN/mm and 3.9 kN/mm, respectively. They are assumed as the stiffness of the cross-laminated timber of the shearwall, computed from finite element modeling [16]. The result is shown in Fig. 3 as the function of θw, the angle between the impact velocity vector and the wall-normal. The figure shows that a higher impact load of the windborne missile is expected on the vertical surface of the building component. These values are further used to design the critical components of the shelter such as shearwall and should consider the allowable stress design coefficient that is excluded in the figure.

3.2 Seismic Analysis

The two-story shelter building is modeled as MDOF with a two-DOF system and floors are assumed as a rigid system. The stiffness of the 1st floor is estimated as twice the 2nd floor with the inter-story stiffness of 0.4 kN/mm. The eigenvalues and

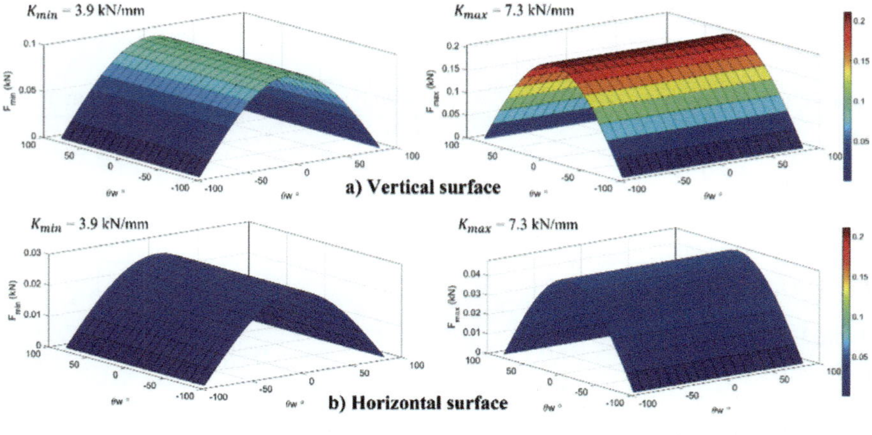

Fig. 3 Maximum impact load of the windborne missile on vertical and horizontal surfaces

eigenvector are computed as $\omega_1 = 12.6$ rad/sec and $\omega_2 = 24.6$ rad/sec, respectively. The associated periods are computed as $T_1 = 0.49$ s and $T_2 = 0.25$ s. Therefore, the 1st and 2nd modes are calculated as $f_1 = 2.04$ Hz and $f_2 = 4$ Hz.

The modal participation factors as the function of eigenvectors and the maximum response of the associated modes are calculated as $\Gamma_1 = 1.3$ and $\Gamma_1 = -0.3$. Using the parameters of seismic design load in Table 1 and 5% damping, $\xi = 5\%$, the displacement of the top floor is computed as 0.0214 m associated with the base shear of 42.16 kN. Considering story drift for a seismic building shall not exceed 3%, the top floor displacement is still within the acceptable value.

3.3 Stability of Structural Member from Glulam

An example of a structural member design using glulam as described in this study is a roof header. A roof header is a girder over an opening such as a window that should be carefully designed to maintain the integrity of the structure as it disperses the continuation of the load outside the opening. As shown in Fig. 4, there are two types of roof header following the roof plan, i.e. roof header with distributed load and with point load. For a typical header, i.e. header with uniformly distributed load, a 24F-V4, DFL No. 14 is designed as roof header, and the support is assumed as simply supported beam. The trial header size is $130 \times 300\ mm^2$ with moisture content $MC = 0.12$.

Each square meter of roof system delivers a total of 20.5 kN/m load from the roof to the structural support and they are uniformly distributed over the entire surface of the roof. Using static analysis, the support reaction of the header R is computed as 26.52 kN and equal to shear, V. The maximum moment, M_{max} of the header

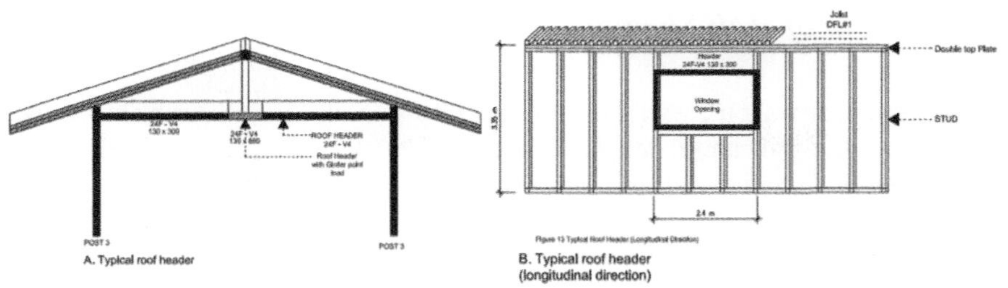

Fig. 4 Typical roof header in transverse and longitudinal directions

is calculated as 17.18 kN-m. The allowable deflection due to live load, $\Delta_{L,allow}$ is obtained as 15.24 mm and due to total load, $\Delta_{TOT,allow}$ is calculated as 21.59 mm.

The header stability factor and volume factor are computed with taking into consideration the slenderness ratio, R_B and the critical buckling design for bending, F_{bE}. The slenderness ratio considers effective span length for a simple span beam with a uniformly distributed load as 16. The critical buckling design load is calculated as 26.83 MPa. The reference design values and the adjustment factors are taken from the provisions with the values as shown in Table 4. The reference design values and the adjustment factors are used to normalize bending, $F_b^{\prime *}$ and shear $F'v$ values as 19.03 MPa and 2.10 MPa, respectively.

Considering the section properties, the allowable bending stress, f_b and shear stress f_v are computed as 1.68 MPa and 0.45 MPa, respectively. The deflection due to live load, Δ_L is calculated as 0.28 mm and due to total load, Δ_{TOT}, 0.25 mm. By checking the adequacy of header based on the stress values, it is observed that for bending check, $f_b < F_b$; 1.68 MPa < 19.03 MPa, for shear, $f_v < F_v$; 0.45 MPa < 2.10 MPa. As for the deflection, it is also shown that deflections due to total load and live load are still below the allowable values, that $\Delta_{TOT} < \Delta_{allow}$; 0.25 mm < 21.59 mm and $\Delta_L < \Delta_{allow}$; 0.28 mm < 15.24 mm. Therefore, glulam 24 V-F4 DFL No.1 with the size of 130 × 300 mm^2 is safe to be used as a roof header with a uniformly distributed load.

Table 4 Reference design values and adjustment factor

Reference design values			
Bending, F_{bx}	16.55 MPa	Elastic modulus, E	12,410.56 MPa
Shear, F_{vx}	1.83 MPa	Minimum elastic modulus, E_{ymin}	5722.65 MPa
Adjustment factor			
Load duration factor, C_D	1.15	Flat use factor, C_{fu}	1
Wet service factor, C_M	1	Curvature factor, C_c	1
Temperature factor, C_t	1	Repetitive member factor, C_r	1
Use volume factor, C_V	1		

3.4 Design of Wall Sheathing

Besides the shearwall, the wall sheathing should also perform well during hurricanes and earthquakes and careful detail is necessary to ensure that there will be no possibility of debris from the wall sheathing. The shearwall frame is designed as CLT and the sheathing uses plywood as the most common material to layer the frame. The design of wall sheathing is given for the upper and lower level of the shelter following the requirement from IBC 2305.3.4 for the transverse and longitudinal direction. The requirement states that the ratio of h/w, shear panel height over width, both in transverse and longitudinal directions, should be less than 3.5 for wind load and 2 for the seismic load. The h/w based on the shelter design for the upper level are 1 for shear panel 1 (transverse), 0.91 for so shear panel 2 (longitudinal), and 1.375 for shear panel 3 (longitudinal). Therefore, all shear panels on the upper level meet the IBC requirement for the h/w ratio.

The transverse reaction to the diaphragm, R_u is computed as 51.33 kN. The seismic coefficient is used to reduce the seismic force on the wall as the function of the wall dead load. The seismic force on the wall, R_{ru}, is computed as 1.83 kN; therefore, the total strength level design seismic force, $V_{u,}$ is equal to 53.16 kN. The unit shear on the upper wall is then calculated as 0.52 kN/m. The wall design for the upper level is based on IBC Table 2306.4.1, 15/32 in (381/812 mm) Struct. I plywood with 10d@4" (100 mm) O.C that at panel edges, 10d@12" (305 mm) field, blocked capacity = 0.69 kN/m is higher than 0.52 kN/m. Therefore, the selected design of upper-level wall sheathing is adequate.

A similar procedure is conducted in designing the wall sheathing in the lower level. A higher seismic force is expected that the seismic force on the 1st-floor wall is computed as 3.66 kN. The unit shear on the lower wall is computed as 0.78 kN/m. The lower-level wall is designed using similar wall sheathing of 15/32 in (381/812 mm) Struct. I plywood but with closer spacing using 10d@3" (75 mm) O.C. Checking the adequacy at panel edges, 10d@12" (305 mm) field, the blocked capacity is 0.90 kN/m that is higher than 0.78 kN/m and 3by framing is required at all edges including wall plates.

4 Conclusion

This study provides the design example of several structural members of the shelter as post-disaster mitigation effort using glulam with a site-specific location at Lombok, NTB. The MWFRS and component and cladding loads for building envelopes are calculated based on the maximum wind speed associated with the severe damages of the tropical cyclone. The hurricane load is estimated through a simplified windborne missile model subjected to the horizontal and vertical faces of the building. The seismic load is based on the parameter generated from the response spectra of the specified location.

The design examples are given for roof header and wall sheathing considering the utilization of glulam and cross-laminated timber as structural components. The stability check is conducted on the roof header analysis and the selected dimension is safe to support the design load. The capacity check is conducted on the wall sheathing analysis for the upper and lower level of the shelter and the selected details are also adequate to support the design load.

References

1. Stoner M, Pang W (2020) Simulated performance of cross-laminated timber residential structures subject to tornadoes. Front Built Environ 6:88
2. Falk RH, Bridwell JJ, Williamson T, Black T (2019) Development of a ready-to-assemble tornado shelter from cross-laminated timber (CLT): impact and wind pressure testing. Gen Tech Rep FPL-GTR-266 266:1–27
3. Falk RH, Bridwell JJ (2016) Development of a tornado safe room door from wood products: door design and impact testing. Research Paper FPL–RP–686. Madison, WI: U.S. Department of Agriculture, Forest Service, Forest Products Laboratory. p 17
4. Falk RH, Bridwell J (2018) A residential severe weather safe room from commodity wood products: design and development. General technical report FPL–GTR–253. Madison, WI: U.S. Department of Agriculture, Forest Service, Forest Products Laboratory
5. Xiong H, Liu Y (2016) Experimental study of the lateral resistance of bolted glulam timber post and beam structural systems. J Struct Eng 142(4):E4014002
6. Sumini V, Chesi C, Zedda R (2016) An innovative system for post-emergency timber housing. In: 41st IAHS World congress on sustainability and innovation for the future pp 1–10
7. Porcu MC (2017) Ductile behavior of timber structures under strong dynamic loads. Wood Civil Eng 2017:173–196
8. American Wood Council (2018) National design specification for wood construction with commentary 2018 edition. Leesburg, VA 20175
9. American Society of Civil Engineers (2010) Minimum design loads for buildings and other structures. ASCE/SEI 7, Reston, VA
10. Western Wood Products Association (1979) Western woods use book: structural data and design tables
11. Kementerian PUPR (2021), Desain Spektra Indonesia homepage, http://rsa.ciptakarya.pu.go.id/2021/, last accessed April 10
12. International Code Council (2006) International building code. ICC, New York
13. ICC/NSSA (2008) Standard for the design and construction of storm shelters, ICC 500
14. Richards P, Pe K, Milne I (2009) Impact dynamics of rod type windborne debris seventh asia-pacific conference on wind engineering. Taipei, Taiwan
15. Florida Building Code (FBC) (2001) State of Florida, Tallahassee
16. Shahnewaz M, Alam S, Tannert T (2018) In-plane strength and stiffness of cross-laminated timber shear walls. Buildings 8(8):100

Floristic Diversities in Konawe Utara District, Southeast Sulawesi

Albasri and Laode Alhamd

Abstract Study of floristic diversities has been done in Konawe Utara, Southeast Sulawesi. The aim was to determine the species composition and vegetation structure. The current study was conducted by using several sampling plots. The results showed that the 62 species in 49 genera and 30 families were recorded. The species richness of trees (DBH or Diameter at Breast Height $\geq$ 20 cm) was 35 species in 31 genera and 24 families, dominated by *Manilkara fasciculata*. For poles (10 < DBH < 20) reaching 37 species in 34 genera and 24 families were dominated by *Helicia serrata* and *Ixonanthes petiolaris*. For saplings (1 < DBH $\leq$ 10) consisting of 47 species in 41 genera and 26 families were dominated by *I.* petiolaris and *Lasjia hildebrandii*, while the seedlings were covered by *Castanopsis buruana* and *Agathis dammara*. Tree density of trees, poles, saplings and seedlings ranged from 268 to 39,667 individuals ha^{-1}, while the basal areas without seedling were around 13.15–29.6 m^2 ha^{-1}. Indexes of species diversity (*H*'), dominance (*C*), evenness (*E*) and richness (*R*) ranged from 3.15–3.52; 0.04–0.06; 0.89–0.95, and 6.03–8.88, respectively.

Keywords Floristic · Konawe Utara · Endemic species · Species index

1 Introduction

Human activities have strongly influenced in deforestation by utilizing forest resources to increase its economy along with regional development. The need for land is increasing, so forest area land will be an option for exploitation. However,

Albasri
Faculty of Forestry and Environment, University of Halu Oleo, Kendari, Southeast Sulawesi, Indonesia

L. Alhamd (✉)
Reacearch Center for Ecology and Ethnobiology, National Research and Innovation Agency (BRIN), Cibinong, West Java, Indonesia
e-mail: trialbgr@gmail.com

M. A. Rahandi Lubis (ed.), *Proceedings of the 10th International Symposium for Sustainable Humanosphere*, Lecture Notes in Civil Engineering 255,
https://doi.org/10.1007/978-981-19-1422-5_25

these activities sometimes do not pay attention to the sustainability of biodiversity so that it has an indirectly impact on high rates of deforestation [1].

The rate of deforestation in Southeast Sulawesi was approximately 63.7 thousand ha year-1 in the period 2000–2009 and the area of forested land was approximately 1.4 million ha (Directorate General of Planology 2009). This will affect the loss of forest area, such in Konawe Utara as one of the regencies in Southeast Sulawesi, whose forest area is 374,784 ha out of 500,339 ha of Konawe Utara area [2]. The forest area in Konawe Utara gradually continues to decline with increasing of mining locations and/or plantation areas.

The loss of forest area also affects the reduction in the biodiversity potential of forest areas, especially in North Konawe. In general, Southeast Sulawesi has a high level endemicity of biodiversity because the Sulawesi area is bordered by the Wallacea Line. The biodiversity in the form of vegetation potential plays a very important role in an ecosystem balance.

Vegetation, in the forest, plays a very important role in an ecosystem related to the regulation of the balance of carbon dioxide and oxygen in the air, improvement of physical, chemical and biological properties of soil, regulation of groundwater management, and others. Although in general the presence of vegetation in an area has a significant impact, the vegetation varies depending on their structure and composition.

Therefore, the objectives of the current study were to observe the existence of floristic species, and to find out also the diversity type in Konawe Utara, Southeast Sulawesi.

2 Study Site

The research location is at coordinates between 122°8′22, 80″–122° 14′57, 6″ BT and 3° 20′9, 60″–3° 25′ 9, 60″ LS. Administratively, it is located in Lasolo and Oheo Districts, Konawe Utara Regency, Southeast Sulawesi Province. The soil types at the site are generally Dystropepts and Eutropepts including Inceptisol soil, according to the USDA Soil Taxonomy classification (1990).

The annual rainfall during seven years (2011–2020) was ranging from 1,320 to 2913 mm. The high rainfall occurred from May to July, while the low rainfall within September to October. The average monthly rainfall is moderate reaching 174.4 mm (Rainfall data from the Ministry of Public Works, Directorate General of Water Resources, Sulawesi IV River Basin Center, Southeast Sulawesi Province in 2012).

3 Methods

The characteristics of species growth were obtained by species composition, and structure of vegetation. Within six sampling plots, a 20 m × 20 m (0.28 ha) plot size was constructed in representative areas. All vegetation in the plot were inventoried and recorded according to growth rate, namely trees, poles, saplings and seedlings. Trees were defined as having a DBH (Diameter at Breast Height) $\geq$ 10 cm and tree height larger than 1.3 m, poles with DBH ranging from 5 to 10 cm and larger than 1.3 m, sapling with 2 to < 5 cm in DBH and seedlings with DBH < 1 cm.

The present study was conducted in Konawe Utara District, Southeast Sulawesi, as shown in the Figure 1. There are 15 sampling plot, 20 × 20 m^2 of each plot, within the plot was subdivided into measurement of tree (20 × 20 m^2), pole (10 × 10 m^2), sapling (5 × 5 m^2) and seedling (2 × 2 m^2).

Tree measurement were in size 20 × 20m^2 of sampling plot, while the plots of pole, sapling and seedling were, respectively, in size 10 × 10m^2, 5 × 5m^2 and 2 × 2m^2 (Fig. 2). All voucher specimens were taken for further species identification in Research Center for Biology.

Tree structure described the distribution of tree species number (individuals ha^{-1}) based on the tree diameter in plot. Important Value Index (IVI) is utilized to figure

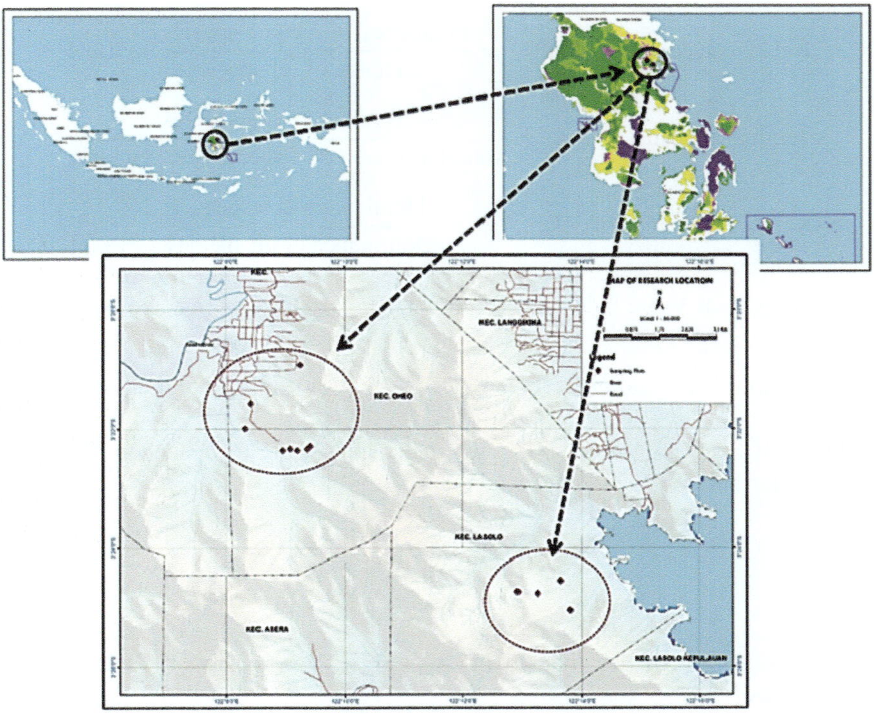

Fig. 1 Map of study site of Konawe Utara District, Southeast Sulawesi

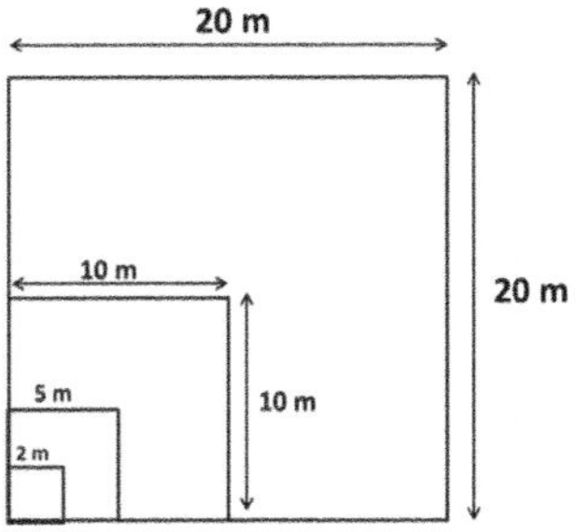

Fig. 2 Sampling plot sizes for tree, pole, sapling and seedling measurements

out the dominance of a species in a community. The calculation equation of IVI, as follows:

$$\text{Density (D; individualsha}^{-1}) = \frac{\text{Number of individuals}}{\text{Unit area in plot}}$$

$$\text{Relative Density (RD; \%)} = \frac{\text{Number of individuals of a species}}{\text{Total number of individuals of all species}} \times 100\%$$

$$\text{Frequency(F)} = \frac{\text{Points occupied of a given species}}{\text{Distribution a species in plot}}$$

$$\text{Relative Frequency (RF; \%)} = \frac{\text{Frequency of a given species}}{\text{Frequency of all species}} \times 100\%$$

$$\text{Dominance}\left(\text{Dm; m}^2/\text{ha}\right) = \frac{\text{Basal area of a species}}{\text{Total basal area of a species per unit area}}$$

$$\text{Relative Dominance (RDm; \%)} = \frac{\text{Dominance of a species}}{\text{Dominance of all species}} \times 100\%$$

$$\text{Important Value Index(IVI)} = \text{RD} + \text{RF} + \text{RDm, for tree and pole}$$

$$\text{Important Value Index(IVI)} = \text{RD} + \text{RF, for sapling and seedling}$$

Index of Species Diversity (*H'*) was calculated using the Shannon–Whienner diversity formula [3], as follows:

$$\boldsymbol{H'} = -\sum_{i=1}^{n} \boldsymbol{Pi} \ \mathbf{ln} \ \boldsymbol{Pi}$$

In where: H' = Species diversity index, $Pi = \frac{ni}{N}$, ni = density of i species, and N = total density.

There are three criteria of the species diversity index, namely if the value of $H' < 2$, then it is included in the low category, the value of $2 < H' < 3$, then it is included in the medium category and will be included in the good category if $H' > 3$ [3].

Index of dominant species (C) was calculated using the formula:

$$C = \sum_{i=1}^{n} \left[\frac{n_i}{N}\right]^2$$

where C is index of dominant species, ni is density of i species, and N is total density

The value of the dominance index (C) ranges from $0 \leq C \leq 1$, if the value of C is close to 1 indicating a certain area only one species is controlled. In contrast, if the value of C is close to 0 it means uncontrolled of a species or there are several dominating species in the area.

The evenness index of species (E) shows the level of individual evenness of species. If the value of E is getting closer to 1, the value of evenness is getting higher. The value of E is calculated using the following formula [3]:

$$E = \left[\frac{H'}{\ln S}\right]$$

where E is evenness index of species, H' is species diversity index, and S is total number of species

The magnitude of $E < 0.3$ indicates a low evenness of species, $0.3 < E < 0.6$ indicates a moderate level of species evenness and $E > 0.6$ indicates a high level of evenness [3].

The species richness index (R) was calculated using the Margallef formula [3] with the following calculations:

$$R = \frac{(S - 1)}{\ln(N)}$$

The value of $R < 3.5$ indicates a low species richness, a value of $3.5 < R < 5.0$ indicates a moderate species richness and $R > 5.0$ indicates a high species richness.

4 Results

A total of 62 species were found for tree, pole, sapling and seedling levels, consisting of 49 genera and 30 families.

4.1 Tree Level

There are 24 families which were dominated to Myrtaceae (consisting of seven species) and Moraceae / Rubiaceae (each five species). The species of *Manilkara fasciculate* (49.06), *Diospyros foliosa* (27.53), *Helicia serrata* (18.36) and *D. ellipticifolia* (18.24) were four dominant species out of 35 species according to Important Value Index (IVI), described in Table 1. Although the density of *D. foliosa* was highest at tree level, the *M. fasciculata* were highest in dominance and frequency. The number of density was 268 individual ha^{-1}, and total basal area (TBA) reached 16.2 m^2 ha^{-1}.

Table 1 Fifteen species of tree according to important value index (IVI) of Konawe Utara, Southeast Sulawesi

No	Family	Species name	Local name	RD	RDm	RF	IVI
1	Sapotaceae	*Manilkara fasciculata*	Kumialu	10.63	29.82	8.62	49.06
2	Ebenaceae	*Diospyros foliosa*	Kayu Hitam	11.25	8.52	7.76	27.53
3	Proteaceae	*Helicia serrata*	–	6.25	4.35	7.76	18.36
4	Ebenaceae	*Diospyros ellipticifolia*	Lawaru	8.13	5.81	4.31	18.24
5	Elaeocarpaceae	*Elaeocarpus sphaericus*	Koleuhu	7.50	5.08	5.17	17.75
6	Leguminosae	*Archidendron pauciflorum*	–	6.25	4.39	6.03	16.67
7	Ebenaceae	*Diospyros* sp.	–	5.00	5.83	5.17	16.00
8	Lamiaceae	*Gmelina palawensis*	–	3.75	2.59	5.17	11.52
9	Ixonanthaceae	*Ixonanthes petiolaris*	Kasu Lare	3.75	2.08	4.31	10.14
10	Fagaceae	*Castanopsis buruana*	Eha	3.13	3.49	3.45	10.07
11	Phyllanthaceae	*Cleistanthus oblongifolius*	Nogoai	2.50	4.58	2.59	9.67
12	Anacardiaceae	*Buchanania arborescens*	Mangga hutan	3.13	2.06	4.31	9.50
13	Lamiaceae	*Premna* sp.	Uroko	3.13	1.70	4.31	9.14
14	Penaeaceae	Crypteronia cumingii	–	2.50	1.49	3.45	7.44
15	Malvaceae	*Sterculia heterophylla*	Olimbute	2.50	1.96	2.59	7.05

Note RD, RDm, and RF represented as Relative Density, Relative Dominance and Relative Frequency, respectively

4.2 Pole Level

There are 24 families which were dominated to Lamiaceae (consisting of three species) and Moraceae (three species) in the sampling plot. The species of *Helicia serrate* (22.03), *Ixonanthes petiolaris* (21.66), *Diospyros ellipticifolia* (19.96) and *Elaeocarpus sphaericus* (18.30) were four abundant species out of 37 species according to IVI, as listed in Table 2. One species, *H. serrata*, dominates at the tree level is also found to dominate at this level. Also *H. serrata* was highest in dominance and frequency. The number of density was 1667 individual ha^{-1} and TBA reached 29.6 m^2 ha^{-1}.

Table 2 Fifteen species of pole according to important value index (IVI) of Konawe Utara, Southeast Sulawesi

No	Family	Species name	Local Name	RD	RDm	RF	IVI
1	Proteaceae	*Helicia serrata*	–	7.60	8.50	5.93	22.03
2	Ixonanthaceae	*Ixonanthes petiolaris*	Kasu Lare	8.00	7.74	5.93	21.66
3	Ebenaceae	*Diospyros ellipticifolia*	Lawaru	6.80	7.98	5.19	19.96
4	Elaeocarpaceae	*Elaeocarpus sphaericus*	Koleuhu	7.20	6.66	4.44	18.30
5	Leguninosae	*Archidendron pauciflorum*	–	6.80	6.22	5.19	18.20
6	Phyllanthaceae	*Baccaurea* sp.	–	6.40	6.15	5.19	17.73
7	Rutaceae	*Melicope trichopetala*	–	6.80	5.87	4.44	17.11
8	Lamiaceae	*Premna* sp.	Uroko	5.20	5.47	5.93	16.60
9	Lamiaceae	*Gmelina palawensis*	–	5.20	5.79	5.19	16.17
10	Ebenaceae	*Diospyros foliosa*	Kayu Hitam	4.00	5.25	4.44	13.70
11	Burseraceae	*Cararium maluense*	Dama Mea	4.40	4.14	3.70	12.24
12	Anacardiaceae	*Buchanania arborescens*	Mangga hutan	3.20	3.24	5.19	11.63
13	Penaeaceae	*Crypteronia cumingii*	–	2.40	2.58	3.70	8.69
14	Myristicaceae	*Knema latericia*	–	2.40	2.54	3.70	8.64
15	Cannabaceae	*Gironniera subaequalis*	Apu	2.00	2.99	2.96	7.95

Note RD, RDm and RF Represented as relative density, Relative dominance and relative frequency, respectively

4.3 Sapling Level

We recorded 27 families which were dominated by Myrtaceae (consisting of five species), and Rubiaceae/Lamiaceae and Moraceae (each of four species). The species of *Ixonanthes petiolaris* (21.84), *Lasjia hildebrandii* (20.12), *Castanopsis buruana* (17.12) and *Melicope trichopetala* (16.17) were four dominant species out of 47 species according to IVI, as listed in Table 3. The species of kasu lare (*I. petiolaris*) as most dominant species in this level was highly found also in the pole level. The number of density was 4270 individual ha^{-1} and TBA also reached 13, 2 m^2 ha^{-1}.

Table 3 Fifteen species of sapling according to important value index (IVI) of Konawe Utara, Southeast Sulawesi

No	Family	Species name	Local name	RD	RDm	RF	IVI
1	Ixonanthaceae	*Ixonanthes petiolaris*	Kasu Lare	4.52	11.62	5.69	21.84
2	Proteaceae	*Lasjia hildebrandii*	Makadamia	9.60	3.20	7.32	20.12
3	Fagaceae	*Castanopsis buruana*	Eha	8.47	3.77	4.88	17.12
4	Rutaceae	*Melicope trichopetala*	–	3.39	8.72	4.07	16.17
5	Malvaceae	*Colona celebica*	–	3.95	7.51	1.63	13.10
6	Myrtaceae	*Syzygium* sp1	Ruruhi	3.39	3.65	4.07	11.11
7	Lamiaceae	*Geunsia cinnamomea*	–	2.82	5.18	2.44	10.44
8	Lauraceae	*Cinnamomum subavenium*	–	3.39	1.87	4.88	10.14
9	Elaeocarpaceae	*Elaeocarpus angustifolius*	–	3.95	2.10	4.07	10.12
10	Araucariaceae	*Agathis dammara*	Damar	3.95	3.38	2.44	9.77
11	Moraceae	*Paratocarpus* sp.	–	3.39	2.15	4.07	9.61
12	Phyllanthaceae	*Cleistanthus oblongifolius*	Nogoai	3.39	2.45	3.25	9.09
13	Myrtaceae	*Syzygium* sp2	Jambu-Jambu	3.39	3.25	2.44	9.08
14	Myrtaceae	*Syzygium* sp3	Jambu-Jambu	3.39	1.61	4.07	9.06
15	Phyllanthaceae	*Baccaurea* sp.	–	1.69	4.46	1.63	7.78

Note RD, RDm, RF Represented as relative density, Relative dominance and relative frequency, respectively

4.4 Seedling Level

There are 22 families which were dominated by Myrtaceae (consisting of five species) and Moraceae (four species). The species of *Castanopsis buruana* (12.83), *Agathis dammara* (11.57) and *Ixonanthes petiolaris* (10.31) were three abundant species out of 47 species according to IVI, as listed in Table 4. The number of density was 39,667 and TBA reached 8.9 m^2 ha^{-1}.

Indexes of species diversity (*H*'), dominance (*C*), evenness (*E*) and richness (*R*) for all growth levels were shown in Fig. 3. In which the *H*' ranged from 3.15 (tree) to 3.52 (sapling), while *C* ranged from 0.038 (sapling) to 0.056 (tree), *E* ranging from 0.89 (tree) to 0.95 (seedling) and *R* ranging from 6.03 (seedling) to 8.88 (sapling). The several indexes showed the high categories of *H*', *E* and *R* indicating high species for all growth levels, while *C* was low category describing no main species in the growth rates.

Table 4 Fifteen species of seedling according to important value index (IVI) of Konawe Utara, Southeast Sulawesi

No	Family	Species name	Local name	RD	RF	IVI
1	Fagaceae	*Castanopsis buruana*	Eha	7.56	5.26	12.83
2	Araucariaceae	*Agathis dammara*	Damar	6.30	5.26	11.57
3	Ixonanthaceae	*Ixonanthes petiolaris*	Kasu Lare	5.04	5.26	10.31
4	Asparagaceae	*Dracaena angustifolia*	–	4.62	5.26	9.89
5	Phyllanthaceae	*Cleistanthus oblongifolius*	Nogoai	5.04	4.51	9.55
6	Proteaceae	*Lasjia hildebrandii*	Makadamia	4.20	5.26	9.46
7	Clusiaceae	*Calophyllum Inophyllum*	Bintangur	4.62	4.51	9.13
8	Myrtaceae	*Syzygium* sp1	Ruruhi	4.20	4.51	8.71
9	Casuarinaceae	*Casuarina sumatrana*	Kayu Angin	3.78	4.51	8.29
10	Cannabaceae	*Gironniera subaequalis*	Apu	3.36	4.51	7.87
11	Lauraceae	*Cinnamomum subavenium*	–	3.78	3.76	7.54
12	Ebenaceae	*Diospyros ellipticifolia*	Lawaru	3.36	3.76	7.12
13	Burseraceae	*Canarium maluense*	Dama Mea	3.78	3.01	6.79
14	Leguninosae	*Archidendron pauciflorum*	–	2.94	3.76	6.70
15	Anacardiaceae	*Buchanania arborescens*	Mangga hutan	3.36	3.01	6.37

Note RD and RF Represented as relative density and relative frequency, respectively

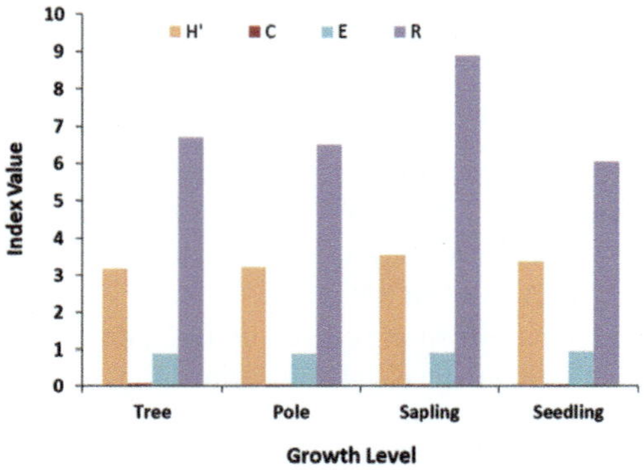

Fig. 3 Indexes of *H'*, *C*, *E* and *R* in different growth level

5 Discussion

The important value index (IVI) describes the high influence by a species in the community. If there is a species with a high IVI, it indicates that the species is the dominant species. Density is a factor that affects tree growth, in case of the density is high then the competition for nutrients and sunlight is getting bigger. A frequency value also describes the distribution pattern of a species in a habitat. In addition, a species has a high frequency, then the species will grow widely, and conversely a species will grow in groups and the low number of densities resulting from the low value of frequency.

The dominance of species is a characteristic of the community that influences of controlling one species in the community to other species so that the population of other species will relatively decrease in number or living ability [4]. The distribution of various plants starting from ground cover plants, seedlings, saplings, poles, and trees in their development will be selected through succession stages resulting in the dominant species that can survive formed standing formation.

The dominant species also indicates that the species can compete in utilizing water resources, nutrients, and growing space, and the dominant species can utilize environmental factors more efficiently than other species in the same place. In this current study, the species with a high IVI indicate that the presence of those species plays a very important role in the balance of the ecosystem [5]. Also, a species has a major role in the community if it has an IVI value of more than 15% for the tree and pole level.

The species with the largest IVI indicates that the species is widespread and dominates a forest area. A low index value indicates that there is high ecological pressure, either from biotic factors (competition between individual plants for each level) or abiotic factors. This high ecological pressure causes not all species to survive in a certain environment [5, 6].

Species composition based on the growth rate, namely trees, poles, saplings and seedlings, is a transitional species composition between lowland forest and sub-mountain forest, with secondary forest land cover characteristics. In some locations, land cover changes to secondary forest caused by massive illegal logging activities by the community around the forest area, causing extensive damage. Forest degradation and species composition of the tree were also influenced by impact logging [7].

The high number of species in the Myrtaceae family is thought to be because the family has good adaptability to environmental conditions where it grows. The Myrtaceae family is one of the most widely distributed families in the tropics [8]. This family has approximately 2,050 species belonging to 137 genera.

Regarding the conservation status of the plant species found at the sampling plot, there were no species included in the list of protected plants of the Minister of Environment and Forestry Regulation No. 106 of 2018, which can be caused by forest conditions that have been disturbed by the closed community such illegal log and open areas. Although, there are still several endemic floristic species of Sulawesi Island recorded, namely ruruhi (*Syzygium* sp.), kayu besi (*Xanthostemon petiolatus*), *Gmelina palawensis*, eha (*Castanopsis buruana*), pololi (*Lithocarpus celebicus*), *Colona celebica*, makadamia (*Lasjia hildebrandii*), *Archidendron pauciflorum* and belimbing bajo (*Sarcotheca celebica*).

6 Conclusion

The family of myrtaceae was dominant and in the tree, sapling and seedling levels, and the that of lamiaceae was dominant in pole level. Almost similar found species for tree and pole was recorded, such *Helicia serrata*, although the species of *Ixonanthes petiolaris* was found from pole to seedling levels. Also, the species number tended to increase from tree to sapling. For species indexes of species diversity (H'), dominance (E) and richness (R) was high, while in contrast to evenness (C).

Acknowledgements We thank to Ministry of Environment and Forestry (PKTL) for opportunity conducting our study in the production forest of Konawe Utara, Southeast Sulawesi. Thus, our special thanks to Forestry Department in Province of Southeast Sulawesi for supporting secondary data, and all field assistants for their guiding us in the field works.

References

1. Muttaqin M, Subarudi Z (2013) Effect on REDD+ implementation in Indonesia: tenure, stakeholders and livelihoods. Cent Clim Change Policy Res Dev For Res Dev Agency-Ministry For Bogor 7(2):150–163
2. Setiawan H, Jaya INS, Puspaningsih N (2015) Spatial model deforestation in Konawe Utara and Konawe, Southeast Sulawesi. Med Konservasi 20(2):166–176
3. Magurran AE (1998) Ecological diversity and its measurement. Australia (AU). Croom Helm
4. Martono DS (2012) Analysis of vegetation and associations between the types of main trees of lowland tropical forests in Mount Rinjani National Park West Nusa Tenggara. Agric Technol 13(2):18–27

5. Depari EK, Susatya A, Wiryono (2013) The potential for standing stock and carbon of Kayu Bawang (*Dysoxylum mollissimum* Blume) in Hutan Rakyat Bengkulu. Final report of Agriculture Faculty, University of Bengkulu
6. Junaedi Y (2006) Species composition of tree in the primary forest and succession forest in HPH PT. Tri Tunggal Eboni Corporation, Central Sulawesi. Thesis. Forest Management of Department Forestry, Agriculture Faculty, University of Tadulako
7. Astiani D (2016) Tropical peatland tree-species diversity altered by forest degradation. Biodiversitas 17(1):102–109
8. Lutfiasari N, Dharmono D (2018) Diversity of family myrtaceae in Tabanio Beach Forest Takisung District, Tanah Laut Regency. Proceed Nat Semin Wetlands 3(1):186–190

Study on the Potential of Carbon Storage in House Buildings with A-Frame Structures Made from Sengon Wood Impregnated with PF Resin

Jamaludin Malik, Basyarudin, Rudi Dungani, and Ali Albirnaqy

Abstract Climate change is a common concern of humankind in the world and every party should take action for addressing this global problem. This paper discussed how A-frame structure of a house made from impregnated sengon wood could contribute to national carbon storage. Assuming all houses from a national backlog in Indonesia would be built using sengon wood with an A-frame structure, the potential of national carbon storage from this part of the construction development sector would be 12.82 bt CO_2.

Keywords Carbon storage · A-frame structure · House · Impregnated sengon wood

1 Introduction

According to Paris Agreement 2015/COP21 of the United Nations Framework Convention on Climate Change (UNFCCC), global community should commit to holding the increase in the global average temperature to well below 2 °C and pursuing efforts to limit the temperature increase to 1.5 °C above pre-industrial levels, recognizing that this would significantly reduce the risks and impacts of climate change [1]. Every country is fostered to implement low greenhouse gas emissions development or—in other words—countries have to reduce emissions in their development. One year later, Indonesia committed to reduce emissions by

---

J. Malik (✉)
Research Centre for Biomass and Bioproducts, BRIN, Cibinong - Bogor District, West Java, Indonesia
e-mail: jamaludin.malik@brin.go.id

Basyarudin · A. Albirnaqy
Civil Engineering, Kalimantan Institute of Technology (ITK), Balikpapan, East Kalimantan, Indonesia

R. Dungani
School of Life Sciences and Technology ITB, Labtek-1 ITB, Jatinangor, West Java, Indonesia

M. A. Rahandi Lubis (ed.), *Proceedings of the 10th International Symposium for Sustainable Humanosphere*, Lecture Notes in Civil Engineering 255,
https://doi.org/10.1007/978-981-19-1422-5_26

29% independently and 41% with international support. As a nationally determined contribution (NDC), this promise was submitted to the UNFCCC secretariat in 2016.

The Environment and Forestry Ministry submitted the latest updated NDC document on July 21st, 2021. The document stated that a large vast of emissions reduction is expected to come from the forestry and land use sectors by 2030, at 24.5% [2]. This comprises around 692 metric tons of carbon dioxide equivalent [3]. The Indonesia's NDC will contribute to achieving the Convention's goal as stated in Article 2 by reducing greenhouse gas emissions and improving climate resilience, leading to long-term economic development. Climate change policies will align with long-term economic development by reducing GHG emissions and improving climate resilience. This policy then place forestry and other land uses as a leading sector together with Industrial processes and product use in major large-scale industries (IPPU) sector by enhancing efficiency in feedstock utilization and CO_2 recovery in upstream. Furthermore, Indonesia has taken significant steps in the land use sector to reduce emissions, including instituting a moratorium on clearing primary forests, reducing deforestation and forest degradation, restoring ecosystem functions, and implementing sustainable forest management [2].

Regarding CO_2 emissions from biomass, Kazulis et al. [4] explained both natural and anthropogenic processes cause these emissions. If biomass is allowed to putrefy in open air, CO_2 and other gases are gradually released into the atmosphere during the natural decomposition. Combustion of biomass also releases CO_2 into the atmosphere, but the energy generated in the combustion process is used. Producing goods from biomass can be considered as carbon storage in bioproducts. The most important questions here are how much CO_2 emissions are generated during the production process of a bioproduct, how long will be the durability of a bioproduct and if a bioproduct is substituting a product which is made from fossil resources. The same way as carbon in biomass has three possible paths—open air putrefaction, combustion (energy source) and being turned into a bioproduct also bioproduct at the end of its life cycle faces these three potential paths: landfill (open air putrefaction), combustion and recycling (being turned into a new product). Carbon storage in bioproducts and recycling can create a circular bioeconomy loop as the bioproducts would be recycled and turned into new products at the end of their life. For the environmental perspectives of production—the circular bioeconomy loop is the most desirable solution to manage CO_2 emissions., including using wood for end product and its application for building material.

Depending on the tree species, wood consists in varying proportions of cellulose (40%–55%), hemicellulose (12%–15%), lignin (15%–30%) and extracts (2%–15%) [5]. As a result, wood is effectively 50% carbon [6]. For European Standard DIN EN 16,449: 2014-06 EN 16,449: 2014 [7], the value of the carbon content of the wood biomass is determined to be 0.5. The other main elements are oxygen (44%) and hydrogen (6%). Some wood species also contain small amounts of mineral components, which are irrelevant in the present case. Trees and wood also contain moisture in different forms and in different amounts. Wood is generally dried down to the moisture content appropriate for the end use. The moisture content of some wood products is proportionate to the manufacturing process. The amount of moisture contained in the wood affects both the dimensions and the bulk density; With moisture removal, the wood shrinks and the bulk density decreases. The moisture content at this point is also required to determine the amount of carbon dioxide absorbed based on the specified values for volume and gross density of the wood to be used and the wood products [8]. This paper presents a study on sengon wood used for house material with A-frame structure in relation with carbon storage therein.

2 Wood Material in House Buildings

The manufacture of cement, steel, and other building materials will be a significant source of greenhouse gas emissions. This potential threat to the global climate system could be turned into a powerful tool for mitigating climate change by investigating the potential of mid-rise urban buildings designed with engineered timber. This may provide long-term carbon storage while avoiding the carbon-intensive production of mineral-based construction materials. For each ton of material, wood has a much higher carbon storage capacity (0.48t) and a much lower emission (0.12t) than cement and steel. Cement and steel have almost no carbon storage capacity and emit 0.16 t and 0.54 t, respectively [9].

Harvested wood products (HWP) include all forest products (sawn timber, veneer, chips, and their derivatives), including paper and paperboard [9]. The first stage of timber processing is the harvesting of wood. Roundwood refers to felled trees with branches removed and trunks cut to length for transportation. The use of roundwood (logs) or solid sawn or shaped timbers was part of a long, preindustrial building history. Although wood is still used for structural purposes today, primarily in the form of light-frame construction associated with low-rise residential buildings, the use of heavy solid timber for industrial and commercial multi-story buildings declined in the late nineteenth and early twentieth centuries.

Ramage et al. [8] consider the broad adoption of engineered timber structures and their potential global substitution for steel and concrete in mid-rise structural systems based on their analysis. Mass timber refers to engineered wood products that are laminated from smaller boards or lamella into larger structural components such as glue-laminated (glulam) beams or cross-laminated timber (CLT) panels. Mass-timber production methods such as finger-jointing, longitudinal and transverse lamination with both liquid adhesive and mechanical fasteners have enabled the re-formulation of large structural timbers [10]. These new approaches address wood's natural inconsistencies and improve the predictability of its mechanical performance in large structural members [5].

Architects, builders, and sustainability advocates are all abuzz about a new building material that has the potential to significantly reduce greenhouse gas (GHG) emissions in the building sector, as well as reduce waste, pollution, and construction costs while creating a more physically, psychologically, and aesthetically healthy built environment. Wood is undeniably the most environmentally friendly material to reduce greenhouse gas emissions when used in place of other materials. Wood

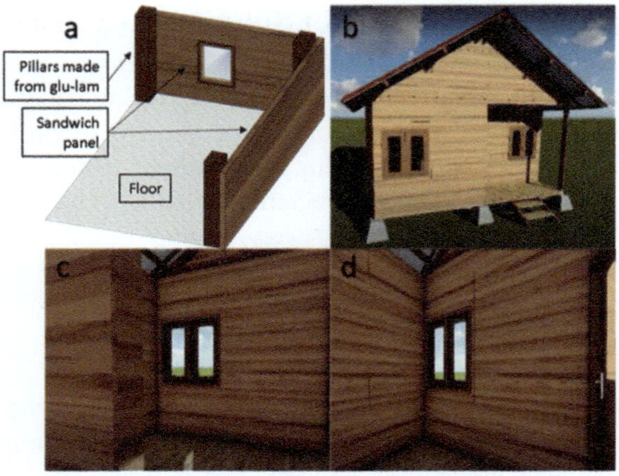

Fig. 1 The design of wood-based sandwich applications: wall and beams structure **a**; wall with fully-wood sandwich panel **b**; interior views (**c** and **d**)

has many advantages over other building materials because it is the only renewable building material. It requires less energy to produce in terms of resource extraction, manufacturing, and transportation [8, 11]. In the modern world, a new way of using wood, known as "mass timber" (short for "massive timber"), has brought the material back into the spotlight. The term "mass timber" refers to a broad range of products with varying sizes and functions, such as glue-laminated (glulam) beams, laminated veneer lumber (LVL), nail-laminated timber (NLT), and dowel-laminated timber (DLT). Malik et al. [12] proposed wood-based sandwich panel for rapid built houses combined together with glulam beams to be used in the construction of a house (Fig. 1).

However, cross-laminated timber (CLT) is the most common and well-known type of mass timber, and it is the one that has opened up the most new architectural possibilities [8]. Glulam beams, sandwich panels, LVLs, NLTs, and CLTs are engineered wood products designed to eliminate some of the weaknesses of wood and improve its quality. Treatments such as heating, densification, impregnation, and compregnation are other ways to improve wood quality. Timber is expected to meet building material requirements as a result of the treatment. In making plans for the design of wooden structures in Indonesia, there is an applicable standard that can be referred to, namely SNI 7973–2013—Design Specifications for Wooden Construction [13]. Furthermore, The Rule of The Minister of Settlement and Regional Infrastructure No. 403/KPTS/M/2002 about Technical Guidelines for Construction of Simple Healthy House guides how to build a house for all types of materials including wooden house.

3 Wood Material in Buildings as Carbon Storage

In the context of the climate change issue, the carbon in HWP is important for at least three reasons. First, the carbon in HWP is sequestered from the atmosphere while the products are being used. Second, after use, a certain fraction of HWP is disposed of in landfills where it continues to sequester carbon. Third, methane released to the atmosphere from HWP in landfills adds to the greenhouse gas levels in the atmosphere. While HWP are being used, the carbon they contain is sequestered from the atmosphere, but the time in use varies considerably from one type of product to the next. In some cases, the products remain in use for very long periods of time. Many building materials are in this category [9].

According to Tonn and Marland [14], there are three alternative approaches for carbon stored dealing with wood products as follows: (1) The flow approach follows the same logic as the fossil fuel approach in tracking carbon emissions; reporting would be of actual carbon fluxes to or from the atmosphere at the time and place where they physically occur; (2) The stock-change approach uses a pure accounting for changes in carbon stocks, reporting stock changes where and when they occur, regardless of whether the inputs and outputs are solid materials or CO_2 exchanges with the atmosphere; (3) The production approach adopts a philosophy of continuity, reporting all stock changes as continuous derivations from the original forest harvest, i.e. the party that harvested the wood would continuously report the changes in the stock of the harvested wood products derived from its forests as this actually (or statistically) occurred, regardless of who physically held the carbon-containing products.

However, there are also some simplifications accounting carbon stored in wood products as proposed by Kazulis et al. [4]. They reported that in order to determine the potential amounts of CO_2 that can be avoided by storing carbon in bioproducts, an analysis evaluating the amounts of avoided CO_2 emissions stored in the product is required. The amount of CO_2 avoided in a tonne of raw material can be calculated by assuming that a tonne of carbon requires 2.67 tonnes of oxygen to be completely transformed into CO_2, and thus a tonne of carbon can produce 3.67 tonnes of CO_2. Another simplification approach for calculating carbon content in wood products presented in DIN EN 16,449:2014 *Wood and wood-based products—Calculation of the biogenic carbon content of wood and conversion to carbon dioxide.* This standard is used as a reference in this study.

4 Carbon Storage in A-Frame House Structure from Impregnated Sengon Wood

An A-frame house should resemble a capital "A". Today's interpretation of the A-frame can be a little wonky, as architects cling to the well-known name and apply it to structures with a slight downward slope on both sides and a symmetrical peak while taking liberties elsewhere. Englishman [15] defines A-frame structure that is a triangular-shaped home with a series of rafters or trusses that are joined at the peak and descend outward to the main floor with no intervening vertical walls. In addition, Cameron [16] state that which rejects modern loosey-goosey interpretations, A-frame has a steep peak and a long, sloping yet symmetrical roof on both sides, which can extend almost all the way to the ground. The main advantages of an A-Frame house are: timeless architectural style, simple/minimalist design, secure and strong structure, requires fewer materials, adaptable and versatile, good insulation, works well in cold and warm climates, scalable build, more affordable than a traditional four wall home [15]. This structure gives a more chance to using lower quality of timber as the materials for the house due to the secure and strong structure of A-frame.

In this study, we used A-frame structure (Fig. 2) as the basic of calculation of a pre-planned house made from impregnated sengon wood (*Falcataria moluccana*) using PF resin. Sengon is one of most familiar wood species planted either by the farmers in community forest or by forest companies in Indonesia. Our previous research demonstrated that impregnation treatment sengon wood using PF resin resulted in improvement on the physical and mechanical characteristics of the impregnated sengon wood by 23% density and 9.1%, respectively [17]. Referring to other research, PF treatment also can increase durability of wood [18, 19].

The A-frame structure as shown in Fig. 2 need the total volume of the impregnated sengon wood for one unit of the house is 3.4 m^3 (Table 1).

Based on the volume above, we calculated carbon storage in impregnated sengon wood referring to DIN EN 16,449:2014-06 EN 16,449:2014 (*D*) [20]. The calculation in the standard is based on the atomic weights of carbon (12) and carbon dioxide (44) following the formula:

$$P_{CO_2} = \frac{44}{12} \times cf \times \frac{\rho_\omega \times V_\omega}{1 + \frac{\omega}{100}} \tag{1}$$

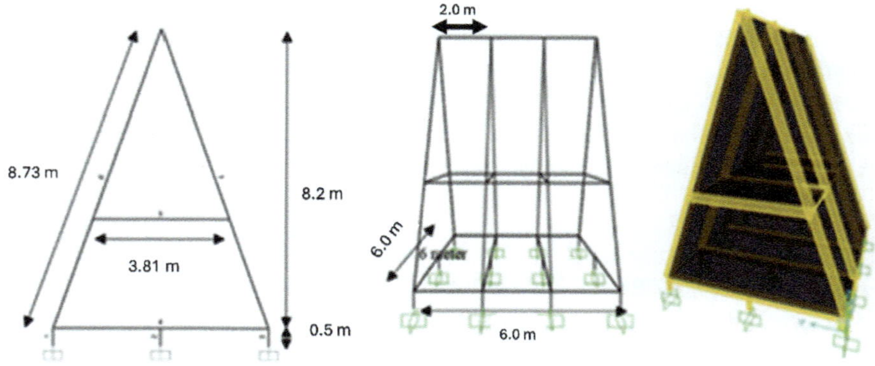

Fig. 2 The structure of A-frame house designed from impregnated sengon wood using PF resin [17]

Table 1 Volume recapitulation of sengon wood frames for house structures

Parts	Width (cm)	Thick (cm)	Length (cm)	Number of pieces	Volume (cm^3)
Column	10	10	50	12	60,000
Tilt column	10	30	873.2	8	2,095,680
Beam L1	10	30	600	4	720,000
Beam L2	10	30	381	4	457,200
Connecting beam	5	5	200	15	75,000
	Total			43	3,407,880

where:

P_{CO2} is the biogenic carbon oxidized as carbon dioxide emission from the product system into the atmosphere (e.g. energy source at the end of life) (kg);

cf is the carbon content of the wood biomass (kiln-dry mass), 0.5 as the standard value;

ω the moisture content of the product (e.g. 12%);

$\boldsymbol{\rho_\omega}$ the gross density of the wood biomass of the product at this moisture content (kg/m3);

V_ω is the volume of the solid wood product at this moisture content (m^3). In the case of wood products, the wood volume content V_ω = VP × percentage of wood. VP is the gross volume of the wood product.

$$Pco_2 = \frac{44}{12} \times cf \times \frac{\rho_\omega \times V_\omega}{1 + \frac{\omega}{100}}$$

$$Pco_2 = \frac{44}{12} \times 0.5 \times \frac{299kg/m^3 \times 3,4m^3}{1 + \frac{8\%}{100\%}}$$

$$Pco_2 = 1725.71kg$$

$$Pco_2 = 1.72571\text{ton } CO_2$$

where 299 kg/cm^3 = density of sengon wood; 3.4 m^3 = total volume of the impregnated sengon wood for one unit of the designed house; 8/100 = moisture content of the wood [17].

The calculation above shows that the carbon storage in sengon wood in 1 unit designed house with A-frame structure is 1.73 bt CO_2. This number is very important and would be a significantly contribution to carbon storage in national level if many houses built using wood as their materials. If we refer to national data, Indonesian 'backlog' until 2019 reached 7.6 million units [21, 22]. If we use the assumption that all houses from the backlog would be built using sengon wood with A-frame structure, the potential of national carbon storage from this part of construction development sector would be 12,82 bt CO_2.

References

1. UN (2015) Paris agreement
2. MOEF (2021) Updated Nationally Determined Contribution Republic of Indonesia, Ministry of Environment and Forestry: Jakarta
3. Aqil AMI (2021) Indonesia aims to transform forests into carbon sink, in The Jakarta Post, PT. Bina Media Tenggara: Jakarta
4. Kazulis V et al (2017) Carbon storage in wood products. Energy Proced 2017(128):558–563
5. Roger M, Rowell, RP, Tshabalala MA (2012) Handbook of wood chemistry and wood composites, in Handbook of Wood Chemistry and Wood Composites, R.M. Rowell, Editor, CRC Press: London
6. Brown S (1997) Estimating biomass and biomass change of tropical forests: a primer, in, FAO: Rome, in FAO forestry paper. FAO, Rome
7. CEN (2014) Wood and wood-based products—calculation of the biogenic carbon content of wood and conversion to carbon dioxide; German version EN 16449:2014, European Committee For Standardization: Brusel
8. Ramage MH et al (2017) The wood from the trees: the use of timber in construction. Renew Sustain Energy Rev 68:333–359
9. Miner R, Lucier A (2003) Item 9c: carbon in harvested wood products, in FAO advisory committee on paper and wood products, 44th session. FAO, Rome
10. Churkina G et al (2020) Buildings as a global carbon sink. Nat Sustain 3:269–276
11. Adhikari S, Ozarska B (2018) Minimizing environmental impacts of timber products through the production process "From Sawmill to Final Products." Environ Syst Res 7(6):1–15
12. Malik J, Trisatya DR, Abdurachman (2020) Potential use of fully wood based sandwich panel for house component in Indonesia for disaster mitigation and rapid built: A review. In: International conference on forest products (ICFP) 2020: 12th International symposium of IWORS 2020. Bogor, Indonesia: IOP Conference series: materials science and engineering
13. BSN (2013) SNI 7973–2013 Spesifikasi Desain untuk Konstruksi Kayu, National Standard Agency: Jakarta
14. Tonn B, Marland G (2007) Carbon sequestration in wood products: a method for attribution to multiple parties. Environ Sci Policy 10(2):162–168
15. Englishman K (2020) A definitive guide to the a-frame house: history, how to build & more, in field mag. Community Garden New York City, USA
16. Cameron S (2019) Advantages and disadvantages of a-frame houses. [Cited 2021 25 July]; Available from: https://www.hunker.com/12585450/advantages-and-disadvantages-of-a-frame-houses
17. Birrnaqiy AA (2021) Analisis Perlakuan Impregnasi Resin Fenol Formaldehida Terhadap Kayu Sengon Untuk Diaplikasikan Pada Rumah Modern A-Frame (Impregnation Analysis of Phenol Formaldehyde Resin on Sengon Wood for Appli-cation in A-Frame Modern House), in Civil Engineering, Institut Teknologi Kalimantan: Balikpapan
18. Bakar ES et al (2013) Durability of phenolic-resin-treated oil palm wood against subterranean termites a white-rot fungus. Int Biodeterior Biodegradation 85(November):126–130
19. Stefanowski BK, Spear MJ, Pitman A (2018) Review of the use of PF and related resins for modification of solid wood, in Timber 2018, Spear M, (ed). 2018, Bangor University: Bangor, Philipine
20. CEN (2014) EN 16449:2014 (D): Wood and wood-based products—calculation of the biogenic carbon content of wood and conversion to carbon dioxide. 2014, European Committee for Standardization: Brüssel.
21. Kuswara K (2019) Flagship Bangunan Tahan Gempa, Tahan Api, Cepat Bangun, dan Murah. Presentation Material for Coordination Meeting of Flagship of Earthquake-ResistantBuilding. Research and Development Centre for Settlemet (Puskim), Ministry of Public Works and Housing (PUPR): Depok 21 August 2019
22. Hutapea E (2019) Backlog Rumah 7,6 Juta Unit, in Kompas, PT Kompas Gramedia: Jakarta

Tropical House Design and Analysis Considering Energy Efficiency and Local Seismicity-A Case Study in Sukabumi, West Java, Indonesia

Dany Perwita Sari, Ananto Nugroho, and Luna Nurdianti Ngeljaratan

Abstract The total energy consumption of a household in temperate countries such as Indonesia is not only counted from the cooling system but also from the lighting equipment. Therefore, it is necessary to design a housing that considers the energy efficiency to reduce the cooling and electricity bills as well as the carbon emission. Besides energy efficiency, seismicity also should be a crucial part of the design as a seismic-prone country like Indonesia should implement the seismic-resistant structure concept in every aspect of the structural design. This study selects Sukabumi, Jawa Barat, Indonesia as a case study in which a tropical house type is planned to be located. The study focuses on improving energy saving and assessing the seismicity of Indonesian housing using solar shading devices and local-site analysis. The preliminary design shows that the proposed housing concept from the architectural and structural perspective can be implemented and requires further analysis to assess the high energy efficiency and structural safety against earthquakes.

Keywords Energy efficiency · Energy saving · House · Seismicity

1 Introduction

Indonesia has been experiencing the rise of average temperature over the past ten years, with the mean annual temperature increase by about 0.3 °C and projected warming from 0.2 to 0.3 °C per decade [1]. The morbidity and mortality rates caused by either the global or local climate change, or by the combination with the rise of

D. P. Sari · L. N. Ngeljaratan (✉)
Research Center for Structural Strength Technology, National Research and Innovation Agency, Tangerang Selatan, Indonesia
e-mail: luna.ngeljaratan@biomaterial.lipi.go.id

A. Nugroho
Research Center for Biomass and Bioproduct, National Research and Innovation Agency, Cibinong, Indonesia

M. A. Rahandi Lubis (ed.), *Proceedings of the 10th International Symposium for Sustainable Humanosphere*, Lecture Notes in Civil Engineering 255,
https://doi.org/10.1007/978-981-19-1422-5_27

pollution level are previously reported and highly alarming [2]. Therefore, households in Indonesia and other temperate countries have a high dependence on air conditioning to achieve thermal comfort. The installation of cooling devices at home is in a high demand as they do not only preserve a cooler temperature but also maintain the health of the members from health issues [3, 4]. However, buildings with air conditioning typically consume more energy and emit more carbon dioxide.

Another source of energy use in an Indonesian household is the use of lighting equipment especially for middle- and low-income families [5, 6] that adds into the total energy consumption of a household. Therefore, it is of interest to design the most-effective and energy efficient house to reduce the cooling and electricity bills. The design should consider the site-specific climate and materials to improve the energy efficiency. For example, by understanding the local climate and sun path diagram, it is possible to reduce the lighting electricity usage. Also, house design should require detailing that meet the performance requirements.

Besides the rising temperature, the design of Indonesian house should also confront the hazard as the country lies in a highly active seismic zone. For example, an earthquake of magnitude 5.0 that struck near Sukabumi on Tuesday, 27 April 2021 reported many casualties and damaged of the houses. Most of the houses are unrepairable so a new construction is preferred. However, most of the house replacement is not for long-term use, they serve more as an immediate or temporary housing responding to the disaster mitigation scenarios. Therefore, it is of importance to not only implement the energy efficient concept in a house design but also to ensure the safety of their occupants during a major seismic event.

The objective of this research is to improve the energy saving of a household and to assess the design safety from the seismicity perspective by maximizing the daylight and reducing heat gain using solar shading devices as well as analyzing the local site seismicity. This study proposes that earth science, climate science, as well as earthquake engineering should be combined to explore an innovate architectural design of a house with a safety assurance from the earthquake perspective. As a case study, the house is assumed to be located at Sukabumi, West Java, Indonesia, with a recent active seismicity and considerably high temperature. Sukabumi was selected due to the region's tectonic and geological conditions, which frequently cause natural disasters such as earthquakes and tsunamis. An innovative concept for houses is presented in this paper. The house design is described first followed by the analysis of energy efficiency and structural safety against earthquake. The preliminary results are reported and the conclusions are drawn based on the findings and observations.

2 Method

2.1 General Description of the House

The architectural model of the house is shown in Fig. 1. Modeling is done using Sketchup for web. The selection of house location considered the previous earthquake in which there were several places in Sukabumi that were affected with severe damages. As shown in Fig. 1, the house only has one floor for a simplified design. The design of the floor plan complies with local culture to ensure the comfortability of the occupants [7]. The cross ventilation is achieved by installing windows in the bedroom, living room, and roof.

The most important building element for tropical climates is window shading especially for the Indonesian housing as it located on the equator. Thermal radiation passing through the opening increases the indoor temperature. In addition to this, opening is crucial for reducing the electricity consumed by lighting appliances. An estimation of the possibility of shading is necessary when designing openings for windows. For shading analysis, it is necessary to know the time of shadow cast per day of the year [8]. Figure 2 shows a simplified graphicamethod that determines the shadow based on the the solar angle for the surface, or the solar profile angle, p that is determined when the surface is normal and the sun's rays are projected on the plane. The solar profile angle, p, can be calculated using (1) in which α is the solar altitude angle, z is the solar azimuth angle, and Z_s. is the surface azimuth angle. This equation is used to design all the shading and opening in Sukabumi's house shown

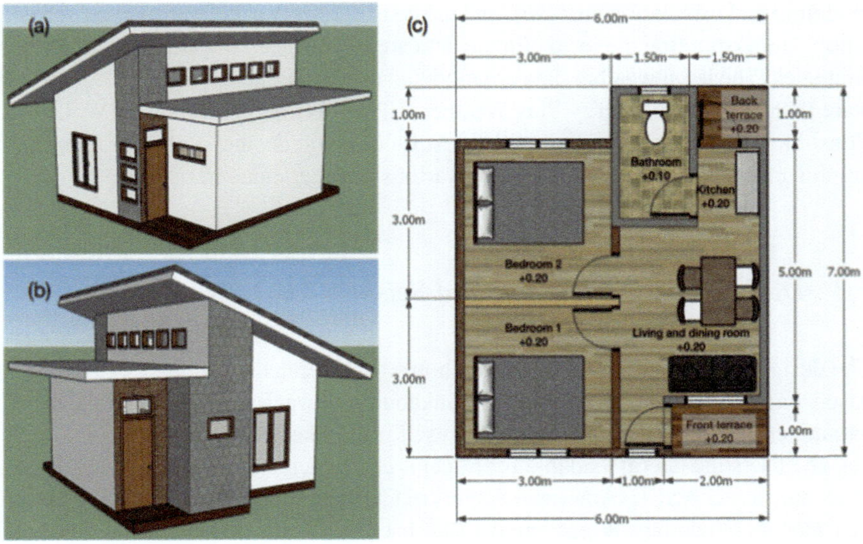

Fig. 1 An architectural model of Sukabumi's residential building; **a** the front façade of the house facing south, **b** the back side of the house, **c** an open floor plan that allows for maximum ventilation

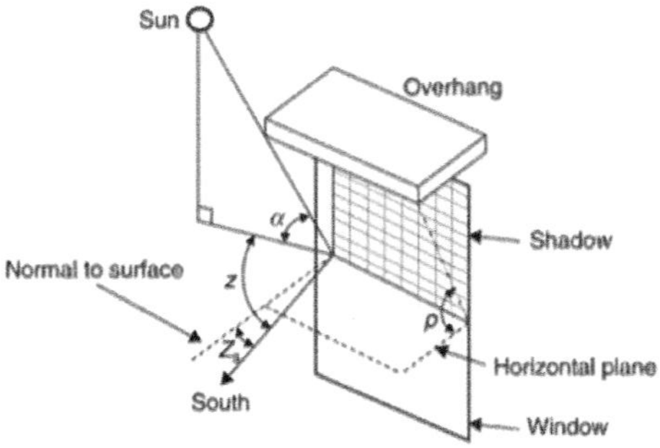

Fig. 2 A simplified graphical method to illustrate shadows from the solar angle [8]

in Fig. 1. With its ceiling-mounted windows the building is designed to maximize daylight in the bedroom. Large overhangs were designed near all of the windows as a means of reducing glare.

$$\tan(p) = \frac{\tan(\alpha)}{\cos(z - Z_s)} \tag{1}$$

2.2 *Analysis of a High-Energy Efficient Housing*

In order to effectively analyze a high-efficiency house, three steps are considered, i.e. the analysis of the local climate and meteorological data, in this case Sukabumi, modeling and orientation analyses, and analysis of energy efficiency and cross ventilation. Understanding the local climate data are the first step toward designing future strategies. The local weather data are modeled using Climate Consultant 6.0.15 software that reads from Energy Plus Weather (EPW) format [9]. Using EPW, Climate Consultant helps designers select the best bioclimatic design strategies. Dozens of graphic charts are generated that show various weather characteristics.

2.3 *Design of Seismic Load and Seismic Analysis*

The designed response spectrum of the house is generated seismic design map [10] based on the location coordinates of Sukabumi, Jawa Barat. The seismic design parameters are given in and the values used in the design are given in Fig. 3. The design of seismic lateral load in this project is conducted using the Simplified Design procedure as per ASCE (American Society of Civil Engineers) 7–10 [11]. The Simplified Design Procedure is permitted to be used following limitations such as: the building category shall qualify for Occupancy Category I or II, the site class shall not be class E or F, the structure shall not exceed three stories in height above grade,

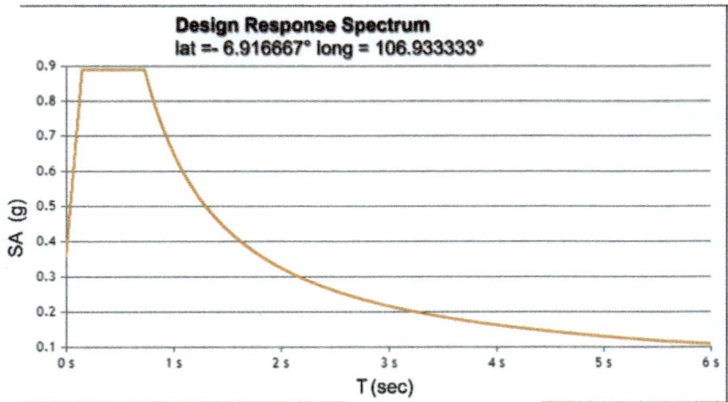

Seismic Design Parameters

Site class = D (soft soil)
PGA (g) = 0.56
S_s (g) = 1.33
S_1 (g) = 0.56
S_{DS} (g) = 0.89
S_{D1} (g) = 0.65
T_0 (sec) = 0.15
T_s (sec) = 0.73

Fig. 3 Design response spectrum and seismic design parameters of tropical house

the seismic-force resisting system shall be either a bearing wall system or building frame system, there are no system irregularities caused by in-plane or out-of-plane offsets of lateral force-resisting elements, and the lateral-load-resistance of any story shall not be less than 80% of the story above.

From the inspection of the building plan and configuration, the building was categorized as simple building; therefore, the Simplified Design for calculating the seismic load could be used. However, a response spectrum analysis is still conducted to ensure safety considering the system as single degree of freedom (SDOF) structure as per (2):

$$m\ddot{x}(t) + c\dot{x}(t) + kx(t) = f(t) \tag{2}$$

To achieve equilibrium, then the sum of all forces acting on the mass should be equal to zero with $f(t)$ an externally applied force as a function of $m\ddot{x}(t)$, the inertial force, $c\dot{x}(t)$, , the (viscous) damping force, and $kx(t)$ the restoring force. In mechanical structures, the damping coefficient c is usually very small resulting in a complex conjugate pole pair as shown in (3):

$$\lambda = -\sigma \pm i\omega_d \tag{3}$$

with $f_d = \omega_d/2\pi$ the damped natural frequency, $f_n = \omega_n/2\pi$ the un-damped natural frequency where $\omega_n = \sqrt{k/m} = |\lambda|$, $\zeta = \frac{c}{2m\omega_n} = \frac{\sigma}{|\lambda|}$ the damping ratio, so that $f_d = f_n\sqrt{1-\zeta^2}$. The natural frequency, ω_n, , is in units of radians per second (rad/s) but it is preferred to present it in Hertz (Hz) especially for dynamic analysis of structures. The damping factor can also be represented as a percent of critical damping that is the damping when the system experiences no oscillation such as modal damping. Even though there are three distinct damping cases, only the underdamped case ($\zeta <$ 1) is important for structural dynamics applications.

3 Result and Discussion

3.1 Weather Analysis

Sukabumi is located between 6° 57 and 7° 25 South Latitude dan 106° 49–107° 00 East Longitude [10]. Sukabumi has a total area of approximately 4.162 square kilometers. The climate here consists of two seasons: the dry season and the wet season. Climate Consultant software was used for more detailed analysis of local weather data. The weather data of Sukabumi is shown in Fig. 4 that illustrates the dry bulb temperature. The chart shows the average temperature for each month in a year. Throughout each month, the dry bulb temperature remains relatively stable. The local dry-bulb temperature is relatively comfortable, around 18 °C in the morning and 25.62 °C at noon. The wind rose in Fig. 4b, describing the predominant wind direction and frequency, as well as wind velocity for Sukabumi. Relative humidity above 70% is indicated by a light green shade. In general, the wind comes predominantly from the north.

A climate consultant can also provide an adaptive comfort strategy, as depicted in Fig. 5. The psychrometric chart illustrated the design strategies. As a powerful visual aid for understanding the relationship between supply air and relative humidity, psychrometric charts are widely used. Throughout the climate data file, a dot represents each hour. Dry bulb temperature is shown on the x-axis, while fresh air humidity is shown on the y-axis. A residential building in Sukabumi was evaluated for its thermal comfort using the adaptive comfort strategies shown in the diagram. As shown in Fig. 5, an adaptive comfort through ventilation is the only proposed strategy for Sukabumi with the estimated values of 32.3%. Therefore, the following strategies can be applied:

Orient the windows with the prevailing breezes and incorporate shaded windows for natural ventilation;

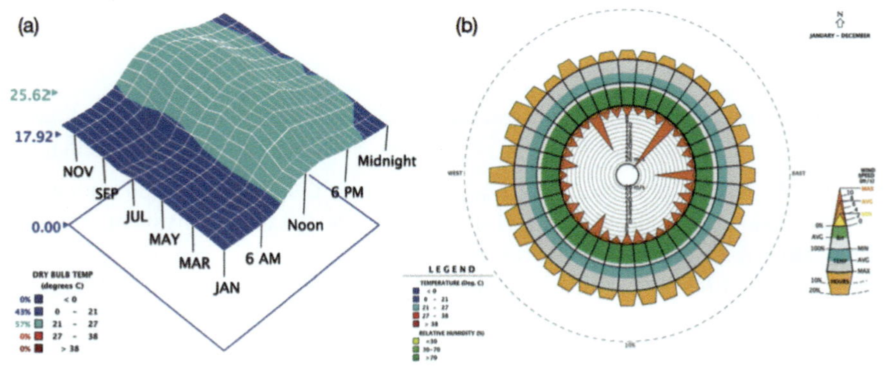

Fig. 4 Sukabumi's weather data: **a** dry bulb temperature; **b** humidity, wind speed, and wind direction

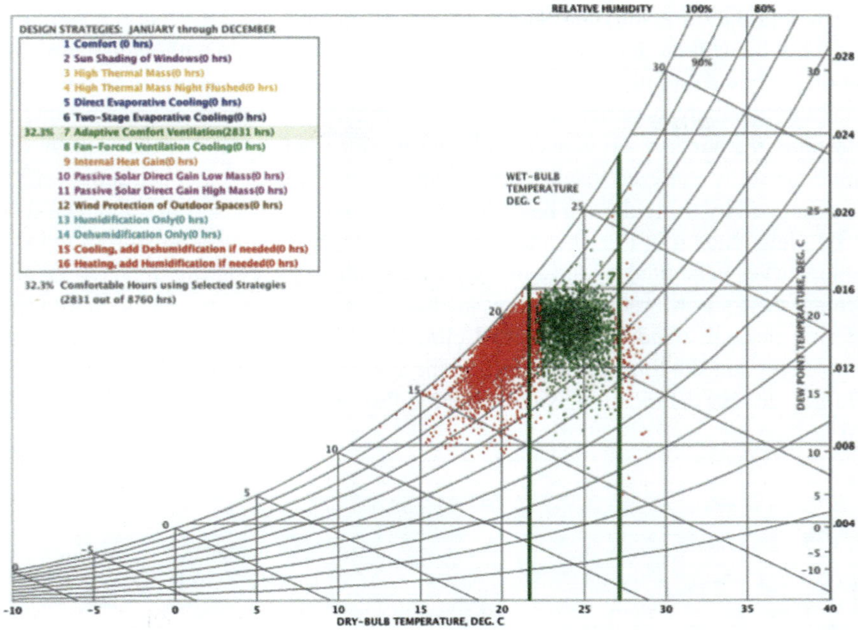

Fig. 5 Psychrometric chart of Sukabumi, West Java, Indonesia

Cross ventilation can be maximized with a narrow, long floor plan;
To reduce glare and heat gain, high ceilings and pitched roofs were used;
Balance daylighting by allowing enough south glazing,; and
Reduce heat gain by minimizing west-facing glazing.

Based on the observation, this passive strategy is sufficient to be proposed for the future designs of house in Sukabumi. Additionally, maintaining thermal comfort while maximizing cross-ventilation is vital during pandemic conditions. Those strategies will be used to reduce the consumption of energy and to provide a healthy living environment.

3.2 *Shading Analysis*

To determine the location of the sun during any given period of time, a sun path diagram is essential. Sun path diagrams reveal how many sunlight hours a location receives at any given location based on each component. The simulation allows for a more detailed examination of the availability of daylight that may be harvested on a given day, which is useful in the calculation of the total daylight extraction on that day. A simulation is conducted to predict how much illumination is available each year.

The simulation results are shown in Fig. 6. In this case, the sun's position is represented one hour at a time as illustrated in Fig. 6a. The sun path map diagram is calculated based on Sukabumi's current latitude and longitude shown in Fig. 6b. Sun path diagram analysis period was varied from month to month in order to be able to determine the sun's position accurately during different seasons of the year. A sun path diagram is analyzed visually using sun vectors, altitudes, and azimuths, which are generated by using the sun path curve for each period. The set curves can be seen in 3D view (shown in Fig. 6c and d) using Sun-Path 3D simulator. Across the sky dome, curves generated by the sun are used to reduce glare during the daytime and increase daylight. Moreover, window shading is a necessity to prevent sunlight from entering through windows in tropical climates. January and June were chosen as samples for sun shading in Sukabumi. In the case of Sukabumi, the selected shading solution reduces the incoming solar radiation flux by more than 80% both in January and in June.

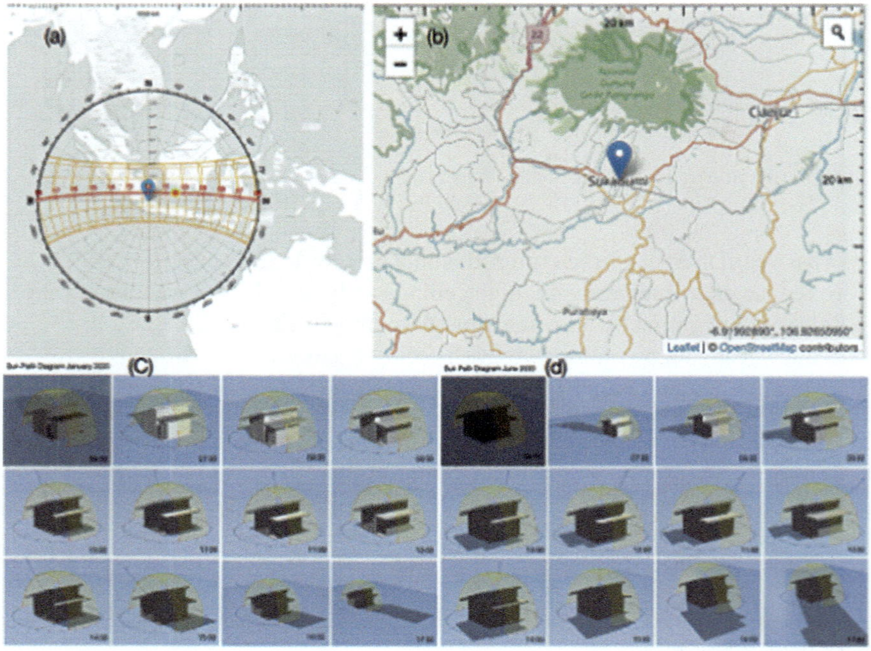

Fig. 6 **a** Sun Path map diagram on Sukabumi **b**, West Java Indonesia; 3D Sun_Path analysis from 3D house model in Sukabumi shows: **c** shading analysis from 6am to 5 pm on 22 January 2021 and **d** shading analysis from 6am to 5 pm on 22 June 2021

3.3 Daylight Analysis

A dynamic daylighting simulator was used to analyze the daylighting analysis. Daylighting analysis is a preliminary simulation to understand and quantify the amount of natural light in a house. This analysis can help reduce lighting loads and create more comfortable spaces. The house was assumed to have a 88 mm single pane glazed window with variative size as shown in Fig. 7. According to Sukabumi's weather data most of the dynamic daylight comes from bedroom 1 and bedroom 2. The average illuminance level (300 lx) is shown in in Fig. 7c.

UDI (Useful Daylight Illuminance) measures the percentage of time that a space receives adequate daylight and also identifies when the light levels are too high or too low. UDI is based on three standard bins: First less than 100 lx is insufficient daylight, second between 100 and 2000 lx is useful daylight and third more than 2000 lx is too much daylight and can result in visual and thermal discomfort. The UDI results from the house are shown in in Fig. 7d. The selected shading solution can provide useful daylight (300–3000%).

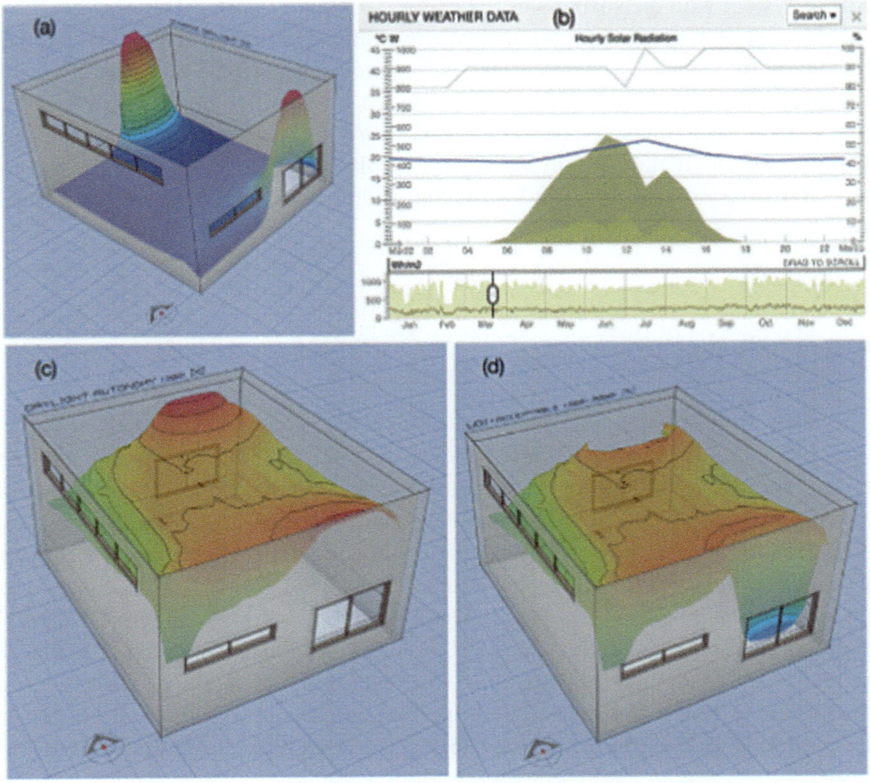

Fig. 7 Average illuminance in Sukabumi's house **a** based on Sukabumi hourly weather data **b** show the daylight autonomy **c** and UDI **d**

Table 1 Storey and diaphragm forces

Seismic weight, w_x (kN)	Storey forces, F_x (kN)	Storey shear, V_x (kN)
5.50	0.87	0.87

3.4 Seismic Analysis

From the inspection of the building plan and configuration, the building was categorized as simple building; therefore, the Simplified Design for calculating the seismic load could be used. The method uses the assumption such that the house is expected to respond to seismic excitation in its fundamental lateral mode. Therefore, the house should be low-rise without plan irregularities to avoid torsional movement under seismic load.

The static seismic load method is started by computing the seismic weight of the house as the function weight and height as well as the seismic coefficient. The results of storey and diaphragm forces using equivalent static analysis are given in Table 1.

A response history analysis is also conducted in which the house is modeled as SDOF system with the stiffness as 0.2 kN/mm. Using mass and stiffness, the modal frequencies of previous step and employing the principles of dynamics compute the modal frequencies, w, and corresponding mode shapes, φ. The modal mass and modal participation factor, Γ, are then computed as a function of seismic weight and mode shape. The fundamental mode of the house is computed as 0.31 s associated with 3.2 Hz of frequency. The maximum displacement at roof level is computed as 0.154 m associated with 2.8% drift that is still below the limit of 3% drift for structural safet assessment.

4 Conclusion

The result from weather analysis reveals that a sun shading device is an important option to minimize heat gain and improve daylighting. Various window configurations influence the lighting distribution in the building model space in significantly different ways. A configuration like this achieves an adequate amount of daylight transmission into the house. Meanwhile the detailed relationship between the dynamic lighting effect of windows and the position of the sun was analyzed based on the 3D simulation model, which should receive more attention when designing architecture. This suggests that designer, architect, and the Government of Indonesia should focus on the implementation of sun shading device as the priority. As from seismic analysis, it is concluded that the preliminary design shows acceptable performance during the design earthquake. The drift is considered below the maximum limit assuring that the house is expected to perform well during seismic event.

References

1. Case M, Ardiansyah F, Spector E (2007) Climate change in Indonesia: implications for humans and nature. Climate change in Indonesia: implications for humans and nature
2. Gosling SN, Lowe JA, McGregor GR, Pelling M, Malamud BD (2009) Associations between elevated atmospheric temperature and human mortality: a critical review of the literature. Clim Change 92(3):299–341
3. Randazzo T, De Cian E, Mistry MN (2020) Air conditioning and electricity expenditure: the role of climate in temperate countries. Econ Modell 90:273–287
4. Santamouris M (2020) Recent progress on urban overheating and heat island research. Integrated assessment of the energy environmental, vulnerability and health impact. Synergies with the global climate change. Energy Build 207:109482
5. Batih H, Sorapipatana C (2016) Characteristics of urban households electrical energy consumption in Indonesia and its saving potentials. Renew Sustain Energy Rev 57:1160–1173
6. Sukarno I, Matsumoto H, Susanti L (2017) Household lifestyle effect on residential electrical energy consumption in Indonesia: on-site measurement methods. Urban Clim 20:20–32
7. Sari DP, Chiou YS (2019) Transformation in architecture and spatial organization at Javanese house. ASONAM'19: Proceedings of the 2019 IEEE/ACM international conference on advances in social network analysis and mining
8. Kalogirou SA (2014) Solar energy engineering (second edition). ISBN 978-0-12-397270-5. https://doi.org/10.1016/C2011-0-07038-2
9. Climate Consultant 6.0.15 (2021). Available online: https://climate-consultant.informer.com/6.0/
10. Kementerian PUPR (2021) Desain Spektra Indonesia homepage, http://rsa.ciptakarya.pu.go.id/2021/. Last accessed April 10
11. American Society of Civil Engineers (2010) Minimum design loads for buildings and other structures. ASCE/SEI 7, Reston, VA

The Natural Disaster Rescue Cube

Martinus B. Susetyarto

Abstract Earthquakes, landslides, and tsunamis are very powerful natural disasters. These disasters are difficult to predict. But, they occur very quickly and take many lives, damage infrastructure, and disrupt economic activities. In response to the impact of the disaster, an initiative emerged to build a strong and safe rescue cube to protect women and children from earthquake shocks, landslides, and tsunami waves, as well as the impact of the collapse of building construction around them. This futuristic idea is a follow-up to my research with experimental test methods in 2019. The natural disaster rescue cube is made of wooden planks with a connection system without nails, similar to the wooden construction system at the *Horyuji* shrine in *Ikaruga*, Japan, which has stood for thousands of years. The cube is designed to be strong enough to accept a load of 10 tons/m2, tested for durability in a credible laboratory, and meets the SNI test. The cube is equipped with a seat belt, emergency buoy, oxygen cylinder, mineral water storage, first aid kit, surveillance binoculars, and an emergency signal that sends the location of the cube. The emergency signal is automatically connected to the IT system of the National Disaster Rescue Agency at the Central and Regional levels, as well as the artificial intelligence network the Ministry of Communication and Information Technology. Now the natural disaster rescue cube is being patented at the DJKI Kemenkumham R.I.

Keywords Disaster · Rescue · Cube

1 Introduction

The Nusantara is located on the emerald equator, which is prone to earthquakes, volcanic eruptions, landslides, hurricanes, and tsunamis. It is difficult to predict when this natural disaster will occur and how big its impact will be. The disaster occurred very quickly and took many lives, damaged infrastructure, and disrupted

M. B. Susetyarto (✉)
Universitas Trisakti, DKI, Jakarta, Indonesia
e-mail: bambang.s@trisakti.ac.id

M. A. Rahandi Lubis (ed.), *Proceedings of the 10th International Symposium for Sustainable Humanosphere*, Lecture Notes in Civil Engineering 255,
https://doi.org/10.1007/978-981-19-1422-5_28

the economic activities of the community. Responding to the impact of the disaster, an initiative emerged to build a strong and safe rescue cube to protect residents, especially women and children, from earthquake shocks, landslides, and tsunami waves, as well as the impact of the collapse of building construction around them. This futuristic idea was a follow-up to my research with experimental test methods in 2019. This disaster rescue cube was made of a strong wooden plank with a nail-free joining system, similar to the wooden construction system at *Horyuji* temple in *Ikaruga*, Japan [1] which was built a thousand years ago. The cube was inspired by the *One* building, a cube of the *Sa'o* traditional house in the *Ngadha* culture, the *Ngada* Regency, the East *Nusa Tenggara* Province, *Flores* Island, has proven to be reliable. There was also the *Bagha* building that looks like the *One* building, representing the presence of the ancestors, which was erected in the courtyard of the traditional kampongs. Many indigenous people believe that living in the *One* and/or the *Bhaga* buildings will be safe from all natural disasters and threats of disease. Their ancestors who lived in the *One* and/or the *Bhaga* will always protect them. They believe that although Flores is located in the ring of fire, and tectonic earthquakes often occur which sometimes pose a risk of tsunami waves, but the number of human casualties recorded in traditional *Ngadha* settlements due to natural disasters was not large. Therefore, the cube-shaped *One* or *Bhaga* buildings measuring $\pm$ 3.60 $\times$ 3.60 square meters or smaller, with ceiling height of about 3.00 m, made of wood and bamboo structures, which was built strong, by means of splicing wooden pegs without using nails or screw, has inspired researchers to develop a natural disaster rescue building cube.

Referring to the experiences and testimonies of indigenous peoples during the shaking due to natural disasters, the most severe damage occurred when the *One* and/or the *Bhaga* buildings were shaken violently that the buildings were separated from their position in the construction of the stage floor, where the construction was only supported by wooden pile foundations, or megalithic stone pillars that can slip from their position on the stone pedestal, shift, and the *One* or the *Bhaga* building falls to the ground in the condition that the cube building remains intact [2]. Parts of the floor and ceiling that were firmly attached to the construction of the *One* or the *Bhaga* including the wooden column ties and wallboards that surround it remain intact. The floor and ceiling construction elements were also not detached, or collapsed due to the impact of the construction roof from above. The interlock system to connect wooden posts and wall boards without threads or threads was made so sturdy, both in terms of workmanship and in terms of material use. Thus, in the event of an earthquake, family can save themselves by entering the *One* or the *Bhaga* cube for a few minutes to avoid accidents and safety threats from a more severe natural disaster. The *One* or the *Bhaga* disaster cube prototype model deserves to be patented, and then tested for its strength, durability, technical feasibility study, and market feasibility for downstream processing. Furthermore, in collaboration with industry, government, and local governments, this disaster rescue cube property will be created on a real scale of 1:1 as a pilot project.

2 The Structure of *Ngadha* Traditional House

2.1 The Sa'o

Traditional houses in *Ngadha* culture are referred to the *Sa'o.* The structure of the *Sa'o,* the *Ngadha* traditional house vertically consists of head—body—feet, or in the *Bajawa* language called *zeta ulu—zale wekki/zeta wekki—zale wa'i* and mentions the parts of the human body. *Ulu* is the head, *wekki* is the body, and *wa'i* is the feet. If referring to the roof space, walls, and pillars of the stilt floor, then the roof space as *zeta ulu,* walls as *zeta wekki,* stage floor as *zale wekki.* The space under the stilt floor as the foundation pillar of the cube with stilt construction is called *zale wa'i.* So, vertically, traditional houses are grouped into four vertical spaces, namely roof (*ubu sa'o*), walls (*kebi ube kedu*), stilt floor (*befa*), and poles (*tere leke/leke*) [2]. The structure of the *Sa'o* traditional house is understood vertically and horizontally as illustrated in Fig. 1 below:

Horizontally, the components of the *Sa'o* traditional house spaces are the *One*, the *Teda one*, and the *Teda wewa.* Particularly in the *One* cube, there are two recitations, namely the women's zoning, it is called *papa bhoko*; and the men's zoning, and it is called *papa lewa.* This enlightenment is not interpreted as segregation of zone, but as a state of mutual complementarities. Whereas, the *Teda one* and the *Teda wewa* are

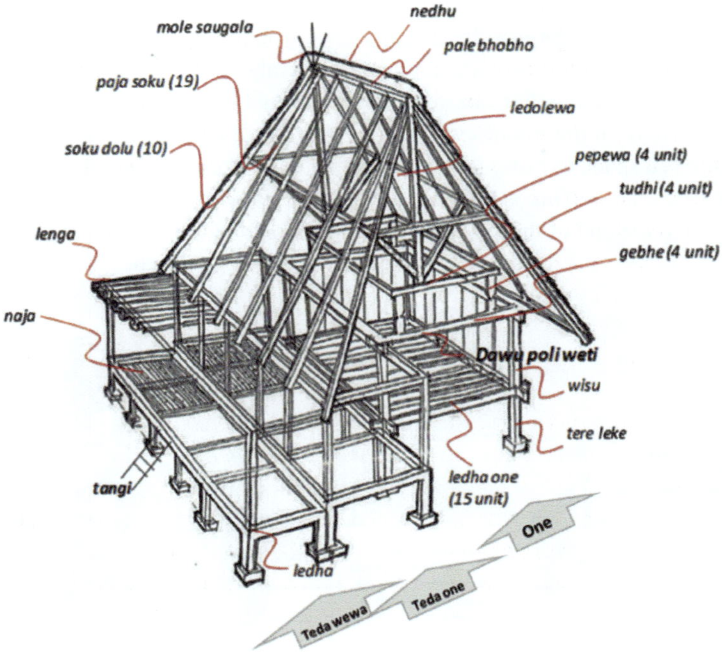

Fig. 1 Structure of the *Sa'o. Source* Susetyarto, 2013

not conceptualized as different zones. However, these two spaces are enlargement of the *One* as the additional rooms, or *Bheli* in the *Bajawa* language, at the right or at left of the *One*, where the floor surface is made not as high as the surface of the *One* floor. The concept of the structure and construction of the *Sa'o* building which is so strong, of which the *One* building is the core, has been experimentally proven to be able to withstand the shock of earthquakes, hurricanes, and other natural disasters.

2.2 *The* One

The *One* is as a core space inside the *Sa'o*, the main building of the "*Sa'o*," or it is called the real residence of the owner. The word of *One* means a space to live in, a place to get togetherness. Multi-use functions of the *One* namely accommodating the independent activities of family members/clans, they are called "*Ana woe*" in the *Bajawa* language; resting, cooking, traditional rituals, traditional meetings, giving birth to babies, laying down bodies, and others. The benefits of the *One* cube are to get protection, security and safety from the ancestors. By living in the *One* cube together in a family/clan entity, then protecting each other, keeping each other safe, and surviving can be realized. Please study carefully the *One* building construction system below [2, 3] (Fig. 2).

In the *One* cube, there are ancestral spirits that guard the spiritual life of the inhabitants of the traditional house, they called *Ana sa'o* in the *Bajawa* language; guarding against interference from evil spirits around the area. The ancestor which is called *ebu nusi* in the *Bajawa* language, is symbolized in the form of *sau uwi*, a piece of bamboo *aur* and *bhoka*, the *tuak* traditional drink bottles, which are stored in the *mataraga* on the middle wall of the *One*, is called as *ube kedu*. The meaning of the *One* is a space to bring together the spiritual needs between *Ana sa'o* and *ebu nusi*. In addition, the *One* is the base of their life, which is called *ulu* in the *Bajawa* language and the end of their life, which is called *eko* in the *Bajawa* language. Thus,

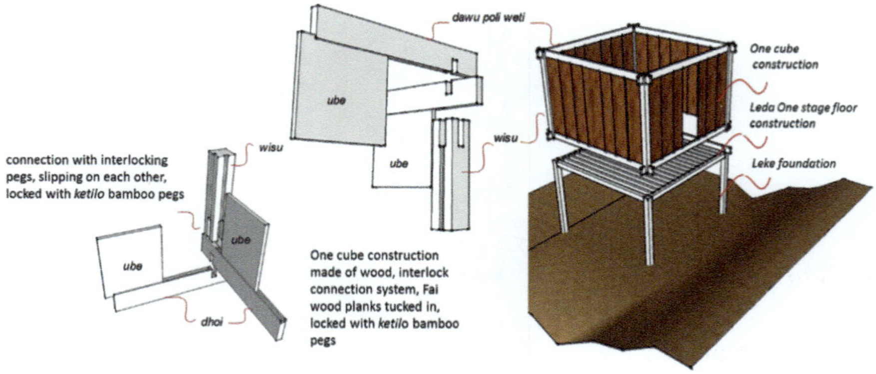

Fig. 2 Structure of the *one*. *Source* Susetyarto, 2013

the *One* is as the smallest world, or micro cosmos, of the life of *woe* and *Ana sa'o* in the indigenous people of the *Ngadha* culture. Therefore, this meaning is articulated more clearly by the *Lima pade*, a name of 'traditional architect' of *Ngadha* culture, by making the floor surface of the *One* has the highest floor surface among the other rooms in the *Sa'o*. The cube of *One* is a single box building without a permanent construction join, without the use of nails and iron hinges, and has a separate structure and construction from the *Teda one* or the *Teda wewa* building. Besides being used for traditional ritual activities, the *One* cube is also used for cooking chicken as a sacrifice for ancestors. Therefore, on the right side near the entrance are three stone stoves as a traditional kitchen, is called *lapu lika* in the *Bajawa* language. The right side of the *One* area, there is a traditional kitchen is called a female area, or *papa bhoko Ana fai* in the *Bajawa* language, and vice versa it is called a male area, or *papa lewa Ana saki* [2–4].

2.3 *The* Bhaga

The structural system and construction of the *Bhaga* building stand in the *loka woe,* which is the courtyard in front of the *Sa'o*, belonging to the tribe/clan of the owner of the traditional house. Therefore, a *loka* is a representation of the ownership of the outdoor space which is named the same as the name of the tribe/clan that owns the outdoor space. The *Bhaga* building is essentially similar to the *One* building. Its function is as a traditional kitchen when traditional rituals are held. Initially, the *Bhaga* building had the same structural system and construction as the *One* building. However, in its development the *Bhaga* building was made with a smaller size, approximately 1.5 × 1.5 m2 [2, 3]. The *Bhaga* building has one door that is always open facing to *Ulu,* the front of the village. In front of the *Bhaga* stands the *Ngadhu* building, an erect thatched roofed structure, which is philosophically tasked with guarding the spirits of the ancestors who are believed to reside in the *Bhaga* cube. The *Ngadhu* building is a representation of the guardians of ancestral spirits, guardians of indigenous people (*woe*). Therefore, the *Ngadhu* building is decorated like a palace guard who is equipped with weapons. The *Bhaga* building was made based on the decision of the *Mosalaki* meeting, with the main material being boards and bamboo, joined without using nails or screws. *Bhaga's* height is relatively the same as that of an adult. Because, one day at a traditional event, the space in the *Bhaga* will be entered by girls, both in adulthood and in their teens (Fig. 3).

3 The Research Method

The research method which be used to understand the characters of traditional houses in the *Ngadha* culture was an inductive method with a scope of problems in the *Sa'o* design in several traditional villages in the *Ngada* Regency. Lesson learnt

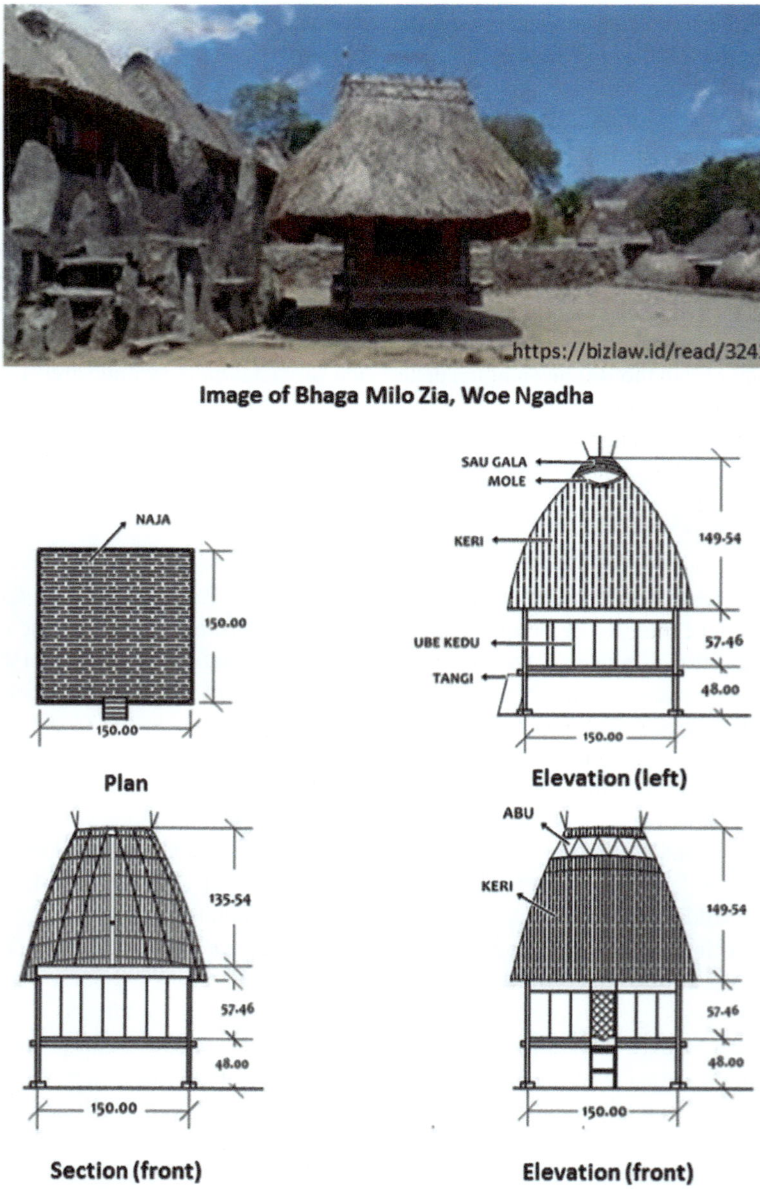

Fig. 3 Structure of the *Bhaga*. *Source* Susetyarto, 2013

about the structural system and construction characters of tradistional buildings, how its strengths and weaknesses in accommodating indigenous people's activities, the impact of security and safety of indigenous people, and specifically understanding its rigidity so that it has advantages as a cube building that can be used as a lifeline for natural disaster. However, in an effort to find a cube building design for natural disaster rescue, experimental design research methods were used which had been taught by Henri Achten [5, 6] and modified by researchers, as shown in Fig. 4 below.

Explanation Fig. 4. Starting from the analysis of researchers (red box) conducting an Alpha evaluation study of several natural disaster rescue building designs that have been carried out by other researchers, to prove the design concept of a natural disaster rescue cube from each cube, which was compared with each others, in order to obtain the best formulation of the prototipe natural disaster rescue cube concept, in accordance with the character of natural disasters in the Nusantara, the earth of ring of fire. This stage was commonly referred to as proof of concept (*PoC*). The results of the *PoC* study were synthesized (maroon box) comprehensively, to lead to the process of making a model (orange box) of the rescue cube as intended by the researcher. The next stage, the finished cube model was simulated (yellow box) in a computerized laboratory, and evaluated (green box) by the internal evaluator team for input, suggestions, and criticism. This was called as the Beta evaluation. The results of the evaluation of the rescue cube design for natural disasters, which were in the form of notes on the performance of the cube, size, materials used, practicality of use, and the method of industrialization, were very important to be used in subsequent analyzes (red box Analysis 2), both in the context of improving existing designs and preparing new designs until acceptable to the market, government, and the engineering industry willing to produce it. This last analysis was mentioned as the Gamma analysis.

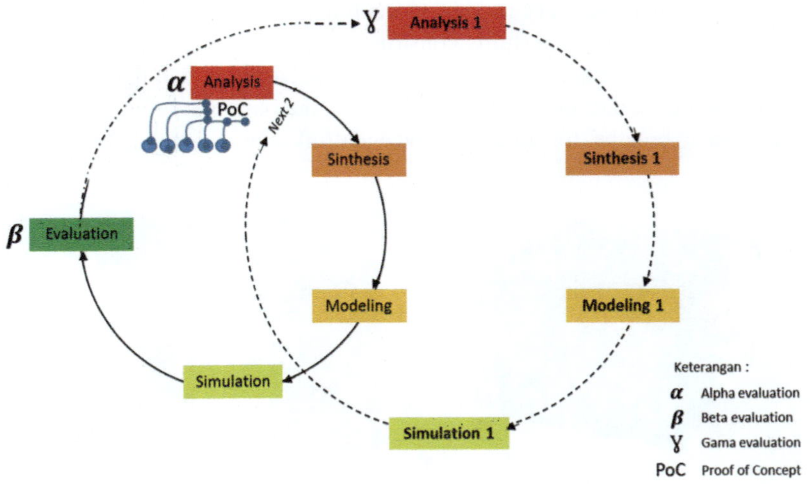

Fig. 4 Modified experimental design research methods. *Sumber* Susetyarto, 2021

4 The Rescue Cube

Inspired by the characteristics of the *One* and the *Bhaga* building from the *Ngadha* traditional houses in Flores, namely as the main residential building, which protects and saves indigenous people from various disasters that threaten, and knows that the *One* and the *Bhaga—Ngadhu* are traditionally believed by indigenous people as a house for their ancestors, and as the center of the orientation of the *ulu* and *eko* axis of their life in the village, their smallest world, then the researchers made an initial model of the disaster rescue cube as a protipe [7, 8], an initial design (V-0) in Fig. 5a. The design of the V-0 was simulated, and tested for its strength against compressive forces caused by disasters and due to construction collapse in the computer laboratory room. The simulation of the model by computer had to be carried out because at that time the atmosphere of the pandemic required the research team to act according to health protocols. The strength test simulation through a computer program [9] was recorded and distributed to the internal evaluator team to get input, suggestions, and criticisms needed for the next analysis. Then, the results of the second evaluation are used to make a synthesis and a second model called the V-1 design, as shown in Fig. 5b.

The rescue cubes have more or less the same meaning with the *One* and the *Bhaga.* The new "*One*" and the new "*Bhaga*" buildings are not meant to replace the *One* and the *Bhaga* current presence in the *Ngadha* traditional houses. However, the new "*One*" and the new "*Bhaga*" buildings are developed for villagers in earthquake, landslides [10, 11], hurricane, and tsunami prone areas throughout Indonesia. Actually the design is an idea development from the sustainability model of the vernacular village, the *Ngada* district, Flores as discussed in some previous article [12, 13]. The V-0 design of the rescue cube building materials were selected from high quality and transparent fiberglass. The structural system and construction which be used an interlocked connection model, very strong in accepting earthquake, landslides, hurricane, and tsunami loads, stable and sturdy construction, tightly closed and waterproof and shockproof. In fact, to maintain tidiness it was added a rubber belt that will automatically lock when the last door was closed. Inside the building, of course, there was enough space to breathe, for adequate food and drink supplies, so that those who live

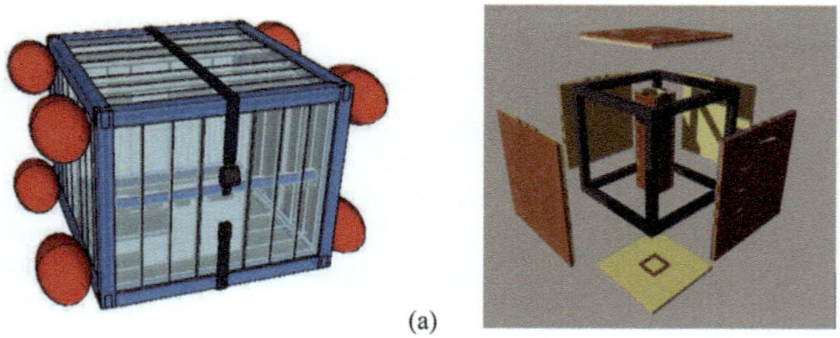

(a) (b)

Fig. 5 The rescue cubes. *Source* Susetyarto, 2019, *Source* Susetyarto, 2021 in it can survive in an emergency. This rescue cube box was also equipped with a number of balloons that automatically inflate when submerged in water, so that it will always lift *"One"* above the water surface in the event of an earthquake, landslides [5, 14] followed by a tsunami hazard. This kind of idea should be introduced and socialized to the Ngadha indigenous people, especially to the *Utu Bhou Dia Sa'o*, the forum for the construction of the *Ngadha* traditional house [2, 13]. Perhaps, there will be wise suggestions from the forums to make the *"One"* design better for the future. Actually, the *Ngadha* indigenous people really understood the character of the earthquake, landslides, hurricane or/and tsunami on the island of Flores. What was more important to convey in the forum was that the *Ngadha* indigenous people must know the idea of *"One"* which was derived from their traditional architecture and was designed for those who urgently need safety equipment from the natural disasters. That way a sense of pride and confidence will grow in the hearts of the *Ngadha* indigenous people to continue to be advanced people wherever they were. People knew that in the *Ngadh*a area, the lack of human resources and the threat of natural disasters were part of a serious problem that must be addressed immediately apart from the problem of poverty. Therefore, academics and the government had time to be responsible for solving these problems.

The structural system and construction of the rescue cube will continue to be tested for strength in civil and water resources engineering laboratories, tested for its durability, technical feasibility, as well as market feasibility, so that the [engineering or construction] industries are ready to produce it. The concept and schematic design of the rescue cube also tested by some students of Architecture, Planology, Information Technology, Computer Technique, and Information system. They gave the important inputs especially in term of its practicality. So, those inputs could be a preleminary examination before be developed by the industries, and be marketed to the government and local governments who need it. The rescue cube building can be placed in the inner room, on the terrace of the building, on the porch of the house, on the yard, in the green open space, on the sports field, on the beach space, and especially on disaster evacuation routes made by architects so that it can be easily seen, can be reached by families, residents of residential areas, or the general public who are facing the dangers of natural disasters. The researchers will continue improve its architectural design [15, 16] towards better product design, more effective and efficient, user friendly, funny and interesting to have it. Of course, as attractive as any of the rescue cube design must be guaranteed to be robust, their reliability highly tested in accepting earthquake. Landslides, hurricane, and tsunami loads.

References

1. Marstein N, Larsen KE (2000) Conservation of historic timber structures: an ecological approach p 21. Elsevier. ISBN 9780750634342. Retrieved 2007-04-03
2. Susetyarto MB (2013) Arsitektur Vernakular, Keberlanjutan Budaya di Kampung Bena Flores, Penerbit: Padepokan Seni Djayabinangun

3. Susetyarto MB (2019) One, The 1st international conference on environmental sciences (ICES2018), OP Conference series: earth and environmental science 314, 012030, https://doi.org/10.1088/1755-1315/314/1/012030
4. Susetyarto MB (2012) Architecture and environmental sustainability: critical issues in Vernacular Kampong of Bena, Flores Appl Mech Mater, 253–255 22–26, https://doi.org/10.4028/www.scientific.net/AMM.253-255.22
5. Rudiyanto R (2010) Analisis Potensi Bahaya Tanah Longsor Menggunakan Sistem Informasi Geografis (Sig) Di Kecamatan Selo, Kabupaten Boyolali, Fakultas Geografi, Universitas Muhammadiyah Surakarta (Skripsi)
6. Cross N (2000) Engineering design methods: strategies for product design. Wiley, Chichester
7. Griffith K, Sass L (2006) Computing & materializing non-uniform shapes: an evolutionary approach to generate and digital fabricate non-uniform masonry walls. CAADRIA 2006 [Proceedings of the 11th international conference on computer aided architectural design research in Asia] Kumamoto (Japan) March 30th—April 2nd 2006, pp 227–235
8. Tang CS (2006) Smart structures; designs with rapid prototyping. In van Leeuwen, Jos P, Timmermans Harry JP (eds) Progress in design & decision support systems in architecture and urban planning—proceedings of the 8th International DDSS conference, Eindhoven: Eindhoven University of Technology, pp 415–429
9. Liu YT (2003) Digital architecture: theory, media and design. Digital design—research and practice—proceedings of the 10th international conference on computer aided architectural design futures, Tainan (Taiwan) 13–15 October 2003, pp 9–18
10. Surendro B, Longsor B, Penanggulangannya C (2004) Februari-Maret (Tahun II) 105–117/17660-ID-bahaya-longsor-cara-penanggulangannya.pdf. Vol 21. No 1
11. Highland LM, Bobrowsky P (2008) The landslide handbook—a guide to understanding landslides U.S. geological survey, Reston, Virginia. World Wide Web: http://www.usgs.gov/pubprod/C1325_508.pdf
12. Djarot P, Benediktus B, Herman FH, Ignatius H, Kristiana B, Reginaldo CL (2016) *Menelusuri Arsitektur Vernakular NTT* (Padang, Indonesia: Rumah Kayu) pp 89–93
13. Susetyarto MB (2018) Model of sustainability of vernacular kampongs within Ngadha culture, Flores IOP conference series earth and environmental science, 106 012076, https://doi.org/10.1088/1755-1315/106/1/012076. Licence CCBY 4.0; Project: Cultural Sustainability (2018)
14. Prasad N, NB (2017) Landslides—causes and mitigation, Sr. Scientist, Centre for Water Resources Development and Management Kozhikode—673571, Kerala, INDIA https://www.researchgate.net/publication/317328970
15. Petric J, Lindsay M (2003) Digital prototyping. CAADRIA 2003—Proceedings of the 8th international conference on computer aided architectural design research in Asia, Bangkok Thailand 18–20 October 2003, pp 837–852
16. Dritsas S, Kashyap S (2005) Scripted mockups: bridging digital and physical through computation. CAADRIA 2005—Proceedings of the 10th international conference on computer aided architectural design research in Asia, New Delhi (India) 28–30 April 2005, vol 1, pp 351–357

Socio-Economy in Sustainable Development Goals

Building a Synergic Commitment to Govern the Plastic Waste Management in the ASEAN Region

Sanusi and Angga Wijaya Holman Fasa

Abstract The increase in plastic waste is alarming, especially since the population growth spike is difficult to control. This issue is, of course, a challenge for heads of government in ASEAN because the problems mentioned above are closely correlated with the level of environmental pollution. The COVID-19 pandemic, in which all countries in the world are still struggling to handle the transmission of the viruses, has added to a long line of home work for each country with the emergence of a new form of pollutants during the COVID-19 crisis. This review of the literature would result in recommendations for policy direction on regional synergic cooperation in plastic waste management. Addressing the challenge of plastic pollution in Southeast Asia would require novel problem solving, innovative ideas and technology, and establishing a supportive innovation ecosystem in which these solutions can thrive. In summary, waste management synergies are essential to preserving the environment.

Keywords Plastic waste management · Policy · Regional cooperation · ASEAN · Technology transfer

1 Introduction

Human activities are increasing from time to time, and the types and amounts of waste produced are growing. One of them is plastic waste. Especially in Southeast Asian (ASEAN) countries, the increase in plastic waste is quite alarming, especially since the population growth spike is difficult to control. This issue is a challenge for heads of government in ASEAN countries because the problems are closely correlated with

A. W. H. Fasa (✉)
Research Center for Population, National Research and Innovation Agency (BRIN), Jakarta, Indonesia
e-mail: angga.wijayafasa@ekraf.go.id

Sanusi
Ministry of Creative Economy/Creative Economy Agency, Jakarta, Indonesia
e-mail: sanu002@brin.go.id

M. A. Rahandi Lubis (ed.), *Proceedings of the 10th International Symposium for Sustainable Humanosphere*, Lecture Notes in Civil Engineering 255,
https://doi.org/10.1007/978-981-19-1422-5_29

the escalation of environmental pollution. However, this issue has not been raised to the top of the priority list for ASEAN leaders. Although the 34th ASEAN Summit in Thailand in June 2019 reached a consensus on the Bangkok Document on Combating Marine Debris in the ASEAN Region, it has emerged that the declaration does not address the issue of plastic waste importation into Southeast Asia. The problem of sending waste from Western countries to the Southeast Asia region after China closed its doors two years ago was a crucial issue during the ASEAN Summit. Based on statistical data between 2016 and 2018, plastic waste imports in the ASEAN region reached a significant increase of 171%, from 836,529 tons to 2,265,962 tons. This volume is equivalent to about 423,544 20-foot containers [1].

The COVID-19 pandemic, in which all countries in the world are still struggling to handle the transmissions of the viruses, has added to a long line of homework for each country with the emergence of a new problem, namely the abundance of disposable mask waste. It has become common in efforts to protect oneself from COVID-19 by wearing disposable face masks [2]. However, the next series of consequences is contamination that causes environmental problems due to current disposal processes, namely incineration and reclamation, due to the release of toxic chemicals. Face masks usually have several layers, especially polypropylene (PP), because polypropylene microfibers are hydrophobic, skin-friendly, and non-allergenic [3]. Elastic bands for fastening, boxes containing filters and transparent covers are all plastic products. All masks and particle filters are made of plastic, and standards set filter performance (at least 80/94/99% efficient) so that they are challenging to recycle [4]. The use of masks as a health protocol, the impact of government policies to prevent the transmission of the COVID-19 virus has even become an obligation. It will be subject to sanctions for refusal or deniable it. However, it is clear that the increased demand for personal protective equipment (PPE) and the obligation to wear masks.

Waste cannot be disposed of irresponsibly because it not only causes unsightly views but also causes health problems because it is unhygienic and can cause disaster for our environment. Besides, finding an appropriate allocation for landfilling waste near settlements is also not an easy task. Community concerns about the impact of odors, pollution, pests can cause a decrease in the value of their homes, apart from the problem of disruption of comfort. As a result, there will be considerable costs to provide a landfill facility that is strategically located and environmentally responsible.

Recently, the idea of integrated waste management, which is implemented to reduce waste, starts from the source, even before it enters the waste stream. This means that the resulting waste material should be processed first to be reused and recycled, then the rest is disposed of at the final waste management location. However, it turns out that disposal cannot be used as a final solution; it needs sustainable management with accompanying policies to realize effective implementation.

2 Method

We examined published research in order to identify and evaluate the research challenges being addressed, most notably those relating to ASEAN-region plastic waste policy and management. Between 2002 and 2021, we assessed and analyze peer-reviewed journal publications, reports, and news stories that contained keywords relating to plastic waste policy and management not just in the ASEAN region but also globally. This process of doing a literature review comprises multiple processes, including defining the research question(s) and objective(s), conducting a literature search, screening for inclusion, evaluating the quality of primary studies, extracting data and information, and analyzing data and information [5]. Based on the analysis of data and information from the selected literature, we synthesize them to answer research issues.

3 Result and Discussion

3.1 The Actual and Potential Impact of Plastic Waste

In recent years, there were many campaigns that plastic should be banned and replaced with other eco-friendly materials because it pollutes the environment. The plastics industry must continuously maintain the argument that plastics do not degrade the environment and that their ecological footprint is one of the most minor compared to other materials in general. It means to pollute the environment; they mix it up by littering and do not know the meaning of an ecological footprint [6].

It is undeniable that plastics have a huge role in the implant and medical aid industry and the pharmaceutical industry, including implants, prostheses, bonding elements, plastic organs, and medical devices. Plastics can help treat medical procedures so that they are faster, more comfortable and less painful. However, the implications arising from the use of plastics have left a severe environmental problem. Quoting from The Guardian, June 8, 2020, regarding the news that the French non-profit organization Operation Mer Proper, whose activities include collecting trash along the Cote d'Azur, started sounding waste warning alarms in the form of dozens of latex gloves, masks and sanitizer bottles below. Waves of the Mediterranean Sea. The waste is floating together with disposable cups and floating aluminum cans, like jellyfish scattered on the seabed [6].

According to the Indonesian Doctors Association (IDI), the negative potential of medical masks that can harm the environment can take hundreds of years to decompose, which takes decades to rot in the soil; studies say it can be more than 100 years [7]. Waste of medical masks and associated objects, such as disposable plastic gloves, is estimated that number is enormous; one day, it will become a threat to the environment if it has been started early childhood is not appropriately handled.

Disposable mask waste is now easy to find people dumping everywhere in its intact condition. Of course, this will be homework for our environment that will never end.

Surgical masks are treated as Hazardous and Toxic Materials in the form of infectious waste, which can cause infection in other people. Thus, this type of waste must be explicitly treated by the medical device handling protocol issued by the WHO. They sorted mask waste into hazardous and toxic waste to anticipate the risk of misuse of mask waste. The dangers of repeated use of disposable masks that can lead to infection are often ignored by the community, especially related to people's concern for doing the right way when disposing of disposable mask waste. Waste management personnel are among the people who have the most significant possibility of exposure to infection due to a lack of responsibility in treating used mask waste [8].

The public awareness dissemination for lifestyle changes to reduce the use of plastic products has long been encouraged but has not had a significant impact. Preemptive action is needed in the form of policies and innovations both in the use and management of this mask waste so that the effects that are generated can be reduced and overcome and not become an additional burden on the environment.

3.2 Plastic Waste Management Issues

Plastics are inexpensive, lightweight, and resilient materials that may be easily molded into a range of shapes and used in a wide variety of human activities. Plastic manufacture has increased considerably over the last 60 years as a result of these advantages. However, with the current rate of use and disposal, a plethora of environmental problems arise. Around 4% of global oil and gas production, a non-renewable resource, is utilized as raw material for plastics, and another 3–4% is required to power their manufacturing [9].

Plastic recycling terminology is varied and, at times, perplexing due to the multiplicity of recycling and recovery procedures. While the majority of thermoplastics may be recycled closed-loop, plastic packaging frequently contains a variety of polymers and other elements, including metals, paper, pigments, dyes, and adhesives, complicating recycling. As a result, recycling procedures must be classified based on the materials included therein [10].

Although the plastics recycling process has been conducted since the 1970s, the amount varies geographically, according to their type and use. The Indonesian government itself has made various efforts to reduce the use of plastic in people's lives. In February 2016, the government launched a paid plastic bag scheme in an attempt to curb the use of plastic. Environmentalists, on the other hand, lament the low cost of this plastic. Additionally, they believe in the importance of transparency about the utilization of funds gathered. In October 2016, the Indonesian Retail Entrepreneurs Association (Aprindo) opted to discontinue the program, citing the absence of a legislative framework for its implementation [7].

3.3 Technology Transfer Urgency

Plastics are critical because they are required for sustainable development and possess unique features not found in other materials. Plastics are critical because they are required for sustainable development and possess unique features not found in other materials. Instead of banning them, it should take responsible and necessary actions to improve quality of life, prevent climate catastrophes and, if needed, play a role in stopping the spread of infectious diseases, including the current coronavirus. Polymers or plastics have been instrumental in practically every key technological advancement over the previous few decades, including 3D printing.

Due to its superior properties and ability to assist in resolving environmental challenges, this technology, called "engineered plastics," is used in a number of industries, from automotive to construction to furnishings. For example, current vehicles are lighter than ever before as a result of polymer innovation, resulting in better fuel efficiency. Nonetheless, the increasing use of plastic items in everyday life is wreaking havoc on the environment, necessitating the creation of eco-friendly substitute materials.

It needs a technology shift so that plastic can continue to be used. There are various plastics, including biodegradable plastics, recyclable plastics, biodegradable plastics, and alternatives to compostable plastics, all of which differ from one another, and each of these differences is important to know. In fact, it is necessary to understand that plastics that are said to be biodegradable are plastics that, when they begin to disintegrate, then triggering the contamination of harmful chemicals into the food chain and endanger the environment and human health [11].

Alternatively, polymers derived from petroleum can be converted to bio-based polymers. This action might result in hundreds of millions of tons of carbon dioxide being removed from the atmosphere each year. Not only are biopolymers more environmentally friendly and sustainable to manufacture, they also contribute to climate change mitigation by functioning as carbon sinks. Regrettably, not all biopolymers are equal. Furthermore, biopolymers are non-biodegradable and non-recyclable.

Recycling remains the most significant choice for the time being. Plastic materials can be recycled in a variety of methods, and the simplicity with which they can be recycled varies according to the polymer type, the packaging design, and the product type [10]. For instance, rigid containers consisting of a single polymer are simpler and more economical to recycle than multi-layer and multi-component packages. Thermoplastics, including PET (polyethene terephthalate), PE (polyethene) and PP (polypropylene), all have a high potential for being recycled mechanically.

Plastics are recyclable in a variety of methods, and their recycling ease varies according to the polymer type, container design, and product kind. The greatest impediment to recycling plastic waste is the incompatibility of the bulk of plastics due to their molecular structure and macroscale processing requirements [11].

3.4 ASEAN Existing Legal Framework

Since its inception in 1967, the ASEAN legal document, notably the ASEAN Charter, has contained environmental restrictions. In Article 1 point 9 of the ASEAN Charter, it is explained that one of the goals of the establishment of ASEAN is to promote sustainable development that ensures protection of the environment in the region, sustainable natural resources, preservation of cultural heritage, and quality people's lives [12]. Based on this regulation, it can be seen that environmental issues have been the main focus of ASEAN member countries since the beginning.

However, in the context of cooperation between member countries, there is also the 'ASEAN Way' principle which contains three fundamental things, namely: (1) the principle of non-intervention, namely that member countries are prohibited from interfering in the domestic problems of other member countries; (2) prioritize building a cooperation consensus based on good faith rather than being bound by a certain legal agreement (this is exempted for cases of extraordinary impact and requiring mutual commitment); and (3) prioritizing program implementation rather than forming a strong regional bureaucracy and regulating cooperative governance [13].

The existence of these principles creates the legal framework of ASEAN, including in the environmental sector, which is based on the soft approach law resulting from the process of consensus among the heads of government through diplomatic channels (quiet diplomacy) [14]. Due to this approach, ASEAN, in contrast to the European Union, has not yet established a regional political body with a mandate based on legal legitimacy that functions to compose and enforce laws but chooses to identify policy directions in a voluntary manner by member states that emphasize informal relations between heads of state and reaching consensus through direct conflict avoidance [15]. Thus, until now, there is no legal framework, in the sense of a legally binding instrument for each member country, which regulates environmental governance, and this issue is left to the legal policies of each member country.

In relation to the legal framework for environmental protection in order to overcome the problem of plastic waste, there is no legal framework that regulates mitigation cooperation. In fact, ASEAN member countries regulate it in their respective national legal frameworks as a form of membership implementation in several global multilateral conventions [16]. However, not all member countries have ratified or signed some of these conventions. Of the several conventions, only one convention, namely, The Basel Convention on the Control of Transboundary Movements and Disposal of Hazardous Wastes (1989), which was ratified and signed by all member countries (Table 1), as noted in the table below [17]:

Table 1 Status by ASEAN countries of international conventions related to plastic waste

International treaties	Number of ASEAN countries being states parties
United Nations Convention on the Law of the Sea (1982) (UNCLOS)	9 (1 Signatory State)
The 1972 Convention on the Prevention of Marine Pollution by the Dumping of Wastes and other Matter (London Convention)	1
London Protocol (1996)	1
Annexes V to the 1973 international convention for the prevention of pollution from ships (MARPOL)	7
United Nations Watercourses Convention (1997)	1
The Convention on Biological Diversity of 1992	10
The Agreement to Implement the UNCLOS Provisions on the Conservation and Management of Straddling and Highly Migratory Fish Stocks (United Nations Fish Stocks Agreement)	4
The Stockholm Convention on Persistent Organic Pollutants (Stockholm Convention) of 2001	8 (2 Signatory States)
The Basel Convention on the Movements of Hazardous Wastes and their Disposal (1989)	10

Source Hisham and Florent [17]

3.5 The Plastic Waste Management in ASEAN Countries

Due to its policies in taking action to deal with plastic waste problems, there are differences in the management of plastic waste among ASEAN member countries. The difference is at least visible in terms of plastics production and trade; manufacturing and trade of plastic goods, including packaging; distribution and use; waste collection and management; recycling and use of recycled plastics; and import of plastic waste for recycling [18]. The innovation and technology capacity factor is also a differentiating factor and the pattern of waste management and the level of waste recycling. According to data from the United Nations Environment Program, based on the method applied, among ASEAN member countries Vietnam, Singapore, Thailand, Malaysia, and Myanmar, the recycling rate of plastic waste is above 50%; while the Philippines, Indonesia, Laos, and Cambodia are below 50%; and Brunei by 15%, whereas in countries with a percentage of plastic waste recycling above 50%, the methods generally applied are composting, sanitary landfill and open dump, as shown in the table below [19] (Table 2).

Table 2 Plastics waste management comparison among ASEAN countries

Countries	Recycling rate (%)	Treatment methods				
		Composting	Incineration	Sanitary landfill	Open dump	Open burning
Vietnam	70	✓			✓	
Myanmar	70		✓	✓	✓	
Malaysia	50–60		✓	✓	✓	
Singapore	> 50		✓	✓	✓	
Philippines	23–42	✓		✓	✓	
Thailand	< 50	✓	✓	✓	✓	
Indonesia	< 50	✓	✓	✓	✓	✓
Lao PDR	< 50	✓		✓	✓	✓
Cambodia	< 50	✓		✓	✓	✓
Brunei	15			✓	✓	

Source UNEP [19]

3.6 Future Policy Direction

Tighter regulations could limit the use of plastics, set standards for reducing waste in the packaging process and oblige producers to take responsibility for waste disposal [20]. Stricter rules could also set out obligations regarding reuse, recycling and disposal [21]. In addition, in this pandemic condition, it is necessary to choose a model that emphasizes the use of resources repeatedly with the aim of reducing waste production, emissions and energy expenditure [21]. According to Rita Helbra Tenrini & Sofia Arie Damayanty [22], switching to a circular economy method can be an alternative. A circular economy system will provide benefits for Indonesia because it is not only absorbing labor but also reducing waste and pollution and preserving nature. This method is found in the 5R concept, namely Reduce, Reuse, Recycle, Recover (restore), and Revalue (provide added value) [23]. From an economic point of view, this concept allows utilizing plastic waste from the previous production process to reduce the cost of processing (recycling) waste globally, and the social side of this concept can create jobs for workers [24]. By utilizing waste to share tools and services, it will ultimately reduce the environmental burden from the exploitation of raw materials and pollution [25].

Numerous policy recommendations to ASEAN member states are possible, including the strengthening of waste management policies, particularly those managing infectious waste [26]. Second, an inclusive plan can be developed by considering all parties at risk of infection, including waste collectors, scavengers, and recycling centre employees [27]. Apart from physical protection, they strive to alleviate stigma by receiving community acknowledgement and appreciation for their critical contributions throughout the pandemic. Another topic to consider, while keeping everyone's safety in mind, is the administration of a market system to recompense scavengers who lost their source of income as a result of the pandemic's decline in recyclable garbage prices [28, 29]. ASEAN member countries can raise awareness about this issue, exchange best practices, and develop more effective regulatory frameworks. Adoption of the Act will not only assist in preventing a pandemic in the short term but will also result in an effective waste management system in the long run.

4 Conclusion

Waste management must start with reduction, while disposal must be the last option. Waste reduction must also be a priority for sustainable waste management. Waste separation at the source is essential to achieve reuse, recycling and recovery of waste. Waste processing and treatment is very important in holistic plastic waste governance. It is essential to critically examined and identifying all potential aspects and options for waste treatment.

Numerous initiatives have been proposed to encourage collaboration among ASEAN member countries in solving Southeast Asia's plastic waste problem: first, strengthening research collaborations in the field of Science and technology related to the environment to bring up new innovations in plastic waste management. Second, strengthening technology intervention by applying for technology transfer among ASEAN member countries. Third, formulation of a joint policy framework and action plan on tackling plastic waste, which contains guidelines for steps to implement circular economic activities, research collaboration, technology intervention, and technology transfer. Another strategy shall be concluded to schedule a mutual collaboration agreement between ASEAN member countries which contains a policy framework and action plan for tackling plastic waste in Southeast Asia to manifest the joint commitment.

References

1. Uhm Y (2020) Plastic waste trade in Southeast Asia after China's import ban: implications of the new basel convention amendment and recommendations for the future. Cal WL Rev 57:1
2. Czigány T, Ronkay F (2020) Editorial corner—a personal view. The coronavirus and plastics. Exp Poly Lett 14:510–511. https://doi.org/10.3144/expresspolymlett.2020.41
3. Fadare OO, Okoffo ED (2020) COVID-19 face masks: a potential source of microplastic fibers in the environment. Sci Total Environ 737:140279
4. Jung S, Lee J, Park YK, Kwon EE (2020) Bioelectrochemical systems for a circular bioeconomy. Biores Technol 300:122748
5. Templier M, Paré G (2015) A framework for guiding and evaluating literature reviews. Commun Assoc Inf Syst 37(1):6
6. Henderson L, Green C (2020) Making sense of microplastics? Public understandings of plastic pollution. Mar Pollut Bull 152:110908

7. Yudhastuti R (2020) The use of masks during the pandemic period in Indonesian people. Kosmas. Nat Publ Health J. 15. https://doi.org/10.21109/kesmas.v15i2.3945
8. Chan AW, Song F, Vickers A, Jefferson T, Dickersin K, Gøtzsche PC, Van Der Worp HB (2014) Increasing value and reducing waste: addressing inaccessible research. The Lancet 383(9913):257–266
9. Wright T, Waddell S (2017) How can Indonesia win against plastic pollution? 5 September. The Conversation. https://theconversation.com/how-can-indonesia-win-against-plastic-pollution-80966
10. Lambert S, Wagner M (2017) Environmental performance of bio-based and biodegradable plastics: the road ahead. Chem Soc Rev 46(22):6855–6871
11. Hopewell J, Dvorak R, Kosior E (2009) Plastics recycling: challenges and opportunities. Philos Trans R Soc B Biol Sci 364(1526):2115–2126
12. Tan SYL, Kamaruddin H (2019) Environmental challenges within ASEAN: contemporary legal issues and future considerations. In ASEAN Post-50 (pp 155–176). Palgrave Macmillan, Singapore
13. Simon S (2008) ASEAN and multilateralism: The long, bumpy road to community. Contemp Southeast Asia: A J Int Strateg Aff 30(2):264–292
14. Lian KK, Robinson NA (2002) Regional environmental governance: examining the association of Southeast Asian nations (ASEAN) model. Glob Environ Gov Options Opportunities, 101–121
15. Lyons Y, Neo ML, Lim A, Tay YL, Dang VH (2020) Status of research, legal and policy efforts on marine plastics in ASEAN+ 3: a gap analysis at the interface of science, Law and policy
16. Hermawan S, Astuti W (2021) Analysing several ASEAN countries' policy for combating marine plastic litter. Environ Law Rev 23(1):9–22
17. Hisham MM, Florent MZ (2019) Overview of plastic issues in ASEAN, focusing on marine debris and microplastics in the region, https://pic.org.kh/images/2019Research/20191227_Overview%20of%20plastic%20issues%20in%20ASEAN,%20focusing%20on%20marine%20debris%20and%20microplastics%20in%20the%20region-Eng.pdf
18. Akenji L, Bengtsson M, Kato M, Hengesbaugh M, Hotta Y, Aoki-Suzuki C, Gamaralalage PJD, Liu C (2019) Circular economy and plastics: a gap-analysis in ASEAN member states. Brussels: European commission directorate-general for environment and directorate-general for international cooperation and development, Jakarta: Association of Southeast Asian Nations (ASEAN)
19. UNEP (2017). Regional Study on Waste Management in the ASEAN Countries. United Nations Environment Programme, Nairobi, Kenya.
20. Syberg K, Nielsen MB, Clausen LPW, van Calster G, van Wezel A, Rochman C, Hansen SF (2021) Regulation of plastic from a circular economy perspective. Curr Opin Green Sustain Chem, 100462
21. Fletcher CA, Clair RS, Sharmina M (2021) A framework for assessing the circularity and technological maturity of plastic waste management strategies in hospitals. J Clean Prod, 127169
22. Tenrini RH, Damayanty SA (2020) A circular economy during a pandemic: creating jobs and reducing waste. Published October 16 2020. Click to read: https://theconversation.com/ekonomi-sirkular-saat-pandemi-mencipta-lapangan-kerja-dan-mengurangi-sampah-142663
23. Hahladakis JN, Iacovidou E, Gerassimidou S (2020) Plastic waste in a circular economy. In: Plastic waste and recycling (pp 481–512). Academic Press
24. Meys R, Frick F, Westhues S, Sternberg A, Klankermayer J, Bardow A (2020) Towards a circular economy for plastic packaging wastes–the environmental potential of chemical recycling. Resour Conserv Recycl 162:105010
25. Bellou N, Gambardella C, Karantzalos K, Monteiro JG, Canning-Clode J, Kemna S, Lemmen C (2021) Global assessment of innovative solutions to tackle marine litter. Nat Sustain 4(6):516–524
26. Alpizar F, Carlsson F, Lanza G, Carney B, Daniels RC, Jaime M, Wahdera S (2020) A framework for selecting and designing policies to reduce marine plastic pollution in developing countries. Environ Sci Policy 109:25–35

27. Xuan Son NT (2021) Policy on marine plastic waste in ASEAN and Vietnam. Environ Claims J 33(1):41–53
28. Akenji L, Bengtsson M, Hotta Y, Kato M, Hengesbaugh M (2020) Policy responses to plastic pollution in Asia: summary of a regional gap analysis. Plast Waste Recycl, 531–567
29. Vanapalli KR, Sharma HB, Ranjan VP, Samal B, Bhattacharya J, Dubey BK, Goel S (2021) Challenges and strategies for effective plastic waste management during and post COVID-19 pandemic. Sci Total Environ 750:141514

GPSR Compliance

The European Union's (EU) General Product Safety Regulation (GPSR) is a set of rules that requires consumer products to be safe and our obligations to ensure this.

If you have any concerns about our products, you can contact us on ProductSafety@springernature.com

In case Publisher is established outside the EU, the EU authorized representative is:

Springer Nature Customer Service Center GmbH
Europaplatz 3
69115 Heidelberg, Germany

Batch number: 10755466

Printed by Printforce, the Netherlands